Achievement Motivation
Recent Trends in Theory and Research

Achievement Motivation

Recent Trends in Theory and Research

Edited by
LESLIE J. FYANS, JR.

Illinois State Board of Education
Springfield, Illinois

PLENUM PRESS • NEW YORK AND LONDON

Library of Congress Cataloging in Publication Data

Symposium on Achievement Motivation, Toronto, Ont., 1978
  Achievement motivation.

  Based on the proceedings of the Symposium on Achievement Motivation, held at the 1978 American Educational Research Association Convention in Toronto, Ontario, Canada, in April 1978 and on the proceedings of the Motivation in Education Conference held at the University of Michigan, Ann Arbor, Michigan, in October 1978.

  Includes index.
  1. Achievement motivation—Congresses. I. Fyans, Leslie J. II. American Educational Research Association. III. Title.
  BF501.5.S94 1978   153.5   80-19826
  ISBN 0-306-40549-0

Based on the proceedings of the Symposium on Achievement Motivation held at the American Educational Research Association in Toronto, Ontario, Canada in April 1978 and on the proceedings of the Motivation in Education Conference held at the University of Michigan, Ann Arbor, Michigan in October 1978.

© 1980 Plenum Press, New York
A Division of Plenum Publishing Corporation
227 West 17th Street, New York, N.Y. 10011

All rights reserved

No part of this book may be reproduced, stored in a retrieval system, or transmitted, in any form or by any means, electronic, mechanical, photocopying, microfilming, recording, or otherwise, without written permission from the Publisher

Printed in the United States of America

Preface

This book started as a symposium on Achievement Motivation at the 1978 American Educational Research Association Convention. The participants in that symposium were Jack Atkinson, Martin Maehr, Dick De Charms, Joel Raynor, and Dave Hunt. The subsequent response to that symposium indicated a "coming of age" for motivation theory in terms of education. Soon afterward, at a Motivation in Education Conference at University of Michigan, Ann Arbor, it became apparent that due to this emergence of motivation what was needed was a comprehensive perspective as to the state of the art of achievement theory. Achievement theory had by now well surpassed its beginnings in the 1950s and 1960s and was ready for a composite presentation and profile of the recent research and theories of motivation. Thus, this volume was born.

I would like to take this opportunity to thank each contributor to this book as well as Robert L. Linn who critically reviewed several of the manuscripts. Thanks are also due to my former graduate advisors, Martin L. Maehr, Maurice Tatsuoka, and Harry Triandis, for the viewpoints given me in graduate school education which I hope have benefitted this undertaking. Joyce Fitch did a splendid job typing many of these chapters and special gratitude should be given to Judy Cadle of Professional Services, Inc. for the composition and proofing of this book.

Finally, I would like to thank the members of my family; my wife Paula, my mother, my father, and Mike Mast who have been a continual source of motivation and achievement.

Leslie J. Fyans, Jr.
March 5, 1980

Contents

SECTION I
Achievement Motivation: Trends in Theory and Research

Chapter 1   INTRODUCTION
Leslie J. Fyans, Jr. 3

SECTION II
Achievement Motivation Theory

Chapter 2   MOTIVATIONAL EFFECTS IN SO-CALLED TESTS OF ABILITY AND EDUCATIONAL ACHIEVEMENT
John W. Atkinson 9

Chapter 3   THE ORIGINS OF COMPETENCE AND ACHIEVEMENT MOTIVATION IN PERSONAL CAUSATION
Richard deCharms 22

Chapter 4   MOTIVATION, EVALUATION, AND EDUCATIONAL TESTING POLICY
Kennedy T. Hill 34

Chapter 5   EFFECTS OF FAILURE: ALTERNATIVE EXPLANATIONS AND POSSIBLE IMPLICATIONS
Margaret M. Clifford 96

Chapter 6   MOTIVATION AND EDUCATIONAL PRODUCTIVITY: THEORIES, RESULTS, AND IMPLICATIONS
Herbert J. Walberg and
Margaret Uguroglu 114

Chapter 7   A MODEL OF DIRECT AND RELATIONAL ACHIEVING STYLES
Jean Lipman-Blumen, Harold J. Leavitt, Kerry J. Patterson, Robert J. Bies, and Alice Handley-Isaksen 135

SECTION III
Achievement Motivation and Life Span Human Development

Chapter 8   THE GRAYING OF AMERICA: IMPLICATIONS FOR ACHIEVEMENT MOTIVATION THEORY AND RESEARCH
Martin L. Maehr and Douglas A. Kleiber 171

Chapter 9   MOTIVATIONAL DETERMINANTS OF ADULT PERSONALITY FUNCTIONING AND AGING
Joel O. Raynor 190

Chapter 10   A DEVELOPMENTAL PERSPECTIVE ON THEORIES OF ACHIEVEMENT MOTIVATION
Diane N. Ruble 225

SECTION IV
Attributions and Achievement Motivation

Chapter 11   ATTRIBUTIONAL STYLE, TASK SELECTION AND ACHIEVEMENT
Leslie J. Fyans, Jr. and Martin L. Maehr ... 249

Chapter 12   A RE-EXAMINATION OF BOYS' AND GIRLS' CAUSAL ATTRIBUTIONS FOR SUCCESS AND FAILURE BASED ON NEW ZEALAND DATA
John G. Nicholls 266

Chapter 13   MEASURING CAUSAL ATTRIBUTIONS FOR SUCCESS AND FAILURE
Timothy W. Elig and Irene Hanson Frieze 289

Chapter 14   ALLEVIATING LEARNED HELPLESSNESS IN A WILDERNESS SETTING: AN APPLICATION OF ATTRIBUTION THEORY TO OUTWARD BOUND
Richard S. Newman 312

SECTION V
Sex Differences in Achievement Motivation

Chapter 15   ACHIEVEMENT MOTIVATION AND VALUES: AN ALTERNATIVE PERSPECTIVE
Jacquelynne E. Parsons and Susan B. Goff 349

Chapter 16   ACHIEVEMENT AND VOCATIONAL BEHAVIOR OF WOMEN IN IRAN: A SOCIAL AND PSYCHOLOGICAL STUDY
Farideh Salili 374

Chapter 17   WOMEN'S ACHIEVEMENT AND CAREER MOTIVATION: THEIR RISK TAKING PATTERNS, HOME-CAREER CONFLICT, SEX ROLE ORIENTATION, FEAR OF SUCCESS, AND SELF-CONCEPT
Helen S. Farmer and Leslie J. Fyans, Jr. 390

SECTION VI
Teacher Expectations and Achievement Motivation

Chapter 18   TEACHER EXPECTATION AND STUDENT LEARNING
Margaret C. Wang and Warren J. Weisstein 417

SECTION VII
Achievement Motivation: A Look Toward the Future

Chapter 19   FROM SINGLE-VARIABLE TO PERSONS-IN-RELATION
David E. Hunt 447

Author Index 457
Subject Index 465

I

Achievement Motivation: Trends in Theory and Research

Chapter 1

Introduction

LESLIE J. FYANS, JR.
*The University of Illinois at Urbana-Champaign
and the Illinois State Board of Education*

This book represents the state of the art of achievement motivation theory and research. The history of achievement motivation stretches back more than a quarter century (McClelland, Atkinson, Clark, & Lowell, 1953). Initally, its research focused primarily on the measurement of a particular motive denoted as "need for achievement" or N-ACH. Soon work with this construct began being related to specific child rearing patterns (Winterbottom, 1953) and to the economic growth of nations (McClelland and Winter, 1969). From these beginnings, research and theory on achievement motivation became quite multifaceted and multidirectional. The area of female achievement became investigated (Horner, 1966), new forms of measurement were proposed (Mehrabian, 1969), new concepts such as risk-taking (Atkinson and Feather, 1966), expectancy and attainment value (Crandall, 1969), evaluation anxiety and negative motivation (Hill, 1972) and attributions (Weiner, 1971) were developed.

Unfortunately, with all of this exponential growth in both theory and research, no one volume housed all of the very recent developments. Several years ago an excellent book by Smith (1969) was written for that purpose. However, it did limit its focus specifically to children and thus could not be sensitive to the expanded horizon of the achievement literature of such as de Charms (1968), Maehr (1974), Raynor (1969), and Walberg (Uguroglu & Walberg, 1979).

This book fills that need in incorporating a wide breadth of chapters on all now researched aspects of achievement motivation. Many varieties and many styles of analysis and theoretical constructs are contained in the mosaic offered here. Thus, factors such as student achievement test performance, teacher expectations, sociocultural change, student self perception, attributional style, future orientation, values systems, vocational achievement, and child development are given. The chapters touching upon these factors have been arranged topologically such that chapters on related variables are adjacent to each other. Thus, the major theoretical advancements for the *construct* of *achievement motivation* have been placed together (Atkinson, de Charms, Hill, Maehr and Kleiber, Clifford, Raynor, and Walberg), the chapters on *attributions and attributional style*, (Nicholls, Fyans and Maehr, Clifford, Ruble, Newman), *sex diffferences in achievement motivation* (Parsons and Goff, Lipmen-Blumen and Leavitt, Salili, Farmer and Fyans) and *teacher expectations* (Wang and Weinstein). A final chapter by David Hunt points to the future achievement theory.

As with past decades, new concepts and directions of achievement motivation will make their appearance in future years. Perhaps in a few years it will be necessary to publish a second edition (or successive editions) of this book. However, these chapters now encapsulate the nature and characteristics of achievement motivation theory and research.

REFERENCES

Atkinson, J. W. and Feather, N. T. (Eds.). *A theory of achievement motivation.* New York: Wiley, 1966.
Crandall, V. C. Sex differences in expectancy of intellectual and academic reinforcement. In C. P. Smith (Ed.) *Achievement-Related motives in children.* New York: Russell Sage Foundation, 1969.
de Charms, R. *Personal causation.* New York: Academic Press, 1968.
Hill, K. T. Anxiety in the evaluative context. *Young Child,* 1972, *2*, 225-263.
Horner, T. The motive to avoid success. Unpublished doctoral dissertation. University of Michigan, 1966.
McClelland, D. C., Atkinson, J. W., Clark, R. W., & Lowell, E. L. *The achievement motive.* New York: Appleton-Century-Crofts, 1953.
McClelland, D. C. & Winter, D. B. *Motivating economic achievement.* New York: The Free Press, 1969.
Maehr, M. L. *Socio-cultural origins of achievement.* Monterey, California: Brooks/Cole Publishing, 1974.
Mehrabian, A. Measures of achieving tendency. *Educational and Psychological Measurement,* 1969, *29*, 445-451.

Raynor, J. O. Future orientation and motivation of immediate activity: An elaboration of the theory of achievement motivation. *Psychological Review*, 1969, *76*, 606-610.

Smith, C. P. (Ed.). *Achievement-Related motives in children.* New York: Russell Sage Foundation, 1969.

Uguroglu, M. E., & Walberg, H. J. Motivation and achievement: A quantitative synthesis. *American Educational Research Journal*, 1979, *16*, 4, 375-389.

Weiner, B., Frieze, I., Kukla, A., Reed, L., Rest, S., & Rosenbaum, R. M. *Perceiving the causes of success and failure.* New York: General Learning Press, 1971.

Winterbottom, M. R. *The relation of childhood training in independence to achievement motivation.* Unpublished doctoral dissertation. University of Michigan, 1953.

II
Achievement Motivation Theory

Chapter 2

Motivational Effects in So-Called Tests of Ability and Educational Achievement

JOHN W. ATKINSON
The University of Michigan

The title identifies a confrontation between two ways of thinking about the very same behavioral phenomena that has been going on for some time. But now motivational psychology has caught up with test theory in the clarity, completeness, and coherence of a mathematical model of motivation and action called the dynamics of action (Atkinson & Birch, 1970, 1974, 1978). The next decade should produce a new understanding of the proper relationship between test theory, which is grounded in statistical considerations, and the new theory of motivation, which is grounded in more than a quarter of a century of research on the problem and psychological considerations.

After a few introductory remarks, my discussion will focus on one aspect of the confrontation: distinguishing the effects of motivation and ability on test performance and educational achievement.

The most lasting contribution of all the work that has been done on achievement motivation is the way in which it is changing our ideas about how motivation influences behavior. The early work helped us to recognize that individuals may react in diametrically opposite ways in situations demanding competent performance. Those who are highly motivated to achieve are challenged. Those who are more fearful of failure are threatened. And we soon came to recognize that the difficulty of the immediate task was an important situational factor controlling the arousal of achievement-related motivation (Atkinson & Feather, 1966). More recently we have come to appreciate the reality of the paradox of fear of success in women (Horner, 1974). And Raynor (1969, 1974) has given us a more general theory of achievement motivation by including, in his broader conception, the motivational significance of more distant future goals anticipated when one views present activity as instrumental to their attainment.

I think much of this is familiar. There remains, however, a deplorable gap between what we are learning at the frontier of research in motivational psychology and the way we are thinking about the role of motivation in mental testing and educational achievement (Atkinson & Raynor, 1974, 1978).

The most recent theoretical development, the dynamics of action, (Atkinson & Birch, 1970, 1974, 1978) has, for example, allowed us to demonstrate by means of computer simulation of the problem that the validity of thematic apperceptive measures of individual differences in motivation does not depend at all on internal consistency or reliability as traditionally conceived (Atkinson, Bongort & Price, 1977). The traditional concept of reliability is founded in a simplistic theory of behavior, vis, obtained score = true score + error. The obtained score is obviously some measurable aspect of behavior. The true score must therefore be the "expected behavior." But the jump from personality traits (such as abilities, motives, conceptions) to behavior requires some kind of coherent theory stating how individual differences in personality are expected to be expressed in behavior. Traditional test theory has never had a theory about the behavioral processses which get us from individual differences in trait (personality) to some observable and measurable aspects of behavior. So for years we have simply gone along assuming that individual differences in some personality trait, presumed constant over some period of time, implied a comparable constancy in the measurable behavioral expressions of the trait. That is the fundamental flaw of traditional test theory. The new theory of motivation provides a very sound theoretical basis for predicting variable behavior from a constant personality trait. And so the long-taken-for-granted idea that internal consistency reliability is logically prerequisite for validity of a test as a whole (e.g., thematic apperceptive f$2n$ Achievement) has been shown to be fallacious. The work cited above answers a persistent criticism of our methodology (Entwisle, 1972). And it defines one confrontation involving contemporary theory of motivation and traditional test theory.

I will now discuss another confrontation of motivational psychology and traditional ideas about tests. We must close the deplorable gap between the frontier of research in motivational psychology and discussion of vital issues in education. Their social significance will not tolerate the normal ten to fifteen year gap in communication from the frontier of science to textbooks concerning applications.

I hope to persuade some educational psychologists that two generalizations based on experimental and conceptual analysis of how motivation influences behavior provide a new basis for comprehending the meaning of declining trends in test scores and the merely modest predictability of educational achievement from tests of "scholastic aptitude," and other problems involving the utter confounding of differences in ability and differences in motivation.

Motivation influences behavior in two ways. First, the strength of motivation for one kind of activity relative to the sum of the strengths of motivation for other, alternative activities within an individual, determines that individual's allocation of time among different activities (Sawusch, 1974).

This generalization concerning the determinants of percentage of time devoted to one kind of activity, such as academic work, as I have stated it non-technically, is the most important theoretical conclusion drawn from the reconstruction of the theory of motivation by David Birch and myself in *The dynamics of action* (1970). We have confidence in it for several reasons. It has been sustained in computer simulations of motivation and behavior based on the new principles of motivation (Atkinson, Bongort, & Price, 1977), and it corresponds in form exactly to the "matching law" proposed by deVilliers and Herrnstein (1976) to summarize what is known about the antecedents of the relative strength of competing activities in many studies of operant conditioning. One merely needs to continue to remember what E.L. Thorndike proposed in the Law of Effect, that the inclination to undertake rewarded actions is strengthened, to imply the dynamics of action as a theory of operant behavior.

The theoretical conclusion is shown in the hatched curve of Figure 1. For simplicity, here, I will treat the relationship as linear as shown in the solid curve. Figure 1 presumes (for simplicity) that the number and strength of tendencies to engage in other competing time consuming activities is constant among the individuals under discussion. If so, the proportion of time spent in the activity of interest to us will be greater when the strength of motivation (TA) for that activity is greater.

In addition to controlling the way an individual distributes the times of his or her life among different activities, the strength of motivation expressed in the performance of some activity influences the efficiency of performance nonmonotonically, in a way that is most simply described by an inverted U-shaped curve (see Figure 2). There has been evidence of this in research with animals since the time of Yerkes and Dodson (1908; e.g., Birch, 1945; Broadhurst, 1959). And it has long been a well-known generalization concerning level of arousal in neurophysiological research (Hebb, 1972). In recent years (Eysenck, 1966; Revell, et al., 1976; Atkinson, 1974) there has been convergent evidence from studies of human motivation

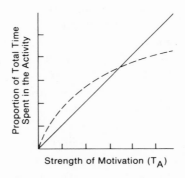

Figure 1. Simplified representation of the effect of strength of motivation for some activity (T_A) on proportion of time spent in the activity. Computer simulations based on current theory show this to be a negatively accelerated function (Atkinson, Bongort, & Price, 1977). Linearity is assumed for simplicity in exposition.

employing both valid measures of individual differences in motivation and experimental manipulation of incentives to be attained as a consequence of good performance.

The implications of this second generalization are shown in

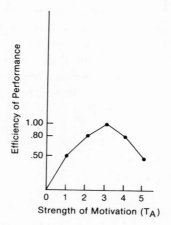

Figure 2. The effect of strength of motivation on efficiency of performance.

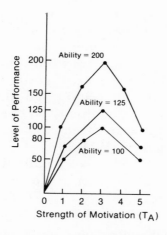

Figure 3. The effect of strength of motivation on level of performance when level of performance = ability × efficiency. The overall strength of motivation is a summation of various components such as intrinsic interest, tendencies to achieve, to gain social approval, to avoid failure, to attain distant future objectives (based on Atkinson, 1974a).

Figure 2. Unless one assumes that individuals do not differ in strength of motivation when taking some test traditionally called an intelligence-ability or scholastic aptitude test, or that strength of motivation has no influence on efficiency of performance, one must consider the several curves in Figure 3 as the most tenable description of the traditional confounding of effects of differences in ability and of strength of motivation on the level of intellective performance as recorded by a test score. One may note that there are three ways to score 100 on the hypothetical test. One may combine a true level of ability of 100 with optimal, that is, moderate strength of motivation on test day which makes for maximal efficiency at the time. One may be more able (level 125) and either undermotivated, e.g., dampened by the dread of failure or deficient in strength of positive motives and consequently less efficient. Or one may combine that same higher level of true ability with such strong motivation to perform well that there is inefficiency attributable to overmotivation, overexcitement.

I would not propose this interpretation of the level of intellective performance on tests such as Scholastic Aptitude tests if we did not have considerable evidence of it in experiments on verbal and arithmetic performance among college students (Atkinson, 1974b). What is needed, of course, are definitive large scale studies using the tests that are conventionally called intelligence-ability or aptitude tests but including the tools available to categorize individuals according to differences in motivation and also employing the techniques available for manipulaiton of the incentives or of intensifying motivation in other ways at the time of performance (see Raynor, 1969, 1974, 1978).

The implications of the two generalizations about motivation can be examined both in terms of cumulative achievement and in terms of performance in a critical ability (testing) situation. Table 1 presents this data. Table 1 puts the two generalizations about motivation into an analysis of how nine students who differ in true ability and in strength of motivation would perform in a critical ability test situation (left side of the table) and how much they would achieve academically in the long run (on the right side of the table).

Table 1

Ability and motivation as determinants of the level of intellective performance and cumulative academic achievement. In this hypothetical numerical illustration it is assumed that all students are more strongly motivated (+1) when taking a test than under normal everyday conditions.

Name of subject[1]	Ability test situation				Normal everyday conditions				
	True ability	Motivation	Efficiency	Level of test performance	Motivation	Efficiency	Level of performance	Time spent in work	Cumulative Achievement
A3	100	3	1.00	100	2	.80	80	2	160
B3	90	3	1.00	90	2	.80	72	2	144
A2	100	2	.80	80	1	.50	50	1	50
A4	100	4	.80	80	3	1.00	100	3	300
C3	80	3	1.00	80	2	.80	64	2	128
B2	90	2	.80	72	1	.50	45	1	45
B4	90	4	.80	72	3	1.00	90	3	270
C2	80	2	.80	64	1	.50	40	1	40
C4	80	4	.80	64	3	1.00	80	3	240

[1]Ss are named according to their true level of ability (A=100, B=90, C=80) and their strength of motivation (TA) in the ability test situation. (Based on Atkinson, 1974a.)

Consider *test performance* first. It is assumed that each student spends all the time available working on the test. The level of performance (or test score) is the product of ability and efficiency. For example, the *true ability* of subject A4 is 100, yet his/her *level of test performance* is only 80 because *motivation* is so strong (4) and *efficiency* is therefore only 80% or .80.

Now consider *cumulative achievement* under normal everyday conditions. Here we assume that intensity of motivation is generally less than in a stressful test for all students (1 unit less for simplicity). This change in motivation has an influence on the efficiency of performance in day to day work. Note that subject A3, the superstar on the big test, is now somewhat less than optimally motivated, less efficient, and the level of performance when at work is therefore lower than when taking a test. Note that A4, the very able student who is overmotivated on the test is now optimally motivated and therefore more efficient. What is more, he or she is still relatively more strongly motivated than others and these individual differences in motivation are expressed in the amount of time spent in academic work (see again Figure 1).

The presumption concerning cumulative achievement, which is evaluated in the grade point average over a long period of time, is that it is the product of the level or quality of performance while at work and the amount of time spent at work. Note that A4 achieves most in the long run, and that A3, the test-taking star, is well down in the distribution on cumulative achievement. One is called an overachiever, the other an underachiever (Thorndike, 1963).

In more elaborate analyses of this sort (Atkinson, 1974a), including more of a range of differences in ability and of motivation and variation in the number and strength of competing interests which also, you recall, will influence percentage of time spent in the critical activity, we have found that the linear correlations between hypothetical so-called ability test scores and cumulative achievement consistently falls between .30 and .50 just where the correlation between college aptitude test and grade point average in college traditionally falls in the studies conducted by the major test agencies (Angoff, 1971; The American College Testing Program, 1973).

In brief, we can deduce one of the best documented facts in mental testing by simply applying what is known about the effects of motivation on behavior problems.

Consider two other current phenomena that are being widely discussed: the decline in test scores of college aspirants in recent years at the very same time that grade averages in college have been on the rise.

The implication of the report of the prestigious commission to explain the decline in test scores (College Board, 1977) produced a number of vague and mouselike conclusions from (presumably) a mountain of considerations. Somehow, and unfortunately because it defines the gap between behavioral science and agencies that should be in contact with it, the report of the commission did not even mention the possibility of decrements in intellective performance attributable to overmotivation. Rather, the general impressions followed conventional wisdom. Students today are probably dumber and less motivated than before; the schools are inept. These are widespread opinions about younger people, in every generation, I might add, and about schools, which seem always to be misusing the hard-earned tax dollars.

The conclusion about the rise of grade averages is called "grade inflation" and attributed to the lack of spine in professors. This is another popular opinion, even among professors—about other professors.

Look back at Table 1 with another hypothesis in mind, one that takes into consideration some factors that might, conceivably, have intensified motivation. The 1960s was marked by liberation of Blacks, of Women, a war that young men could avoid only by getting into college, an increase in the percentage of high school graduates going to college that peaked in 1969 at more than twice what it was in the 1950s, persistent reminders on television that the only way to make it in life was to hang in there and get a college education. Then came the big recession from 1972 to 1975, and the torrent of information about limited future job prospects in this, that, and the other field, you name it.

Is it unreasonable to presume that students taking the college aptitude tests might have confronted that critical situation with a much more explicit sense of the importance of getting a high score, of the instrumental relationship between that test, their subsequent college performance, and achieving their own personal goals in life? And what about a greater awareness of intensified competition for college admission because the pool of aspirants for challenging careers had now been expanded to include all racial groups and Women? I don't find the hypothesis of stronger motivation unreasonable. It was my impression when the movement to pass/fail was initiated and it has been sustained by my day-to-day contact with anguished, grade-conscious college students ever since. Today's students are grinds.

We know from Raynor's program of research that future orientation (that is, perceiving success in the immediate task as instrumental to attainment of future goals) *intensifies* motivation (Raynor, 1969, 1974, 1978).

So look at Table 1 with a single hypothesis in mind—that the difference between yesterday's students and today's is the difference between B3 (a decade ago) and B4 (today). Note the decline in test score from 90 to 72. *Note also the paradoxical increase in cumulative achievement* (that would be expressed in the grade point average in college) from 144 to 270 in our illustration.

I conclude from this application of knowledge about motivation to questions of test performance and educational achievement that today's students *may* be both more able and more highly motivated than any group of students in history. No one can confidently contradict me for the simple reason that ability and motivation have been utterly confounded in test performance and achievement since Binet invented the intelligence test, and today we know more about how motivation influences behavior.

The direction of work in progress having to do with this analysis considers the possibility that the level of performance and time spent in academic work should produce a growth in ability. That would gradually produce a positive correlation between strength of motivation and level of ability (Atkinson, Lens, & O'Malley, 1976). A myriad of new questions concerning the interdependence of ability and motivation arise once we get our semantics straightened out and stop referring to differences in level of intellective test performance by persons who differ in strength of motivation as differences in ability.

If we are to get at those socially significant questions, we must close the gap between the frontier of motivational psychology and educational psychology. Whose responsibility is that?

The first feasible step is to encourage systematic experimental study of how personal and situational determinants of motivation influence the level of intellective performance on the tests traditionally conceived as measures of ability. The test agencies should have been doing this job since their inception.

REFERENCES

American College Testing Program. Assessing students on the way to college. Technical Report for the ACT Assessment Program. Iowa City, Iowa: Author, 1973.

Angoff, W.H. (Ed.). The college board admissions testing program. A technical report on research and development activities relating to the Scholastic Aptitude Test and Achievement Tests. N.Y.: College Entrance Board, 1971.

Atkinson, J.W. Motivational determinants of intellective performance and cumulative achievement. In J.W. Atkinson & J.O. Raynor (Eds.), *Motivation and achievement*. Washington, D.C.: V.H. Winston & Sons, 1974 (a).

Atkinson, J.W. Strength of motivation and efficiency of performance. In J.W. Atkinson & J.O. Raynor (Eds.), *Motivation and achievement*. Washington, D.C.: V.H. Winston & Sons, 1974 (b).

Atkinson, J.W., & Birch, D. *The dynamics of action*. N.Y.: Wiley, 1970.

Atkinson, J.W., & Birch, D. The dynamics of achievement-oriented activity. In J.W. Atkinson & J.O. Raynor (Eds.), *Motivation and achievement*. Washington, D.C.: V.H. Winston & Sons, 1974.

Atkinson, J.W., & Birch, D. *An introduction to motivation: Revised edition*. N.Y.: Van Nostrand, 1978.

Atkinson, J.W., Bongort, K., & Price, L.H. Explorations using computer simulation to comprehend "thematic apperceptive" measurement of motivation. *Motivation and Emotion*, 1977, *1*, 1-27.

Atkinson, J.W., & Feather, N.T. (Eds.). *A theory of achievement motivation*. N.Y.: Wiley, 1966.

Atkinson, J.W., Lens, W., & O'Malley, P.M. Motivation and ability: Interactive psychological determinants of intellective performance, educational achievement and each other. In W.H. Sewell, R.M. Hauser, & D.L. Featherman (Eds.), *Schooling and achievement in American society.* N.Y.: Academic Press, 1976.

Atkinson, J.W., & Raynor, J.O. *Motivation and achievement.* Washington, D.C.: V.H. Winston (Halsted Press/Wiley), 1974.

Atkinson, J.W., & Raynor, J.O. *Personality, motivation, and achievement.* Washington, D.C.: Hemisphere (Halsted Press/Wiley), 1978.

Birch, H.G. The role of motivational factors in insightful problem-solving. *Journal of Comparative Psychology.* 1945, *38*, 317-395.

Broadhurst, P.L. The interaction of task difficulty and motivation: The Yerkes-Dodson Law revived. *Acta Psychologica*, 1959, *15*, 321-338.

College Board. *On further examination: Report of the advisory panel on the scholastic aptitude test score decline.* N.Y.: College Entrance Examination Board, 1977, p. 75.

deVilliers, P.A., & Herrnstein, R.J. Toward a law of response strength. *Psychological Bulletin*, 1976, *83*, 1131-1153.

Entwisle, D.R. To dispel fantasies about fantasy-based measures of achievement motivation. *Psychological Bulletin*, 1972, *77*, 377-391.

Eysenck, H.J. Personality and experimental psychology. *British Psychological Society Bulletin*, 1966, *19*(62), 1-28.

Hebb, D.O. *Textbook of psychology* (3rd ed.). Philadelphia: Saunders, 1972.

Horner, M.S. The measurement and behavioral implications of fear of success in women. In J.W. Atkinson, & J.O. Raynor (Eds.), *Motivation and achievement.* Washington, D.C.: V.H. Winston (Halsted Press/Wiley), 1974.

Raynor, J.W. Future orientation and motivation of immediate activity: An elaboration of the theory of achievement motivation. *Psychological Review*, 1969, *76*, 606-610.

Raynor, J.W. Future orientation and motivation of immediate activity: An elaboration of the theory of achievement motivation. In J.W. Atkinson & J.O. Raynor (Eds.), *Motivation and achievement.* Washington, D.C.: V.H. Winston & Sons, 1974.

Raynor, J.W. Future orientation and motivation of immediate activity: An elaboration of the theory of achievement motivation. In J.W. Atkinson & J.O. Raynor (Eds.), *Personality, motivation, and achievement.* Washington, D.C.: Hemisphere (Halsted Press/Wiley), 1978.

Revelle, W., Amaral, P., & Turiff, S. Introversion/extroversion, time stress, and caffeine: The effect on verbal performance. *Science*, 1976, *192*, 149-150.

Sawusch, J.R. Computer simulation of the influence of ability and motivation on test performance and cumulative achievement. In J.W. Atkinson & J.O. Raynor (Eds.), *Motivation and achievement.* Washington, D.C.: V.H. Winston (Halsted Press/Wiley), 1974.

Thorndike, R.L. *The concepts of over- and under-achievement.* N.Y.: Columbia University Bureau of Publications, Teachers College, 1963.

Yerkes, R.M., & Dodson, J.D. The relation of strength of stimulus to rapidity of habit formation. *Journal of Comparative and Neurological Psychology,* 1908, *18,* 459-482.

Chapter 3

The Origins of Competence and Achievement Motivation in Personal Causation

RICHARD deCHARMS
Washington University

Motivation is Only Part of the Egg

I am going to take the position of the loyal opposition. I maintain that the state of the art suggests to me that it is time that we stop studying achievement motivation. Like all other psychological variables, achievement motivation is a part of a whole—the whole is a person. It is time that we stopped studying variables and started studying *persons in action*. Let me explain with a crude analogy about the state of the art of psychology.

In the past hundred years psychology has been and still is under the influence of logical empiricism, reductionism, operationism and several other outmoded philosophies that were tried in the physical sciences and found wanting. The basic strategy that developed was to reduce the subject matter to its elemental component parts, study each part intensively, and try to see how it was related to the nearest other parts. When we intercept an enemy airplane this strategy works tolerably well. First we get an adequate description of the whole airplane so that after we take it apart we can get it back together again. Then we take it apart and study the *motive* parts (the engine), the *perceptual* parts (the latest radar gear), the *communication* parts (the radio transmitter), and the *learning* parts (the programmed computer components). What if we applied this dissection technique to the pilot? The first thing that would happen is that we would kill him and then no matter how hard we tried "all the kings horses and all the kings men couldn't put Humpty together again."

My major point is that we have failed to get an adequate description of what is uniquely human in persons before we started taking them apart. An adequate description would make it clear that a person in action *is more than the sum of his component physical or even psychological parts* (Ossorio, 1966). To hold up for study one part like achievement motivation may eventually give a more or less complete description of that part but it can never give more than a partial description of a person in action. Partial descriptions are dangerous because as we become more and more specialized we lose sight of the whole that was supposed to be the subject of study. We no longer study persons, we study motives or in the broadest sense we study behavior. A person in action is more than the sum of his component behaviors.

This is not a plea for humanist psychology or an attempt to avoid the rigors of scientific thinking. What I am suggesting is that we apply scientific thinking to our own subject matter—namely *persons in action*—and stop trying to apply an inadequate physicalist analogy to human action. Persons are more than physical objects; persons are more than machines. The "more" is a Gestalt-like quality—the best of human action when a person "gets it all together."

Despite their great contribution to the history of psychology, the classical Gestalt psychologists failed to give us an alternative paradigm to the reductionist approach and it turned out to be practically impossible to deal with the complexity of the whole person without some guidelines for checking whether or how our discussion of more comprehensible parts was related to the totality. Peter Ossorio (1966, 1973) has recently sketched the parameters of an adequate description of human intentional action which allows us to work with comprehensible elements without at the same time divorcing them from the totality of which they are a part. In order to move toward this more complete paradigm we need to analyze first what an adequate description is and then to apply it to a person in action.

What is an Adequate Description?

Suppose that a hit-and-run accident has occurred and there

is only one witness. Asked to describe the offending car he says that it had four wheels, bumpers, a windshield, and a roof. Clearly this is an inadequate description because it cannot help us to find the particular car. It is known to be inadequate immediately by inspection, and the criterion for adequacy is the practical problem of finding the car. If the witness had said that the car had very large wheels, wooden bumpers, a broken windshield, and a red and purple roof we know immediately that the description is more adequate to the practical problem. Note the importance of the practical problem in determining the adequacy of the description.

Or suppose an intricately contrived murder has occurred. The witness says that the suspect had a motive, he knew a lot, he knew how to do things, he tried, and he succeeded. Again this is an inadequate description. A more adequate description from the practical point of view of finding the murderer would be: He hated the dead man, he knew a lot about guns, he was a skilled marksman, he shot with one bullet from a great distance. Again the relevance to the practical problem distinguishes the two descriptions.

But we must note that the structure of the two descriptions is similar. But the first desciption simply names relevant formal elements. These four or five elements are the architecture of a possible adequate description of an act.

Let me now be more formal and present what might be called the architecture of an adequate description of *intentional* action on the part of a human being. At the very least the description needs 5 parameters:

(1) the *identity* of the actor.
(2) what the person *wants* or has reason to do,
(3) what the person *knows* that is *relevant* to the act,
(4) what the person *knows how* to do,
(5) what the person is *trying* to do.

Figure 1 presents these five parameters schematically. The circle itself represents the identity, i.e. the person. Within the circle are *want, know, know how,* and *try*.

Please note that this is not a new theory or even a set of theoretical propositions. Rather it is an attempt to remind ourselves of the elements that we take for granted are important

Figure 1. Components of a competent act.

in any human action. In a sense you can think of it as a game, the diagram as a game board. The basic rules are that we must account for all of the parameters and that the rules must apply to all persons including the investigating psychologist. These rules are like rules in the game of chess; any individual game must conform to the rules (the architecture) but within the rules there are an infinite number of different games. The rules are the formal element, the individual games or actions are what is interesting about the game. They are the problematic, the empirical element. They form the basis of empirical investigation. In all cases the individual elements constitute a unitary whole, a gestalt. All are interdependent.

The Psychologist Doing Research

Now let us apply our game board and its parameters to the psychologist engaged in research. (See Figure 2).
(1) S/he wants to know something.
(2) S/he knows previous research, theory, etc. and the stress is on extending knowledge.
(3) What s/he knows how to do takes the form of methodology, laboratory skills, statistics, computer know-how, etc.
(4) What s/he tries to do is;
 (a) formulate hypothesis,
 (b) operationalize concepts,
 (c) run an experiment, and
 (d) verify the hypothesis.

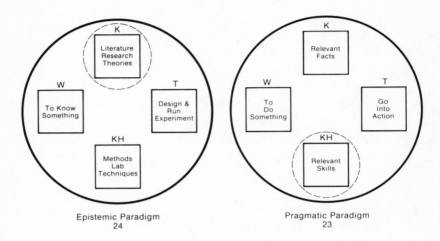

Figure 2. Two paradigms.

This stream of events I shall refer to as the *epistemic paradigm* (Ossorio, 1966, calls it the semantic paradigm) where the goal is to "know that" something is the case (Ryle, 1949), with priority on "knowing that." The basic *value is* knowledge for knowledge's sake. The basic question asked is "How do I know *what I know?*"—the basic question of epistemology.

Psychologist as a Person in Action

The game board for a psychologist in action and interaction with others perhaps trying to change something looks a bit different. (See Figure 2).

(1) The psychologist wants to do something (like enhance motivation in children (deCharms, 1976).
(2) The emphasis now shifts to the "know how" square.
(3) The "know that" square is still important but secondary.
(4) "Try" here will involve planning and engaging in a *skilled performance* in interaction with the other persons involved.
(5) Vindication of the attempt will come in terms of whether what s/he did "made a difference," whether s/he succeeded

in doing what s/he wanted to do and it had the desired effect.

In the *pragmatic paradigm*, as this stream of events is called (Ossorio, 1966), the basic value is action. The basic question is "How do I know *what to do?*" The emphasis is on *knowing how* rather than *knowing that*.

We must not make the mistake of assuming that the epistemic paradigm and the pragmatic paradigm are entirely separate and mutually exclusive. They are not. The diagrams are meant to show that:

> You must KNOW in order to DO
> and
> You must DO in order to KNOW

Personal Causation and the Pragmatic Paradigm

As I look back on my last ten years of research I can see two things. First, my own behavior moved more and more away from the epistemic paradigm and toward the pragmatic paradigm—away from laboratory manipulation of variables toward studying (and even engaging in) skilled performances of persons in interaction in the classroom. Second, I concentrated on the want parameter in students early on and later looked at the know how parameter in the teachers. These are the two points that I want to illustrate.

Personal Causation and the Want Parameter

We started out in the small groups laboratory under the epistemic paradigm to study the motivation variable that we called personal causation. At first (deCharms, 1968) we defined personal causation in terms of perceived locus of causality following Heider (1958) and we distinguished between people who perceived themselves to be pushed around (Pawns) and people who perceived themselves as originating their own behavior (Origins). As we moved into the full pragmatic paradigm it became clear that this model was too restricted. We have since broadened the concept of personal causation and the origin-pawn conception to include all of the parameters and speak now of the *experience* of being an origin or a pawn rather than the *perception* of it.

Briefly it seemed to us that a person who (1) wants something that has meaning to him or her, (2) knows all s/he needs to know about the relevant situation, (3) knows how to do what s/he wants to do, and (4) tries or engages in a skilled performance, will often be successful. S/he will feel personal causation: s/he will feel that s/he originated his or her own action: s/he is an origin. His or her game board makes sense and, in ordinary language, "s/he's got it all together."

But what if the diagram does not comprise a complete gestalt? What about degenerate cases? With our emphasis on personal causation we concentrated first on a potential difference in the WANT parameter and hypothesized that when the "want" was imposed from outside arbitrarily, the "try" would be different from a situation where the "want" was intrinsic. We were able to demonstrate this in the laboratory (see deCharms, Dougherty & Wurtz in deCharms, 1968, Ch. 10) but I won't give the details here.

Armed with epistemic knowledge from the laboratory, we took a bold step and asked ourselves if we could, with the help of teachers in the inner city, help to enhance motivation of elementary school children. The idea was to help the teachers to experience the feelings of being both a pawn and an origin and show them the effects on their own behavior and then to cooperate with them in designing ways to help their students to be more like origins. The pragmatic problem was to improve academic achievement through enhancing motivation in the students.

Since this longitudinal study is completely reported in my book *Enhancing Motivation* (deCharms, 1976) I will describe it very briefly, and concentrate on some follow-up data that we now have approximately ten years after the original study.

Very briefly, the project started with many measures of children at the end of their fifth grade in an inner-city school district. Having randomly assigned one half (16) of the teachers who would teach these children in 6th grade to an experimental group, we conducted a week long residential motivation workshop for the experimental teachers (see deCharms, 1976, Ch. 4 for details) and then worked with them throughout the year in monthly workshops and in their classes to develop origin training techniques for their students (see deCharms, 1976, Ch. 5). It was here that we began to see the power of, and the indespensability of, engaging in skilled performances ourselves with the teachers rather than simply trying to manipulate variables.

We followed the children (both experimental and control) and trained another set of experimental teachers for the year when the children were in 7th grade. The design was basically a longitudinal treatment-control design with pre-measures (5th grade) and post-training measures in both 6th and 7th grades, and a follow-up in 8th and 11th grades. I will show here results on two measures: our measure of motivation (the origin-pawn measure) and the measure of academic achievement.

The origin-pawn measure that was administered to the children in their classrooms was a form of thought sampling where the children were asked to write six short stories to short verbal cues. Their stories were content analyzed according to the scoring manual developed by Plimpton (see deCharms, 1976, Appendix A). The stories were coded for presence or absence of six categories, namely: (a) internal goal setting, (b) internal determination of instrumental activity, (c) reality perception, (d) personal responsibility, (e) self-confidence, and (f) internal control. The academic achievement measure was the Iowa Test of Basic Skills (Lindquist & Hieronymous, 1955).

Figure 3 shows the data on the origin-pawn score for subjects before training in the 5th grade, after training in 6th and 7th, after a follow-up year of no training in 8th grade and again after 3 years of no training in the 11th grade. All of the differences between trained and untrained groups are highly significant ($p < .01$, in every case) except in the 5th grade before training. These data make it clear that the score on the origin-pawn measure increased as a function of training and that the change had long-term effects.

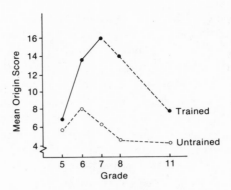

Figure 3. Mean origin score, trained and untrained students.

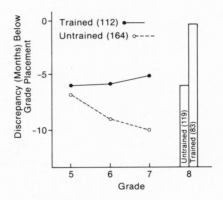

Figure 4. Mean discrepancy from grade placement (ITBS).

Iowa Test of Basic skills scores were available on the children through the eighth grade only. Figure 4 shows longitudinal data for trained and untrained children from 5th through 7th grade, and on a smaller group of children in the 8th grade (bar graph). These means are in terms of months decrement from normal grade placement, since by 5th grade these children are typically more than one-half year behind national norms. Again the effects of the origin training are evident. While the control children continue to lose ground, the trend is reversed in the experimental children. Again the means (which are adjusted for minor differences in I.Q.) are all highly statistically significant ($p < .01$) after training. The results indicate that origin training affects the academic achievement of students in a positive way and that even a year after the training the effects are still measurable.

In our most recent follow-up study we were able to show that origin training in 6th and 7th grades even affected the probability of these children graduating from high school 5 years after training. In our files we had data on 755 children some of whom had received 2 years of training (6th and 7th grades), some who had training one year (either 6th or 7th) and some who had no training. After the year that these children should have graduated from high school we checked yearbooks for the high school for which these children were destined and for four surrounding high schools, since we knew that family mobility was high. Only about 35% of the children who attended 6th grade actually went on to graduate according to school records, and the percentage was even lower for boys (about 26%). When we computed the percentage of graduates that we found in the five school districts of these in our original sample who had been trained 2, 1 or 0 years we found a significant relationship for boys but not for girls. Figure 5

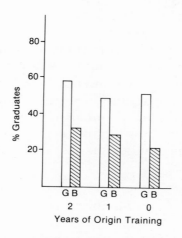

Figure 5. Years of origin training.

shows the results. Significantly more trained boys went on to graduate than did untrained boys.

Frankly, we spent months trying to convince ourselves that these results were not attributable to some unknown artifact because clearly there must have been children who left the city entirely and graduated elsewhere and we have no record of those who didn't graduate. For instance, we did find many children who had moved to a middle class suburb and graduated there. Because there was no way to find all of the children, we decided that the above data would be more convincing if we could devise some direct relationship between the effects of origin training and graduating from high school. Fortunately, we had in our data file pre- and post-training origin scores for a small sample of students who could be classified as having graduated from high school or not (N = 124). Using change in origin score as an indicator of amount gained from the training we looked to see if those who gained most from the training were those on average who graduated from high school. Consistent with the data presented above, there was no difference between mean change on the origin score for the graduating compared to non-graduating girls, *but there was a large and significant difference for the boys (p< .003)*. If pre-post change in origin score is taken as the indicator, apparently those boys who gained more from origin training had a higher probability of graduating from high school 5 years after the training.

There is no time in this paper to discuss our recent work with the "know that" parameter and the "know how" parameter. Suffice it to say that Cohen, Emrich and deCharms (1976/1977) have demonstrated that training teachers to "know that" certain actions in the classroom are effective is no guarantee that they will "know how" to use them in the classroom (see also Koenigs, Feidler and deCharms, 1977).

Summary

We have argued for a broadened view of our task that aimed at the study of person in relation rather than variables. The emphasis is on attempting to use an adequate description of *persons in action* as the basis for research in settings in the real world such as the school. Our data show that when we set out to "make a difference" (the pragmatic paradigm) in the feelings of personal causation of inner-city children in late elementary school, we could show long-term effects on origin scores, academic achievement, and even the probability of graduating from high school.

Finally, let me summarize by suggesting two maxims for research. First, if you start with the pragmatic paradigm, you learn to know how as well as to know that, but if you start with the epistemic paradigm you may learn to know that but never learn how to do something practical.

Second, to understand a complex subject like persons in action it is useful to consider all the parameters that enter into our everyday use of the idea. The complexity precludes simple manipulation of variables. If you think you are beginning to understand something, try to change it by skilled performances with real people in a real situation.

REFERENCES

Cohen, M.W., Emrich, A.M., & deCharms, R. Training teachers to enhance personal causation in students. *Interchange*, 1976/77, 7, 34-39.
deCharms, R. *Personal causation.* New York: Academic Press, 1968.
deCharms, R. *Enhancing motivation: Change in the classroom.* New York: Irvington Publishers. 1976. (Distributed by Halstead).
Heider, F., *The psychology of interpersonal relations.* New York: Wiley, 1958.
Koenigs, S.S., Fiedler, M.L., & deCharms, R. Teachers' beliefs, classroom interaction and personal causation. *J. Appl. Soc. Psychol.* 1977, 7, 95-114.
Lindquist, E.F. & Hieronymous, A.N. Iowa test of basic skills. Boston: Houghton-Mifflin, 1955.
Ossorio, P.G. *Persons.* Boulder, Colorado: Linguistic Research Institute. 1966.
Ossorio, P.G. Never smile at a crocodile. *J. Theory Soc. Behav.* 1973, 3, 121-40.
Ryle, G. *The concept of mind.* New York: Barnes & Noble, 1949.

Chapter 4

Motivation, Evaluation, and Educational Testing Policy

KENNEDY T. HILL*
*Institute for Child Behavior and Development
University of Illinois, Urbana-Champaign*

Prologue

Mark is a 4th grade student in the Jefferson Elementary School in a middle-size American community. His school is integrated and due to changing neighborhood residential patterns has a good mix of children from various socioeconomic backgrounds. The Jefferson Elementary School student population in many ways mirrors the larger community.

Mark's teacher is about to administer a standardized achievement test. It is September and school has been under way for several weeks. In previous years Mark has done almost average school work but has done much worse on similar achievement tests, scoring near the 15th percentile. Based on his school work, his teachers had thought he would score nearer the 50th percentile. But, Mark gets nervous during such testing and has some test-taking problems which keep him from doing his best, from performing optimally. The reasons begin to become clear as we watch Mark take this fall's test.

Mark's teacher explains that they will be taking tests over the next three mornings to find out how much each student knows. She emphasizes that it is important for him to do his

---

*Preparation of this paper and research reported here were supported in part by a research grant from the National Institute of Education HEW-NIE/G-76-0086 to the author and by United States Public Health Service training grant HD-00244 from the National Institute of Child Health and Human Development to the Developmental Psychology Program, Department of Psychology, University of Illinois, Urbana-Champaign. The author wishes to express appreciation to Leslie J. Fyans, Jr., Robert L. Linn, and Martin L. Maehr for their careful reading of an earlier draft of this manuscript and thoughtful comments and suggestions.

best. As she reads the instructions she is somewhat formal, not as warm and friendly as usual. The instructions are fairly long and involve a number of points, and Mark has some difficulty concentrating and understanding them. Several children do raise their hands and ask questions, which helps.

The teacher explains how to use a new kind of answer sheet (a computerized scoring sheet) which Mark has never seen before. In earlier years he could answer right in the booklet during such tests. She also mentions that there will be a time limit for each test. This makes Mark more nervous because he doesn't always finish such tests unless he rushes through faster than he would like. The teacher passes out the first test, asks if there are any more questions, and says "Start!"

Mark does fairly well on the first couple of problems of a vocabulary subtest but then runs into increasingly harder problems. Mark does not know it, but 5th and 6th graders also take the very same test his 4th grade is taking, making it much harder for his group (the testing company instructions inform the teacher of this but not the students). As Mark moves through the test he finds he has to read new instructions for different kinds of word problems, each set of which has its own question and answer format. He soon notices some of the children around him seem to be going faster than he is. He begins to work through the test much more quickly. He is worried about the time limits, and he doesn't take as much time to read each new set of instructions and figure out the answer sheet as he would like. He finds it more and more difficult to concentrate. He is not aware of it, but he is following a "fast but inaccurate" test-taking strategy which often results in poor performance. He finishes the test before the teacher says "Stop!" but does not get many of the problems correct.

The testing goes on for two more days. Mark takes seven different subtests and encounters much the same test-taking problems as before. The problems are difficult, he finds himself rushing through to finish, and he becomes increasingly nervous and has more and more trouble concentrating on the problems. By the third day he feels like giving up and almost quits in the middle of the test. But he knows the teacher wants him to finish and so he keeps working at the test.

Mark is just one hypothetical child taking one representative achievement test in a 4th grade classroom in Anywhere, U.S.A. But there are thousands of high-anxious, poor performing students like Mark at *each* grade level in this country. They come from many sociocultural and socioeconomic backgrounds, and their test-taking problems become worse as they move through the public school years. Aptitude, achievement, and other educational tests given under highly evaluative conditions will increasingly underestimate these students' academic skills, competencies, and knowledge as they become older. For many students who do poorly in school, negative motivation is a significant problem both in the classroom and in the testing situation. This chapter reviews a long-term program of research directed at eliminating such motivational test bias through the development of new testing practices and teaching activities that foster test-taking skills and positive motivation.

General Background

Overview

The role of negative motivational factors has generally been given little attention in this country's testing movement. Motivational factors, however, may be a very important source of test bias (Dweck, 1977; Fyans, 1979a; Hill, 1972; Maehr, 1977). The long-term program of research reviewed here is directed at identifying and eliminating specific motivational causes of test bias for children of all ethnic and socioeconomic backgrounds. Earlier research revealed increasingly strong interfering effects of negative motivational variables such as test anxiety on school and achievement test performance across the public school years (see Hill & Sarason, 1966; Ruebush, 1963). More recent experimental and field research has suggested ways of preparing children for and giving achievement tests which minimize negative motivation and optimize test performance (Hill, 1972, 1977). Ongoing research is developing an integrated elementary school classroom curriculum and school testing program including: (1) the teaching of test-taking skills as part of the regular classroom program; and (2) the phasing in of evaluative aspects of educational testing across the elementary school years as children of all backgrounds gain the necessary skills, strategies, and positive motivation needed to do their best in test situations. The goal of such an educational program is long lasting student gains in both positive motivation and test performance (Hill, 1979a).

The Role of Motivation

There is increasing evidence that the performance of children on achievement and other school tests is strongly influenced by motivational and other test-taking factors (Fyans, 1979a; Hill, 1977; Maehr, 1979). It has been found, for example, that by the end of the elementary school grades high test-anxious children are, compared to low test-anxious children, over two years behind in reading and arithmetic basic skills test performance (see Hill & Sarason, 1966). Ongoing research indicates that the interfering effects of negative motivation on (achievement) test performance are particularly strong for low income/minority and other highly test-anxious children. These students often become anxious and do not perform well under test pressure and have difficulty showing what they have learned in standard achievement test situations (Fyans, 1979; Hill, 1979b; Maehr, 1979). The student's attributions of his successes and failures and the presence or absence of continuing motivation also strongly affect achievement test performance (see Maehr, 1977).

Motivational test bias should operate in any highly evaluative educational test situation (Hill, 1977). Data to be reviewed shortly suggest that the way many standardized tests are composed and given do, in fact, elicit or at least allow strong debilitating motivational dynamics such as test anxiety to operate (Hill, 1979a; Zigler, Abelson, & Seitz, 1973). Such motivational test bias will cause many children to perform well below their optimal level of functioning in the test situation, thereby invalidating their results if one is interested in what the children have learned, as opposed to whether they can demonstrate that learning under heavy testing pressure.

The Importance of Testing

Recently, there has been an ever increasing interest in the nature and assessment of school achievement. Notably a national debate (see for example, Wirtz et al., 1977) has arisen over a prevalent decline in achievement test scores and the reasons for such. Parallel with this, greater stress is being placed on the necessity to assess patterns of school achievement in better ways. Throughout the country there are expressions of concern that school children can't read, and there is a back-to-basics clamor. Requirements that children demonstrate "minimum competencies" for graduation and/or promotion from one grade to the next are closely related to these concerns. Public concerns about education have moved the assessment of achievement into center stage. And a student's performance on standardized tests has become an increasingly important factor in his or her educational progress and future.

Concurrent with demands which have placed increased reliance on measurement and assessment, have been vigorous criticisms of present testing practices (see for example, Houts, 1977; National School Boards Association, 1977; Quinto & McKenna, 1977). Although public controversy over testing is not new (Cronbach, 1975), the criticism in recent years has become ever harsher. Moreover, the debates in the past decade or so have not been limited to academics and journalists. State and national legislatures, the general public, and the courts have all been involved.

The emotion laden atmosphere in which many discussions of assessment issues take place, has too often led to the creation of more heat than light. Several conclusions seem fairly evident, however. Current demands go beyond current measurement capabilities. The best present tests are imperfect assessment devices. Far too little is known about the variables that influence performance on achievement tests. There is a great need for research that will contribute to a better understanding of factors (such as motivation) that influence achievement patterns and to the development of better ways of assessing them.

The recent, renewed emphasis on achievement testing and

the introduction of new evaluation programs such as minimal competency testing should make motivational test bias an *even stronger* factor in the assessment of student's achievement (Hastings, 1979; Hill, 1979a). As the consequences of doing poorly or failing a test become more serious (not being promoted, not graduating, etc.) and more public, test pressure and negative motivational dynamics should become stronger. Such test bias should be strongest for students who do not perform well on standardized tests as now given and who suffer the consequences of test failure the most: the anxious, often low income/minority student. It becomes increasingly important to identify and eliminate motivational causes of test bias through the development of new teaching and testing programs.

A Model for School-University Research

Our recent and current studies of motivation and evaluation involve a new kind of research approach which we refer to as *collaborative school-university research*. The research focuses on the development of specific new educational evaluation practices in the areas of (achievement) testing and report cards which optimize children's motivation, learning, and performance. The general approach, however, is a model for better integration of psychological/educational research and educational practice (see Hill, 1977).

Recent and ongoing research is a collaborative effort among school staff and university personnel in several ways. The areas to be studied are identified by both school and our project staff as being worthy of extensive study. Sometimes school staff take the initiative in developing the research as in the case of several report card projects and development of new school testing programs, and sometimes we take the initiative, as in the case of several projects developing optimal ways of giving achievement tests. Projects tend to evolve over time with reciprocal involvement at *all* phases of the research, from identifying the problems to be studied, to developing the methods and procedures, to actually carrying out and reporting the research together. Often we are helping school staff develop and improve their evaluation procedures and the new approaches become a part of the ongoing school program. Thus classroom teachers give achievement tests under optimizing testing procedures as a part of their student/program evaluation and implement new report card procedures as a part of the school reporting-to-parents system.

As would be expected from the strong mutual interest of school and university staff in such research activities, most projects involve elements of what have traditionally been identified as basic and applied research, which in our case have been woven together into collaborative school-university research.

The research is basic in a number of ways. Projects draw heavily upon both theory and findings in psychological and educational research and are directed at clarifying processes mediating effects of evaluation. Collaborative projects also draw heavily upon the procedures, measuring instruments, design, and analysis techniques from basic research literatures. The research in general draws upon experimental, systematic assessment, and field research paradigms and procedures as appropriate.

The collaborative school-university projects are also applied in nature. Most importantly, they focus on educational issues and problems of direct relevance to the school staff—for example, how to give school or standardized achievement tests so that positive motivation and test-taking strategies facilitate optimal test performance. As noted, portions of the projects are specifically directed at helping schools develop and document new school evaluative practices which help optimize students' learning and motivation. These collaborative endeavors serve as one model for how public school staff and university researchers can work together on projects which at the same time advance our basic knowledge while developing specific new educational practices which help solve basic problems in American education—in this case, motivational test bias.

Research Goals

As alluded to earlier, the program of research has four

general goals. First, the relations between motivation and achievement performance are systematically studied: (1) for children of various sociocultural backgrounds and ages, including lower- and middle-class white, black, and hispanic children, and (2) across various kinds of educational testing situations, including full-scale achievement and aptitude tests, reading and arithmetic skill subtests, and occasionally other forms of problem-solving situations related to achievement (test) behaviors.

A second thrust of the research is a series of studies directed at developing new procedures for giving diagnostic and achievement tests that reduce or eliminate motivational aspects of test bias and thereby optimize children's performance. Emphasis in these studies is placed on removing time pressure, systematic failure, and unnecessary cognitive demands of the test independent of its content. These parameters can all be varied in completely standardized ways in administering achievement and other tests, so that procedures found to minimize motivational bias and optimize performance can be incorporated into testing programs without undue difficulty.

A third goal of the project is to help children learn to cope with and *function effectively in highly evaluative formal testing situations* through the development of additional new kinds of instructional procedures and special teaching programs. One goal here is to teach children test-taking skills, for example, how to balance demands in many testing situations for both speed and accuracy. Performance of anxious, failure-prone children may be further optimized by modifying and making more realistic children's expectancies for success-failure and for later adult evaluation. Modifying children's perceptions and performance strategies is thus a complex process. Some of these test-taking skills can be modified during the (achievement) test situation while others need to be taught in the classroom over time.

A final thrust of the program of research is to develop an integrated set of teaching activities and classroom and formal testing experiences which phase in highly evaluative aspects of testing as students have acquired the test-taking skills and positive motivation needed to perform their best on tests.

These latter qualities will help children cope effectively in whatever testing or other formal evaluative situations they encounter in later years.

Theoretical Background

The research reviewed here is most related to the research literature on achievement-related motives in children (Smith, 1969; Spielberger, 1972). Research in this area was largely stimulated by the work of Sarason and Atkinson. Sarason and his colleagues (Sarason et al., 1960) developed considerable theory and data relating test anxiety in children to achievement and other test performance. Emphasis was placed on parent-child interactions in the preschool and elementary school years and evaluative aspects of the school situation. High test-anxious children develop a strong fear of failure, become very cautious in certain situations, and become overly dependent on adult approval (see also Crandall, 1967; Zigler & Harter, 1969).

Atkinson (1964) developed a theory of anxiety and achievement behaviors in which test anxiety is equated with the motive to avoid failure and achievement motivation is equated with the motive to approach success. Research has indicated that achievement-oriented individuals with a stronger motive to approach success than to avoid failure approach tasks of intermediate risk, persist in evaluative situations, and do better on achievement tests and other problem-solving situations (Atkinson & Feather, 1966; Atkinson & Raynor, 1974). High anxious children and adults avoid achievement situations, do poorly in test situations, and quit in the face of failure (Dweck, 1975).

The theory (Hill, 1972, 1977) which guides the program of research described here combines elements of the Sarason and Atkinson formulations with additional assumptions and emphases. It is assumed that low-anxious children *have stronger motives to approach success* and *obtain approval than to avoid failure and avoid disapproval*. These children should do quite well in evaluative test situations. It is assumed that both approach and avoidance motives increase at higher levels of anxiety but with greater increases for avoidance motives. Thus high-anxious children *have stronger motives to avoid failure and disapproval and weaker* (but still fairly strong) *motives to approach success and approval.* High anxious children should do poorly in testing situations in which failure is likely or quite possible, but may do quite well in achievement situations in which the possibility of failure and adult disapproval are minimized (see Dusek & Hill, 1970). In general, anxious children should rely more on adult evaluation, and less anxious children should be more willing to evaluate the outcome of their achievement activities on their own (see Hill, 1976; Sarason et al., ,1960; Zigler & Harter, 1969). Both success/failure and approval/disapproval are seen as critical to the effects of achievement related motives in children, consistent with Crandall's (1967) formulations. Emphasis is also placed on the effects of social aspects of evaluation from adults, including who the person is, what kinds of feedback are given, and whether comparisons among children are made (Hill, 1976; see also Masters, 1972; Veroff, 1969), and on children's attributions for their successes and failures (Weiner et al., 1971; see also Dweck, 1975; Fyans, 1979a; Hill, 1979b; Maehr, 1979; Weiner, 1977).

Measurement

In the general research literature and the program of research reported here, children's motivation in achievement test situations has been assessed by self-report measures used successfully in a number of studies (see Atkinson & Raynor, 1974; Hill, 1977; Ruebush, 1963). Student's test anxiety is measured by the 30-item Test Anxiety Scale for Children (TASC) and the Defensiveness Scales (primarily the 11-item Lie Scale for children, LSC) developed by Sarason and his colleagues (see Sarason, Hill, & Zimbardo, 1964). Both scales are used because research has shown that the self-reports of highly defensive children, who usually report low test anxiety, are of doubtful validity (see Hill & Sarason, 1966). The effects of defensiveness on self-report of test anxiety is controlled for either by grouping highly defensive children separately in group designs or partialling out defensiveness in correlational analyses following standard procedures in this literature (see Hill, 1972).

The major measure of anxiety is the TASC. Both a nervous (Do you worry a lot while you are taking a test?) and a relaxed (Do you feel relaxed before you take a test?) form of the Test Anxiety Scale are available with predominant use being made of the nervous form (see Feld & Lewis, 1969; Fyans, 1979b; Hill, 1979a). Children's attributing their successes and failures to ability, effort, luck, or task difficulty is assessed by self-report measures developed by Dweck (1977), Weiner (1977), and others (see Fyans, 1979a). The pioneering work of Weiner (1972) suggests that attributions are facilitative when children interpret their successes to ability and failures to lack of effort and most destructive when children attribute their successes to luck or task ease and their failures to lack of ability. Dweck (1975) and others (Fyans, 1979a; Hill, 1972) have found high anxiety to be related to debilitating success-failure attributions and Fyans (1979a) has reported statewide data from Illinois indicating that both test anxiety and success-failure attributions independently are correlated with achievement test performance.

The Program of Research

Let us turn now to see how our collaborative school-university research approach has been applied in the development of new evaluation practices in two areas: achievement testing and report cards. This review of completed and ongoing projects will present in some detail the specific nature and content of the research on changes in evaluative practices. These studies, in turn (along with other research), are a basis for many of the suggestions concerning educational testing policy discussed later in this chapter.

Relation of Anxiety to Achievement

It has been known for some time that there are strong relationships between test anxiety/achievement motivation and performance on achievement and other educational tests

(Atkinson & Raynor, 1974; Ruebush, 1963; Sarason et al., 1960; Spielberger, 1972). In one of the most comprehensive studies, Hill and Sarason (1966) followed some 750 children across the elementary school years in a five-year longitudinal project. Anxiety increased steadily for all groups between first and sixth grade. The negative correlation between anxiety and achievement test performance also increased steadily, approximating −.50 by the end of the elementary school years. By this time, low-anxious, low-defensive children (presumed to experience least anxiety) were over a year ahead of average in basic reading skill development. In contrast, high-anxious, defensive children (presumed to experience the most anxiety) were over a year behind in basic reading skill functioning, the two groups being 2½ years apart on standard reading achievement tests. High-anxious children were also twice as likely to repeat a grade and received much lower grades on school report cards in the basic skill and other achievement areas. There is considerable evidence, then, that anxious, failure-prone children do quite poorly both in school and on standardized tests of achievement and ability.

Much of the early research relating motivation to achievement focused on white and middle-class students (see Ruebush, 1963). There have been studies examining the general relationships between sociocultural variables and school achievement but without a major focus on motivation (e.g., Coleman et al., 1966; Jencks et al., 1973; cf. Shea, 1976). These studies have provided strong evidence that sociocultural factors are of major importance in the determination of school achievement.

Although there has been relatively little research on the role of motivational factors in determining the test performance of minority students, the possibility has been given greater attention recently. Research has indicated, for example, that Spanish-speaking children who do not do well on standard tests are not necessarily lacking in ability (Gezi, 1974; Lazarus, 1977) or in any general motivation to achieve (Ramirez & Price-Williams, 1976). They may, however, be inhibited in demonstrating their competence in standard achievement test situations due to evaluation anxiety which is exacerbated by the ways test are given. This possibility receives some support from the research of Holtzman (Holtzman, Diaz-Guerrero, & Swartz, 1975) and some of our ongoing research to be reviewed shortly. Zigler and his colleagues (e.g., Zigler & Butterfield, 1968) have demonstrated that preschool low-income black and white children also perform poorly on ability tests given under standard conditions due to motivational problems. These children perform much better when testing conditions are modified to reduce debilitating motivational factors such as fear of failure and wariness of strange, aloof examiners.

Hill and Maehr (see Hill, 1979a; Maehr, 1979) have recently studied the effects of test anxiety on achievement test performance for 365 mostly low-income black, white, and hispanic students in grades 4 through 8 in a school system with a bilingual program. In general, highly significant ($p<.001$) differences were found among black, white, and hispanic students in reported test anxiety. At both the elementary (grades 4-6) and junior high school (grades 7-8) levels, test anxiety was found to be highest among hispanic students (in both general education and bilingual programs), intermediate for black students, and lowest for white students. These results were consistent with teacher judgments that minority (both black and hispanic) students often experienced anxiety and frustration in standardized test situations in which they often did not perform well and did not display their knowledge.

Hill and Maehr also found strong negative correlations between test anxiety and performance on a standard arithmetic computation test for black, white, and hispanic students, especially for hispanic children in the bilingual education program whether they were born in Illinois or elsewhere and had moved to Illinois. As can be seen in Table 1, at the late elementary school level, white students in the general education program and hispanic students in the bilingual (all measures given in Spanish) but not the general education programs (all test given in English) show significant negative correlations between test anxiety and test performance. By the junior high school level (grades 7-8), all five groups of black, white, and hispanic students in the general and the bilingual educational programs show *strong* (median correlation of -.45) negative correlations. Brennan and Maehr (1979), Fyans

Table 1

Correlations between test anxiety and arithmetic test performance with defensiveness partialled out for black, white, and hispanic students in the general education and bilingual programs in the ongoing Hill-Maehr project (Total N = 365 students)

Grade Level	General Education Program			Bilingual Program	
	White	Black	Hispanic	Born in Illinois	Other Students
Grades 4-6	-.33** (N=69)	-.14 (N=59)	-.08 (N=26)	-.58** (N=21)	-.36** (N=62)
Grades 7-8	-.36* (N=33)	-.41* (N=33)	-.93***[a] (N=8)	-.51* (N=18)	-.45** (N=36)

*$p<.05$
**$p<.01$
***$p<.001$

[a] -.93 is correct partial correlation; with N of only 8, grade 7-8 results for hispanic students in general education program are highly tentative although in the same direction as other (hispanic) groups

(1979a), and Hill (1979b) have reported additional findings for a sample of some 5000 white, 650 black, and 850 other minority students from low and middle incomes from 4th, 8th, and 11th grade classrooms drawn randomly from schools throughout the State of Illinois. Here correlations between motivation and test performance were in the predicted direction, and while somewhat weaker in the 4th grade, increasingly strong at the 8th and 11th grades. These recent studies clearly suggest increasingly strong interfering effects of negative motivation, in this case, test anxiety, on test performance across the public school years for low-income minority and white students (as well as for middle-class white students).

It is recognized that minority and low-income children may not show their knowledge and learning on standardized tests for many cognitive, language, and cultural-experiential reasons. The fact that these children have a history of doing poorly on such tests and develop test anxiety and other debilitating kinds of negative motivation may place them in double jeopardy. Interfering motivational variables may make the mismatch of cognitive, language, and experiential factors more serious, and in turn, these other problems may worsen the effects of negative motivation. A more thorough understanding of debilitating motivational processes and development of new ways of testing which minimize this form of test bias, then, may not only eliminate motivational test bias but may also lessen effects of some of the other forms of test bias minority/low-income children face.

Developing Optimal Testing Conditions

It is clear that negative motivation is related to low test performance for students of many sociocultural backgrounds on most (highly evaluative) achievement and other educational tests. The critical question immediately arises as to what is the causal underpinning of this correlation? Is it the case that negative motivation (such as test anxiety) modifies students' behavior in ways which lower school and test performance, so that anxiety is the causal factor? Or does a history of failure in school and on previous tests make students anxious, fearful, and experience other negative motivational tendencies in highly evaluative situations such as the achievement test? Here low achievement would be the causal factor in the motivation-achievement test performance correlations. It is also possible, of course, that other variables influence both motivation and achievement test performance.

One of the major reasons to develop optimal testing conditions designed to minimize any effects of negative motivation and thereby produce optimal performance is, in fact, to test the hypothesis that test anxiety (and other forms of negative motivation) cause low performance in evaluative situations. Figure 1 shows four theoretical curves depicting possible results from the research paradigm used in much of our research on optimal testing conditions. In a typical study, students are tested either in a Standard Condition representative of many aptitude/achievement tests or in one (or more) Optimizing Conditions. In the latter conditions, the instructions, procedures, and test conditions are modified to reduce testing pressure and resulting test anxiety, e.g., through the removal of time pressure, the simplifying of instructions, the avoidance of massive failure, etc. Often students are grouped into low, middle, and high test anxious groups and randomly assigned to either the Standard or Optimizing Testing Condition(s).

All four "theoretical curves" in Figure 1 show the same results for the Standard Testing Condition representative of

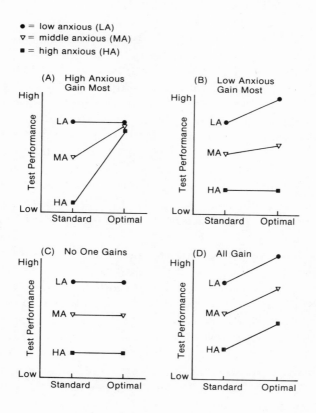

Figure 1. Theoretical curves depicting four possible effects of optimal testing conditions on (achievement) test performance compared to that under standard testing conditions

many achievement tests: low-anxious (LA) students show superior performance, middle-anxious (MA) students do intermediate work, and high-anxious (HA) students perform poorly. This difference simply translates the well established linear negative relationship between test anxiety and performance into the motivation group research design being used.

The important question, then, is what happens in the Optimizing Conditions designed to eliminate negative motivational dynamics? At the top left Figure 1A (labeled "High anxious gain most") depicts the kinds of results which would suggest that motivational bias operates in standard testing conditions such that anxiety causes lower performance. Here, low-anxious students show no gains under optimizing conditions while middle- and especially high-anxious students show marked gains and in fact catch up to the low-anxious group. These results would suggest that high-anxious students know the material being tested but are unable to show that knowledge under standard testing conditions due to motivational and test-taking factors. When time pressure and the like are removed from the testing situation, then negative motivational dynamics are minimized and high-anxious (and other) students should show optimal performance. Low-anxious students are shown performing well in both the standard or optimizing conditions since negative motivation should not be a significant problem for them. Such positively motivated, low-anxious students have, in fact, been found, if anything, to do better in standard than special testing conditions at the secondary and college level (see e.g., I. G. Sarason, 1972).

An alternative possible effect of optimizing conditions, depicted in Figure 1B (labeled "Low anxious gain most"), is that the less anxious children show the greatest gains with high-anxious children showing smaller or even no gains. Here there not only would be no evidence that standardized tests are given in ways which promote or at least allow for interfering effects of anxiety, but just the opposite is true. That is, if more time is given, instructions are simplified, etc., less anxious children do better than ever, and it is their performance which is underestimated by standard testing techniques.

In Figure 1C (labeled "No one gains"), a third possibility is graphed, that nothing happens. That is, all groups continue to perform at the same level as they did in the standard testing condition. Such results would strongly support the hypothesis that low achievement is causal to the experiencing and reporting of anxiety, since there are no gains by high anxious (or other) students in the optimizing condition.

Figure 1D (labeled "All gain") shows the situation in which all three anxiety groups in the Optimizing Condition show significant gains over their respective performance in the Standard Condition. Once again there would be no evidence to support the notion that test anxiety lowers performance since less anxious groups improved as much as the high-anxious group; low performance would still be concluded to be the causal factor. But the situation in which all groups gained significantly albeit equally in Optimizing Conditions would have very different implications for educational testing policy, particularly with regard to diagnostic testing and to minimal competency testing programs. In this situation optimizing conditions would help students at all anxiety levels better show what they have learned and what skills they possess than standard testing conditions. Achievement measures gathered under optimal testing conditions would be more useful for teachers seeking to diagnose students' learning accomplishments and needs. They also would be more appropriate for minimal competency testing programs. It is assumed that what is of primary importance in the assessment of achievement is what students have in fact achieved, and not whether they can necessarily demonstrate that knowledge under highly evaluative testing conditions involving time pressure, social comparison processes, fairly frequent failure, etc. We view the ability to perform well under highly evaluative conditions as a separate matter involving the need for both positive motivation and test-taking skills. The latter skills are, however, important in their own right in our test-oriented culture, and their teaching and development is indeed a goal of portions of the program of research to be discussed later in this chapter.

The development of optimal testing conditions does more than allow for a test of the causality in the anxiety-performance correlations. A second major basic research function of optimal testing conditions is to determine which specific motivational factors interact with which specific testing parameters to determine (achievement) test performance. As in other aptitude by treatment interaction research (Bowers, 1973; Ekehammar, 1974; Endler & Magnusson, 1976), we are finding that the interaction between test anxiety and testing conditions accounts for test performance better than either factor separately (see Hill & Eaton, 1977).

A third critical reason for developing optimal testing conditions is to be able not only to spell out the implications of research for educational testing policy but also to test these implications directly. One of the long term goals of the research program is to develop specific recommendations for educational testing policy based on empirical research which result in more valid and useful educational testing and assessment programs. Suggestions based on data collected to date will be discussed later in this chapter. In the meantime let us turn now to research suggesting which testing parameters might be changed to minimize motivational test bias and provide an optimal estimate of a student's knowledge and skills.

Time Limits, Time Pressure, and Test-Taking Strategies

One of the hallmarks of most achievement tests is that they are given under time limits which generally do not allow all children to finish. Even when most children or older students "finish" a test by trying most problems, it does not necessarily mean that they had sufficient time to do all of the problems adequately. Time pressure is greater at some ages than others. In the elementary school years, for example, children in grades 4 through 6 are often given the same test with the same time limits. This obviously will make the test more difficult for 4th graders than 6th graders and enhance the time pressure on the younger children. Some of the aptitude and achievement tests at the high school level have even stricter time limits so that a fair proportion of students taking the test fail to finish. Time limits may create a strong, pervasive pressure that provokes strong negative motivational dynamics in the test situation, particularly for anxious students with a history of performing poorly on tests.

The position we are coming to take is that achievement tests with stringent time limits do not just measure what skills children have acquired or what they know but in addition whether or not they can demonstrate this knowledge quickly and under time pressure and testing stress. The result is that tests given under time pressure can systematically underestimate performance of anxious low-performing students.

A series of studies have examined effects of time pressure on elementary school childrens' test performance. Hill and Eaton (1977), for example, studied the effect of time pressure on 5th and 6th grade children's performance on basic arithmetic computational tasks which presumably had been mastered several years earlier. Children performed individually either in a "Mixed Success-Failure Condition" involving time limits that allowed for completion of two-thirds of the problems attempted (while a bell indicated time was up on the other third) or in a "Success Only Condition" in which children were allowed to finish all of the problems. The mixed success-failure condition thus involved stringent time pressure, both as perceived and experienced by the child.

The major results from this study are presented in Figure 2, which shows the completion time (rate) and percentage of problems without errors (accuracy) for high (HA), middle (MA), and low (LA) anxious children tested in each of the two conditions. High performance, then, is at the top of the Y-axis, consistent with the theoretical curves presented in Figure 1 earlier. The data have thus been replotted from the original reports (see Hill & Eaton, 1977, Figure 1, p. 208) to facilitate comparisons with other data presented here.

As can be seen in Figure 2, in the mixed Success/Failure Condition involving time pressure and one-third failure, high-anxious children showed weaker performance than their less anxious counterparts, making errors on three times as many problems, even though they were taking twice as much time on each problem attempted. In contrast, in the Success Only Condition the high-anxious children caught up to the less anxious children, going almost as fast and getting just about as

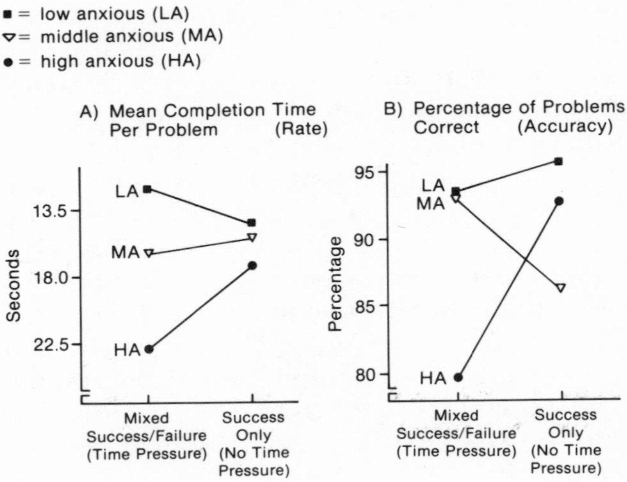

■ = low anxious (LA)
▽ = middle anxious (MA)
● = high anxious (HA)

A) Mean Completion Time Per Problem (Rate)

B) Percentage of Problems Correct (Accuracy)

Figure 2. Performance as a function of test anxiety and time pressure induced success/failure experiences

(Based on Figure 1, Hill and Eaton [1977], with the Y-axis inverted so that superior performance is at the top of the axis, consistent with other figures in this chapter.)

many correct. It is noteworthy that the high-anxious children improved in *both* accuracy and rate in the optimizing Success Only Condition. If the anxious children had increased their speed but made more errors or had improved their accuracy but maintained a slow performance rate, implications for achievement testing and classroom learning would be less striking. The fact that high anxious children *both* went as fast *and* were as accurate as the less anxious children in the Success Only Condition lacking time pressure indicates that anxious children have mastered the specific skills necessary to perform well on basic addition problems *and* can perform them quickly. They don't actually need extra time to do the problems if test pressure is absent. But high-anxious children are unable to demonstrate these skills effectively when the test items are given under time pressure which insures at least moderate failure. Such pressure occurs for many anxious-low performing children on subtests of many standard achievement/aptitude tests. The results indicate that motivational and test-taking difficulties rather than learning or ability deficiencies may underlie and bias the poor test performance of high-anxious children on standardized tests given under stringent time limits. Middle-anxious children, of course, would be expected to show smaller debilitating effects of negative motivation and hence smaller gains under optimal testing conditions than high-anxious children.

Eaton (1979) has extended the Hill and Eaton (1977) findings to a new sample of children for whom test anxiety measures were available for the two years prior to testing as well as the current year. Using similar testing conditions, Eaton found that *change in anxiety* over three years related significantly to performance in the standard and optimizing condition in the direction predicted from the earlier study. Children increasing in anxiety performed poorly in the mixed success-failure condition involving time pressure but much better in the success-only optimizing condition. Conversely, children decreasing in anxiety performed quite well in the mixed success-failure condition but somewhat less well in the success-only condition. Thus children becoming high or low test anxious in the Eaton (1979) study performed in much the same way with and without time pressure as the high and low anxious children in the Hill and Eaton (1977) experiment.

Plass and Hill (1979) have further extended the Hill and Eaton work toward actual educational testing practice in a study of time pressure with several new features. First, third and fourth grade children were tested in small groups of 6-8 students to achieve group dynamics more similar to classroom testing. Second, the actual problems used were developed with the help of classroom teachers based both on algorithms used in classroom teaching materials and on problems found in several currently-used achievement tests appropriate for middle elementary school students. The particular problems used were verbal mathematics problems which tested the four basic mathematical operations of addition, subtraction, multiplication, and division. The correlation between the percentage correct of subjects in the standard time pressure condition and the performance of these same subjects on a full-scale achievement test was .75 ($p < .001$), uncorrected for reliability, suggesting that the set of problems and standard testing condition is a highly comparable analog of an actual standardized achievement test.

The major result of the Plass and Hill (1979) study involved the highly significant triple interaction (F = 7.15, $df = 2/135$, $p < .001$) among test anxiety, time pressure condition, and sex, the means for which are presented in Figure 3. There are two major components of this interaction. First, under time pressure, low test-anxious children perform better, as expected, than middle- and high-anxious children for both boys and girls, but with somewhat stronger differences for boys. Second, in the condition removing time pressure, strong optimizing effects are obtained for boys but not for girls. For boys, high test-anxious (as well as middle test-anxious) students catch up completely and perform as well as low test-anxious boys (who show a small decrement), which indicates that high anxious boys are capable of performing well on achievement test arithmetic problems under group testing if time pressure is

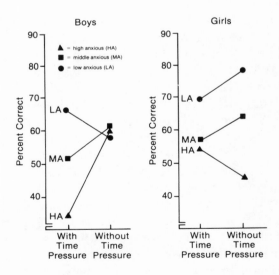

Figure 3. Means determining significant ($p<.001$) triple interaction among test anxiety, time pressure condition, and sex in Plass and Hill Study

removed from the situation. This is the most advanced study in our work showing strong optimizing effects for test anxious, failure-prone elementary school children (boys) in a testing procedure very closely approximating the full scale achievement test situation. As can be seen in Figure 3, low- and middle-anxious girls show some optimizing effects without time pressure but high test-anxious girls show a non-significant decrement in performance.

It is not clear why girls show smaller interfering effects of anxiety (than boys) under time pressure nor why girls fail to show the expected optimizing effect when time pressure is removed. It should be noted that both the interfering effects of anxiety and the effectiveness of optimizing conditions have generally been strong for both boys and girls across a variety of tests and evaluative situations (Hill & Sarason, 1966; Hill, 1977). Additional research is needed to determine the reliability of the present sex difference.

In order to better understand how anxiety interferes with performance in standard testing conditions and how the removal of time pressure changes these dynamics, Plass and Hill examined children's test-taking rate by accuracy strategies in each condition. Plass (1979) has identified four major maladaptive strategies and attributions which research suggests high-anxious children may use to avoid failure and disapproval in highly evaluative situations such as the achievement test (see also Hill, 1972, 1977).

First, while complete task refusal is rare in standardized testing situations, beginning a test and quitting after a few problems is somewhat more common. Dweck (1975) has shown that when children are given the option of stopping in performance situations, children with high negative motivation will stop. Presumably by doing so, they avoid continued evaluation as well as additional possible (to them, probable) failure. Some high-anxious children probably do quit trying in achievement test situations although the presence of their teacher probably prevents them from giving up altogether.

A second alternative strategy which allows considerable task avoidance while presenting some facade of continued effort is slow progress on the test (but not necessarily accurate work) interspersed with considerable off-task behavior. Thus, some high-anxious children might try the task, but then stop to stare around the room for awhile, work some more, glance up, etc. Significantly more of exactly this kind of off-task behavior by high-anxious children than by low-anxious children has been documented in a recent study of off-task glancing during a performance task (Nottelman & Hill, 1977; see also Dusek, Mergler & Kermis, 1976; Wine, 1971). Such a strategy, of course, almost guarantees a low overall test score. If it were true that anxious children could obtain a high performance score by exerting effort throughout the task, and thus obtain approval for good performance, many might choose to do so. Unfortunately, most high-anxious children have a history of doing poorly both in school and on tests. The negative motivation these children develop in evaluative situations comes from numerous classroom experiences as well as the occasional formal testing situations they encounter. Anxious children have little reason to believe that considerable effort during the task will result in anything more than one more devastating critique of low ability.

Plass (1979) notes that for those high-anxious children who, in contrast, for whatever reasons have some history of success (moderate performance) in test situations, a third type of strategy, that of slow, cautious on-task work, appears to be attractive. Williams (1976) found on a timed, achievement test-like arithmetic task, that high-anxious children in advanced (as opposed to regular) math classes who had a history of some success at mathematics were slower and more accurate than their regular math, high-anxious counterparts. Unfortunately, this slow cautious strategy may prove maladaptive on a (stringently) timed achievement test. Here anxious children can avoid immediate failure on specific problems but not do well on the overall test.

For high-anxious children with a history of test failure (consistently low performance), in contrast, such a slow, cautious strategy probably appears unlikely to pay off since such children may think they are likely to fail in spite of their efforts. Such high-anxious, failure-prone children appear more likely to follow one of two strategies: a slow, off-task strategy, as previously described, or perhaps a *fourth* strategy—fast, inaccurate performance.

A fast, inaccurate performance strategy can provide a high-anxious child with several benefits: task avoidance by spending less time on an aversive task, a less threatening "I didn't try hard" attribution for failure, and positive attention from peers for keeping up (at least as perceived by the child).

The rate by accuracy test-taking strategy results obtained in the Plass and Hill (1979) study were quite clear cut and helped explain performance of the various anxiety groups tested with or without time pressure. Table 2 presents the number of children of each anxiety level in each condition who showed a fast, medium, or slow response rate and a high, middle, low level of performance accuracy (based on 20%-60%-20%, approximately one-standard deviation, cut-offs for each dimension). As can be seen, different anxiety groups show quite different strategies.

Looking first at children tested under Time Pressure, nearly all of the low-anxious children (19 of 25 children) show an intermediate rate and either a middle or high accuracy level. These children appear to get right on task and proceed through at a steady but not too rapid rate and do quite well on the test. Middle-anxious children, in contrast, tend to perform at an intermediate level of accuracy (16 of 23 children) and show a slight tendency to go at a slow (6) or medium (6) rather than a fast (4) rate. Many high-anxious children (18) also show moderate performance accuracy and either a slow (7) or intermediate (10) rate, while other high-anxious children (8) perform poorly (boys in particular) but while showing a rapid or intermediate rate (see Plass & Hill, 1979).

In summary, low-anxious children perform at an intermediate rate and do quite well, middle-anxious children perform more slowly (and cautiously?) and do fairly well, and (some) high-anxious children perform quite rapidly and do least well as a group. It is possible that middle-anxious children are more concerned about making mistakes and hence follow a slow, cautious strategy, whereas high anxious children may be trying to keep up with other children in testing situations and hence adopt the fast, inaccurate strategy referred

Table 2

Rate × Accuracy Distributions for each Test Anxiety Group
in each Time Limit Condition in Plass and Hill Study

Time Pressure

% Correct	Low Anxious N=25 (Rate)			Middle Anxious N=23 (Rate)			High Anxious N=27 (Rate)		
	Fast	Med.	Slow	Fast	Med.	Slow	Fast	Med.	Slow
High	1	9	1	1	2	0	0	1	0
Middle	1	10	1	4	6	6	1	10	7
Low	1	0	1	2	2	0	4	4	0

Without Time Pressure

% Correct	Low Anxious N=27 (Rate)			Middle Anxious N=25 (Rate)			High Anxious N=28 (Rate)		
	Fast	Med.	Slow	Fast	Med.	Slow	Fast	Med.	Slow
High	0	12	3	0	4	2	0	5	1
Middle	0	8	1	0	6	11	1	8	5
Low	2	1	0	1	1	0	2	6	0

to earlier. As can also be seen in Table 2, the removal of time pressure does not affect the strategy distributions greatly but if anything seems to have improved performance by reducing somewhat use of the fast-less accurate strategy, especially for boys (see Plass & Hill, 1979).

It should be noted that the general overall similar pattern of results for the low-, middle-, and high-anxious students in the two conditions represents a replication of the rate by accuracy strategy usage of these groups since different subjects were randomly assigned to the two conditions. In other words, the findings suggest that middle-anxious children take tests too slowly, high-anxious children too quickly, and low-anxious children perform at an optimal rate; the removal of time pressure only partially corrects for the maladaptive strategies of the more anxious groups. These strategy results are particularly important because they suggest both what maladaptive strategies underlie the low performance of anxious children in standard testing conditions and what kinds of new strategies might be taught these children to help them perform optimally in any testing situation.

The research reviewed in this section suggests that time limits that induce time pressure is a feature of standardized testing that introduces motivational bias that prevents anxious students from showing what they have learned and the skills they have acquired. This work is particularly relevant to educational testing policy since most tests involve time limits of varying degrees of stringency and increasing time limits to reduce time pressure is a relatively easy and straightforward change to make. We will turn to specific issues involved in changing time limits on standardized tests in the section on implications for educational testing policy.

Success-Failure Experiences During the Test

Theoretical research on achievement motivation (Atkinson & Feather, 1966; Atkinson & Raynor, 1974), success-failure attributions (Weiner, 1972; 1977), and test anxiety (Hill & Sarason, 1966; Hill, 1967, 1972; Spielberger, 1972), has emphasized the role of fear of failure in negatively motivated children's low achievement test performance. Relevant to the focus of this chapter, highly test-anxious children, who typically perform poorly in achievement test situations, would be expected to show strong fear of failure and a tendency to give up when experiencing frequent failure.

Most standardized achievement tests are given to children of a given grade, e.g., third graders receive Form 3 of a given test, or several adjacent grades. Most tests are grade-norm referenced, and children's performance is reported in terms of both grade norms (two years behind norm) and standing relative to other children (15th percentile, etc.). The tests are designed to have a bell shaped distribution with some students way above average, some way below average, and many children near norm. Unlike classroom teaching materials, most tests are not tailored to the current achievement level of the child.

While there may be good reasons for giving the same test to all children at a given grade level, the practice has highly undesirable motivational consequences for the anxious, low performing student. When one examines the actual test performance of children doing poorly on achievement tests, perhaps performing two or more years below age norm, one finds they are often experiencing frequent or almost total failure each time they take such a test. In mathematics computation subtests common to many achievement tests in the late elementary and early junior high school years, for example, these children might get the first, typically easy problem in each sub-area (addition, subtraction, multiplication, division) and then miss most or all of the subsequent problems involving fractions, decimals, etc.

No wonder children who perform well below norm year after year are fearful of test situations, fear failure, and give up quickly on tests. Anxiety and pressure could not be higher, failure could hardly be worse, and any sense of competence or positive motivation almost surely would be lacking. Low-income, minority, and other highly test-anxious students, of course, bring exactly this kind of history of test failure to the standard achievement test situation.

Efforts to develop "optimizing" procedures for giving ability and achievement tests to children which involve greater success (in order to avoid bias due to debilitating motivational processes such as anxiety) date back to the classic study of Hutt (1947). Hutt administered the Stanford Binet IQ test under standard and "adaptive" testing procedures. In the standard testing procedure, the items were presented in order of increasing difficulty, going from the easiest to the hardest. Here failure would be most frequent (and anxiety strong) just as the student reached the part of the test assessing the limits of his problem-solving ability. In the adaptive procedure, easy and difficult items were mixed throughout the test. In this condition, the student experienced frequent success as well as failure throughout the testing session. Children rated as being poorly adjusted, who presumably would include a good number of high test-anxious children, obtained higher IQ scores when tested in the adaptive (mixed) order (mean IQ of 103) than the standard (easy to hard order) condition (mean IQ of 92).

Zigler and his colleagues (Zigler & Butterfield, 1968; Zigler, Abelson, & Seitz, 1973) found similar increases in IQ test performance of black and white low-income preschool and elementary school children. Zigler and his group improved performance both by adding success in the testing situation and positive interactions between the child and the examiner before or during the test situation. In the optimizing condition of the Zigler and Butterfield (1968) study, for example, some easy items came first, and the child was given an easy item each time the child missed two in a row. The examiner also encouraged the child to respond to each problem if the child failed to do so. Lekarczyk and Hill (1969) also found improved performance by high-anxious elementary school children only verbal learning tasks under less evaluative conditions but poorer performance when failure and evaluative aspects of the testing situation became predominant.

Dweck (1975, 1977) has found further that children showing a tendency toward learned helplessness in achievement/test situations, many of whom report high test anxiety, give up quickly in the face of failure in such situations. Dweck (1975) placed such children in an attribution retraining program in which they were taught and shown that their failure to complete problems in evaluative situations was due to their lack of effort and not due to a lack of ability. Children in the training program learned to overcome their learned helplessness and to persist at and complete difficult problems they had consistently given up on earlier. Other children exposed only to success on easier problems quickly reverted to learned helplessness in the face of repeated failure at more difficult problems requiring persistence and continued effort.

These findings suggest that children showing learned helplessness need attribution retraining as well as some success in order to do well in a test situation. Consistent with Dweck's work, emphasis is placed in parts of our work on the child's staying on task, showing continued effort, and doing whatever problems they are capable of doing ("their best"), as will be discussed later under teaching activities.

Research by the author and his students has taken several directions in terms of helping anxious, failure-prone children deal with frequent failure on standardized tests. One approach is to change students' perception of how well they should be doing on a test so that they will realize that some failure during a test is quite common and normal and even considerable failure does not necessarily reflect poorly on the student.

Williams' doctoral dissertation research (1976) reflects this approach (see also Sarason, 1972). Williams presented under time limits arithmetic computation and word problems similar to those on standardized achievement tests to fourth- and fifth-grade children in one of four ways: (1) In the Standard Condition, representative of many achievement tests, the problems were introduced as a test of the child's ability to solve arithmetic problems; (2) In the Diagnostic Condition, the problems were introduced as a test of ability, but the child was told the test results would be used to learn what things the child needed to work on in math; (3) In the Expectancy-Reassurance Condition, problems were again introduced as a test, but the child was told that some of the problems were very hard and that no one gets all of them right. In addition, the child was told that most children get quite a few wrong and not to worry if he couldn't do some of the problems. The Expectancy-Reassurance Condition was designed to lower the level of performance perceived as being successful and lessen the threat of disapproval from the examiner; (4) In the Normative Condition the children were told the examiner wasn't interested in how well individual children did but rather in how interesting the problems were to do and the child needn't put his name on the test booklet. This condition minimized evaluation in general.

It was assumed that the Standard Condition invoked the motives to avoid failure and disapproval, so that high-anxious children would do less well than middle- and low-anxious children. It was expected that high-anxious children would do better in the other three conditions since the motives to avoid failure and disapproval should be diminished in these conditions.

Findings for the significant interaction between test anxiety and instructional condition for proportion correct on the arithmetic *computation problems* are presented in Figure 4. As predicted, high-anxious children were less accurate than middle- and low-anxious children under the Standard test instructions. Performance accuracy of high-anxious children increased significantly in the Expectancy-Reassurance and the Normative Conditions, while less anxious children showed smaller decreases in accuracy in these optimizing conditions.

Williams was also able to group children on the basis of their school assignment to "regular" or "advanced" math classes. A significant triple interaction among Test Anxiety, Test Presentation, and Math Class revealed additional findings, in this case for the arithmetic *word problems* measure (see Table 3 which depicts the proportion correct for each group in each testing condition for this task). In general, children from the advanced math classes were quite accurate and did quite well in terms of proportion of word problems correct in the time limit allowed. In the Standard Condition, however, the high-anxious advanced math children, while accurate, went so

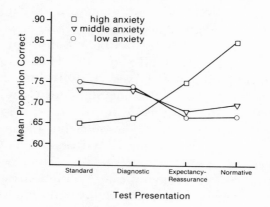

Figure 4. Mean proportion correct on computation problems as a function of test presentation and anxiety (from Williams, 1976, dissertation)

Table 3

Mean Proportion Correct for Word Problems
as a Function of Test Presentation,
Test Anxiety and Math Class (from Williams, 1976, dissertation)

Math Class	Anxiety	Test Presentation			
		Standard	Diagnostic	Expectancy-Reassurance	Normative
Advanced	Low	.84	.77	.85	.78
	Middle	.80	.77	.76	.84
	High	.79	.81	.69	.76
Regular	Low	.77	.60	.60	.61
	Middle	.67	.54	.56	.55
	High	.38	.43	.69	.49

slowly that they did not get as many problems correct as other children. In the optimizing conditions, advanced high-anxious children went faster while maintaining their accuracy, so that they had about as many total correct solutions as other children, especially in the Expectancy-Reassurance Condition.

A different pattern of results emerged for high-anxious children from the regular math classes, children who presumably have experienced more frequent failure in arithmetic work in school and especially on tests. These high-anxious children tended to perform very rapidly but made many mistakes in the highly evaluative Standard Condition (only 38% correct). In the other conditions, high-anxious regular math class children slowed down somewhat and became more accurate, especially in the Expectancy-Reassurance Condition, (69% correct).

The Williams (1976) data indicate that high-anxious, failure-prone children can perform better (including under time pressure) on achievement test problems if instructions are added to weaken the interfering effects of perceived failure and strong failure motives. In addition, the interfering effects of anxiety are mediated by different performance strategies at the task depending on whether the child has done moderately well or poorly in arithmetic in school. The high-anxious, failure-prone students in the "regular" math classes followed the "fast-inaccurate" test-taking strategy Plass and Hill (1979) found for high-anxious students in the study discussed earlier.

Another possible approach to helping children deal with success-failure experiences in the achievement test situation, especially the high test anxious-low performing students who experience heavy failure on such tests, would be to bracket the difficulty level of the test around the child's current level of functioning. It actually makes little sense to expose a low performing child to tests that involve almost total failure year after year. Thus, children three years below norm on a short general achievement test could be given reading-mathematics tests three grade levels below their age, children two years behind in general achievement could be given tests two grade levels below their age, and so forth.

It is recognized that bracketing tests around the child's current level of ability would make norming more difficult and problems would arise from the fact that some children do very well in some areas but not in others. These problems are not insurmountable, however, and indeed some standard achievement test manuals already recommend giving students the test appropriate for one age-level below their grade if there is a history of performing poorly on tests or in the classroom. It is also not unusual for school districts to follow such a practice as part of their general testing policy.

The bracketing procedure should allow all students to achieve at least moderate success during the testing experience. Anxious students should develop more positive motivation and perform optimally once the bracketing procedure takes hold over several testing experiences and the student's tendencies to not try, rush through, and give up are overcome. At the same time, the bracketed tests also would provide somewhat difficult problems that measure the student's most advanced level of achievement, the purpose of the test.

Clearly a good deal of research is needed on the problems as well as the possible advantages of bracketing the difficulty level of an achievement test around the ability level of the student taking it. But both the basic theory and research on motivation and test performance and the educational testing practice discussed above suggest that test bracketing is a feasible and desirable procedure that shows promise of reducing motivational test bias. We will return to the idea of bracketing in the section on implications for educational testing policy.

Cognitive Testing Demands Independent of Content

A third major focus of our current program of research concerns test instructions, response formats, answer forms and other aspects of achievement tests that place cognitive demands on the student completely independent of the achievement content being sampled. Such cognitive test demands can both interfere with the performance directly if the child does not understand how to take the test (and quickly, if under time pressure) and indirectly by increasing interfering motivation dynamics such as test anxiety. At present our empirical work in this area is still in the development stage. Logical analysis from basic research on children's cognitive and motivational development and informal feedback from school personnel, however, suggest that cognitive test demands may be an important source of test bias for the high anxious-low performing student of central interest in this paper. Our increasing interest in this area, in fact, has developed over the past several years directly out of discussions with teachers and school administrators concerned with testing and evaluation. School staff have repeatedly emphasized that aspects of achievement testing which are particularly difficult for elementary school age children include the often complex and changing test instructions and question formats, and the introduction of computerized answer sheets at about the fourth-grade level for many standard achievement tests. Children who have difficulty understanding instructions and using answer sheets should experience more anxiety in the test situation since they don't understand these important aspects of the test. They also should be more likely to give up, have difficulty proceeding effectively on a timed test, and in general, may not be able to answer problems they know because they are either confused about how to do a problem, don't know how to use the computerized answer sheets, or experience strong debilitating motivation. Older students may be better able to understand test instructions and answer sheets but still have trouble dealing with them effectively as time limits become more stringent at later ages.

Inspection of current standardized achievement tests reveals that cognitive and intellectual demands of the typical instructions, response formats, and answer sheets, independent of the material being examined, could readily lower further the performance of the high-anxious already low-performing student. For example, many achievement test instructions are fairly long and difficult to follow. Sample problems are not always given. The response format of the sample problem is not always the same as the response format of the actual problems. Children often have no practice test and are given only one or two practice problems on the computerized answer sheet. Sometimes sample problems are given but are not to be filled in on the computerized sheet. Some subtests have as many as half a dozen instructions introducing portions of the subtest, often with different response formats; at the same time, there may or may not be sample problems which are or are not worked out on the computerized answer sheets. After a brief practice problem, children in grades 4-6 are often expected to read instructions and adapt to different response formats on their own as they proceed through the test under testing pressure. The (younger) fourth graders in particular should have trouble dealing with these demands. Given this myriad of non-content related possible stumbling blocks in many tests, it becomes clear why teachers and members of our project staff feel that difficulty in understanding instructions and answer formats may have a major interfering effect on many children's achievement test performance. We strongly expect that anxious students with a history of scoring far below norm on standardized tests will be most debilitated by these factors.

An ongoing collaborative school-university pilot project in

this area involved nine elementary school teachers giving a standardized achievement test to fourth- through sixth-grade children as part of an evolving school testing program. In this pilot activity, the teachers administered practice tests with the same instructions and actual sample problems (but none of the actual test problems or any other related problems) for up to seven of the subtests in the language arts and mathematics achievement subtests to be given subsequently. Some teachers chose to go over one such practice test carefully, others chose to go over between three and seven. The teachers also talked to the students about the general purposes of the testing, explained that taking tests is one of the things students need to learn how to do, and provided some information on test-taking strategies.

The purpose of this practice testing was to familiarize the students with the test, to be sure they understood the instructions, and to give them experience with question and answer formats as well as the computerized answer sheets while in no way teaching the content of the test itself. This ongoing pilot activity was designed to develop procedures, to obtain teacher judgments, and to obtain some initial information on how well the procedures are working. Teacher feedback indicates the practice session with the instructions and answer sheets seemed to help many children in terms of how to take the test. Children had far fewer questions about the instructions or use of the answer sheets during the test than in previous years, began working the test problems more quickly and with apparently little confusion about how to do the problems, and seemed to progress smoothly through the different subtests. In particular, the anxious students who often are weary and give up in standard achievement tests appeared to have less difficulty than usual with how to do the test and to have worked hard during the tests.

The pilot work suggests that this new kind of classroom activity may be successful in identifying and correcting for another aspect of standard achievement test procedures which allows motivational test bias to operate as well as directly lowering test performance. The school staff is planning to develop and integrate these procedures into the school's teaching and testing programs and to evaluate their effects on motivation, test-taking strategies, and test performance. Gains in positive motivation and performance may become particularly strong for anxious students after their teachers have taught them how to follow and deal with instructions/answer forms across several years and for different tests. The effects of years of test failure take some time, of course, to reverse and eliminate.

The present analysis would suggest that achievement tests themselves might be expanded so that there is an entire practice *session* (not just several practice problems) before the battery of subtests are given, especially for students (e.g., fourth graders) for whom a new aspect of testing is being introduced (e.g., the use of computerized answer sheets). An entire practice session also allows the teacher to talk with students about testing, test-taking strategies, and the like, separate from the testing situation itself. Some current achievement tests involve a practice subtest which can be taken earlier, a useful step in this direction.

Other Testing Parameters

A number of factors have been discussed thus far that influence the effects of motivation on test performance, including *time pressure, test-taking strategies, test difficulty* and *success-failure experiences,* the *supportive/neutral role of the examiner* and *potential later adult approval and disapproval,* and the *cognitive demands of the test independent of content.* Several additional factors will be touched upon briefly here.

Standardized testing often makes *social comparisons* salient to teachers, parents, and older students, which in turn brings competition into play. Social comparisons and competition effect both motivation and performance (see Diane Ruble's chapter in this volume; also see Maehr & Stallings, 1972; Salili, Maehr, Sorensen, & Fyans, 1976; Veroff, 1969). Social comparisons should have their greatest negative motivational and performance effects on children who typically perform poorly on standardized tests and who thus suffer in the social comparison process, the high test-anxious student. In addition, evaluations from others should further enhance the negative consequences of social comparisons (and other forms of evaluation) for these children, due to their sensitivity to failure and negative evaluations and reactions from others (see Atkinson & Feather, 1966; Hill, 1977; Maehr, 1976). Efforts to develop new evaluative practices designed to minimize negative effects of social comparison and to help chidren learn to cope with such comparisons are described in the next section concerning reporting-to-parents.

Consideration of social comparisons draws into attention another important testing parameter, *individual versus group testing*. The Plass and Hill (1979) results discussed earlier suggest that anxious, poor performing students may rush quickly through (and do very poorly at) group-administered achievement test to avoid negative reactions from other children which could result from a slow, careful, alternative approach. In both the Plass and Hill (1979) and the Williams (1976) studies, the fast-inaccurate strategy resulted in a low performance score. Individual testing eliminates situational comparisons among students taking the test, and in addition minimizes distractions and makes it more likely the child can receive appropriate help from the examiner in how to take the test. A major problem with individual testing, of course, is its expense. As will be discussed in more detail in the section on educational policy implications, dual testing programs could be developed in which students doing poorly in group-administered standardized testing conditions could be retested in optimizing conditions (individual testing, no time pressure, etc.) to see if motivational test bias is a problem for different children.

Another pair of related testing factors are what the child *perceives as the purpose of a standardized test* and what kind of feedback the child receives. The instructions of many achievement tests mention that the purpose of the test is to ascertain what the child has learned or whether he knows the material being tested. Elementary school students often receive little or no actual feedback, however, about how well they did on a test or the implications of their performance. Williams' study (1976) referred to earlier examined the effects of explaining the diagnostic value of tests. Teachers can also explain to children the importance of learning how to take tests, that test-taking skills are one of the many things children learn in school. Consistent with earlier discussion, social comparisons can be downplayed and children encouraged to do their individual best. Teachers/examiners might give immediate feedback such as praise for effort and persistence during the test and later feedback in terms of how well the child showed what he knows (not "how well he did on the test" in terms of norms or being compared to others).

The Other Formal Evaluation: Reporting-to-Parents

Let us turn now to a second major area of school practice involving motivation and evaluation feedback, the *report card*. This is a very different form of evaluation than the (achievement) testing situation, lacking the situation dynamics of testing but involving the teacher, parent, and child in assessment of important general qualities. Our work in this area has involved several elementary and several junior high school projects. The work described here involves two ongoing four-year longitudinal studies involving extensive study at all phases of the development and assessment of a new reporting system, one at the elementary school level, the other at the junior high school level.

A general model has developed for such collaborative school-university research on report card change. First, a committee of school and university staff is formed to develop the new evaluative procedures. Often parents serve on this committee, but sometimes parent feedback is obtained through surveys or interviews. The committee begins by thoroughly reviewing the issues and purposes of report cards and other reporting-to-parents systems. Then the school's current reporting procedures are reviewed in terms of how well they are meeting the objectives of such evaluation and their perceived effects on students. Relevant portions of the research literature on motivation and evaluation are reviewed and potential changes and improvements in the development of new procedures are considered. At this point, survey questionnaires are typically sent or given to all teachers, parents, and students (if old enough) in the school to obtain information concerning their perceptions, needs, and priorities concerning reporting to parents.

For example, all three groups are asked to rank order the relative importance of feedback concerning children's achievement, effort, progress, specific strengths and weaknesses, class rank, and national standing in the major academic areas of language arts, mathematics, etc. The impact of the present reporting system on challenging the students to work hard, creating pressure, developing self-confidence and other student qualities is also assessed in the surveys. In addition, preferences and priorities for letter grades, descriptive ratings, individualized written comments, parent-teacher conferences, and other possible reporting procedures are obtained.

On the basis of the committee's review of goals, the research literature, and the present report card, and with careful consideration given to the survey results from parents, teachers, and students, a new reporting system is developed if appropriate. For example, at the elementary school level, A-B-C-D-E letter grades in each subject area may be eliminated for the following reasons: (1) they are too general and lump together teacher evaluation of a number of qualities such as learning, effort, and progress; (2) they essentially compare children and generate unnecessary and potentially harmful social comparison (e.g., if three-quarters of the parents in a school want their children in the top quarter of their class, then half of the children will be seen as failing no matter how hard they work or how well they do). Such unrealistic parental expectations and the harmful social comparisons they generate are related to the work of Ruble reported elsewhere in this volume; (3), "grade inflation" may have occurred such that too many A's and B's are being assigned. In practice "objective" letter grades may convey very little information concerning how well a child is doing or why. In one elementary school, a new card was developed in which individualized written comments were developed in each major academic subject area focusing on the child's achievement, effort, and specific strengths and weaknesses. Comments are designed to provide both positive comments on strengths to be encouraged and constructive feedback on weaknesses to work on. Feedback on social development (getting along with others, self-confidence, etc.) is also provided. Teachers' comments are meant to be as specific and clear (jargon-free, etc.) as possible.

At the junior high school, a different kind of card is emerging from several projects. At this age level, there is a very strong preference for letter grades on report cards by parents, teachers, and students themselves. In one longitudinal project, the school-university committee expanded the present A-B-C-D-E card to include letter grades for both achievement and effort as well as general feedback on areas of strength and concern on a checklist basis. The card thus permits the teacher to differentiate between achievement and effort instead of lumping these two important qualities into one grade. For example, students can be identified who are learning well but not working as hard as they could (A-achievement, B-effort grades), who are not learning well but who are putting in good effort (C-achievement, B-effort grades), etc. Feedback on students' effort represents a major new thrust in the school's reporting system.

Effects of such reporting systems are evaluated in a number of ways. Post-change surveys are again given to all teachers, parents, and students in the school. These surveys test for the general level of satisfaction with the new report card, the perceived effects of the card on the students, and for any changes in priorities, perspectives, and needs assessed in the pre-change surveys. In the elementary school project changing from letter grades to individual written comments, for example, over 90 % of the parents replying, and all of the teachers, were positive in their general evaluation of the new card, and most parents and teachers perceived gains for the students in learning and motivation. The percentage of parents indicating that feedback on children's social develop-ment was important increased from about 50 % before such information was introduced on the new card to over 90 % after it was introduced.

In the junior high school project in which information about areas of strength and concern and a letter grade for effort were added to the one for achievement, the parents, teachers, and students all perceived desirable specific changes in the students, for example, a tendency to work harder, learn more, etc. Parents and teachers showed overwhelming preference for the new card while student reaction was mixed, apparently due to the increased amount of evaluation.

In addition to analysis of the pre-change and post-change questionnaire data, information concerning the students' achievement test performances, achievement-related motives, and perceptions of causes of their successes and failures in school were obtained in several projects before and after the report card change and are undergoing analysis.

The nature of the report cards being developed at the elementary and junior high school levels reflects several general principles for improvements in reporting systems across the public school years. For one thing, feedback should be as specific as possible (single A-B-C-D-E letter grades are not), as broad as possible (including feedback on effort, social development, etc.), and should include suggestions for how students can improve their achievement efforts. Second, certain forms of evaluation such as social comparison should be phased in across the public school years. Consistent with Diane Ruble's findings, social comparison feedback should be avoided in the early elementary school years and phased in gradually during the late elementary or junior high school ages. Transition periods in school evaluative practices deserve careful attention in order to prepare children for the new forms of evaluation coming. Third, and more generally, teachers and other school staff should explain to students the purposes of evaluation and help students develop effective ways of constructively interpreting and coping with report card feedback and other systematic evaluation.

An Integrated Program of Teaching Activities and Achievement Testing

The work review thus far has concerned
(1) the interfering effects of negative motivation on (achievement) test performance, and
(2) ways to modify the ways tests are given to minimize motivational test bias. This section briefly describes

initial work concerning the two other major foci of the longterm program of research:

(3) efforts to teach students how to cope with the pressures and demands of testing (whether highly evaluative or not), and
(4) development of new educational testing programs that phase in evaluative aspects of testing as children have acquired the test-taking skills needed to perform optimally in such situations.

Many of the attitudes and behaviors students bring to the achievement test situation come from somewhat similar evaluative situations in and out of school. The achievement test itself, of course, is designed to measure learning as it has developed in the classroom. From the early elementary school grades on, students function in a variety of formal and informal evaluative situations ranging from formal classroom tests to informal discussions with their teachers. The motivational dynamics found in standardized tests clearly come in part, perhaps good part, from classroom and other school evaluative experiences.

We would suggest further, however, that the ways in which many standardized achievement, aptitude, and other formal educational tests (such as minimal competency exams) are given differ in a number of significant ways from classroom experiences. At the elementary school level, standardized tests even differ markedly in many cases from the tests children receive in schools. We refer to standardized testing conditions as often being "artificial"—not the content of the test or the achievement domains being sampled, but rather the testing parameters, the kinds of conditions and situations standardized tests place children in.

Let us look at the differences between many school tests children are given and the standardized achievement test. To begin with, children may take school tests fairly often, at times daily, often weekly (spelling tests, etc.), and other times monthly. Standardized tests might be taken just once or twice a year and are a much less familiar experience. In terms of basic content of the test, school tests almost always sample the specific kinds of things children are learning in school. On standardized achievement tests, however, it is not unusual for children to take sections (in science, social studies, etc.) or parts of sections (word meanings, kinds of arithmetic problems—use of fractions, etc.) that they are not familiar with. The anxious-low performing child may never have been exposed to materials covered by sections of a test, particularly some of the more advanced materials.

Many standardized tests place greater testing pressure on students than school tests with regard to the three major factors focused on in the present paper: difficulty level, time pressure, and cognitive demands of the test. Most classroom tests focus on the material the students have been learning and are of moderate difficulty, not easy material children learned much earlier or difficult/impossible material they have not yet studied. In contrast, on achievement tests the low anxious-high performing student will find much of the material easy, and the high anxious-low performing student will find much of it very difficult (or impossible).

Although some elementary classroom tests involve time pressure and many involve some kind of time limit, children usually are given a fair amount of time to complete the questions. Standardized tests often involve more stringent time limits placing greater pressure on students, particularly older students. In a similar manner, instructions and answer sheets in classroom tests are often simple, straightforward, and familiar to the students, especially for tests given frequently over the course of the school year. On the other hand, and as noted earlier, test instructions, response formats, and computerized answer sheets can be quite complex, unfamiliar, and difficult for many children to follow and use.

Classroom tests also have advantages in terms of other motivationally relevant testing parameters. The purpose of the test is often constructive, to assess what the child has learned, with the teacher often assigning additional work in the areas not mastered. Teachers usually provide students with feedback fairly soon after the test is taken and often will feel free to help a child if he gets "stuck" during a test (suggesting he move along to the next question, etc.). Individualized feedback to the student also tends to de-emphasize the social comparison/competition aspects of classroom testing. In contrast, students often do not know the purposes of standardized achievement tests, receive less help during the test (due to the way testing procedures are standardized), and often receive little or no feedback as to how well they did on a test.

A final factor worthy of mention is that school tests are often fairly short (10-20 minute spelling or math test, etc.) and given on separate days or weeks. In contrast, standardized achievement tests often include a half dozen or so subtests each of which might take between 20 and 60 minutes at the elementary school level. This means children may be tested for an hour or two each day for two or three days in a row, a testing schedule requiring persistence, attention, and optimal functioning over an unusually long period of time for children of this age.

It seems clear that the standardized achievement test places many demands on students and has many qualities and features that are unfamiliar to most students. It also seems clear that many of these features will continue to be part of standardized testing for some time and that it is desirable that students learn to cope with such testing demands. We would suggest that test-taking skills are one of the more important basic skills which children learn (or fail to learn) in school.

Our own efforts to help school staff develop classroom and other school activities which teach children how to deal with the demands of standardized testing as they are phased in over the elementary school years involves an ongoing collaborative school-university research project on motivation and evaluation now in its sixth year (see Bodine, Hill, & McPhail, 1976). The principal, teachers, and other professional staff of a cooperating elementary school and our university project staff have worked together on a variety of activities directed at enhancing positive motivation and learning. The school has a broad balance of ethnic and socioeconomic groups unusual for a single school, drawing from three major residential areas in the community and reflecting its general population closely.

Children are being taught test-taking skills, coping strategies, and attitudes which facilitate positive motivation and test performance in a variety of evaluative classroom situations ranging from fairly formal school tests (in mathematics, spelling, and other basic skill areas) to workbook exercises to informal class discussions.

The amount and timing of the teaching and classroom testing activities will vary across the three major areas of concern in motivational test bias: (1) time pressure and time limits; (2) success-failure experiences during a test; and (3) understanding instructions, response formats, and how to use computerized answer forms. Teaching children how to cope with time pressure and success-failure experiences in evaluative situations will occur frequently and throughout the school year. Teaching children how to understand and cope with formal test instructions and answer forms, on the other hand, will occur several times each year and be coordinated with the school district's standard testing program in reading, mathematics, and general achievement.

With regard to the first focus, time pressure and time limits wil be gradually introduced by teachers in a series of classroom testing activities moving from the early to later elementary school years and across each school year. Here the teachers will familiarize children with and teach them how to cope with the kind of time pressure they experience on achievement tests as they become older. Children will be taught appropriate test-taking strategies concerning rate and accuracy and the trade-off between these two performance criteria in test situations. They will learn it is not always possible to finish learning exercises and tests in the time allowed. One goal of the teaching in this area will be to help children feel comfortable if they do their best on problems they have time to attempt on any timed test.

The second focus, on success-failure experiences in the test situation, will be approached in a similar manner. Children will be exposed to increasingly difficult materials during parts of their classroom evaluation and testing activities, as well as being exposed to material they are familiar with and have generally mastered in the classroom. Thus, they will be taught how to interpret and cope with failure involving both familiar and unfamiliar materials in test situations, and to persist and perform effectively and to do their best no matter what level of difficulty is encountered in any educational test.

The third area of focus involves teaching students how to master general test instructions, response formats, and computerized answer sheets common to most standard achievement tests during the later elementary (and subsequent) school years. The focus of this intervention will be at
  (1) the second-grade level when formal achievement tests are introduced in the cooperating school's testing program, and
  (2) the fourth-grade level, a time when several new and potentially confusing aspects are added to the test situation, e.g., more complex test instructions and computerized answer sheets.

Consistent with all of the proposed activities, this project does not involve "teaching to the test (content)," but rather with how to effectively take tests and show positive motivation and optimal performance based on what the student has learned.

Both the classroom testing and teaching activities will involve the phasing in of evaluative aspects of testing within and across the elementary school years. During kindergarten and the early part of the first grade, children will work at exercises free of time pressure and any overly difficult or unfamiliar materials. Time limits and difficult materials in workbooks will be gradually introduced over the first grade and into the second grade. The district introduces its formal achievement testing program in the fall of the second grade, by which time children would be taught how to follow general test instructions and how to work effectively on a group test. Moving through the third and by early in the fourth grade, children will have been taught how to deal with more complex test instructions and response formats, computerized answer sheets, and more stringent time limits and difficult test materials. By such time, children must know how to cope with all of these non-content test demands and know how to do their best on achievement and other tests. This teaching will be continued through the sixth grade, with particular emphasis at this age on children having test-taking problems. The goal of the project is to have all children by the end of the sixth grade master these test-taking skills and experience confidence and positive motivation in any test situation. We will refer to providing younger children optimal testing conditions but then phasing in evaluative aspects of testing as they become older and develop good test-taking skills as the "optimal school testing program."

The teaching and optimal testing program under development should not only facilitate performance and provide a more valid measure of high anxious-low performing children's achievement but also enhance the positive motivation, confidence, and test-taking skills of such children in the (achievement) test situation as they come to understand the situation and experience success in it. These gains may be particularly strong for anxious low income-minority students who also face other forms of test bias as discussed earlier. At present, standardized testing is aversive and demoralizing for such children given their inability to show what they know and their subsequent poor performance.

Effects of the new teaching activities and changes in the school's testing program will be assessed by a variety of measures of learning, achievement, and motivation collected by the elementary school and, for some data, in other elementary schools by the school district. It is hoped that the student gains in motivation and performance over the course of the study will be strong and enduring.

Implications for Educational Testing Policy

General Comments

We will focus in this section on major implications of research on motivational test bias for classroom teaching and educational testing policy. Many of the specific implications of the work for teaching and testing have been discussed earlier and will be treated here only for purposes of emphasis or to illustrate general points.

It was noted in the preceding section that the standardized achievement test situation is often artificial in the sense that many aspects of the test and the testing conditions are unfamiliar to the student, being quite different from classroom activities. At the elementary school level, at least, standardized tests often differ markedly even from classroom tests. It is this very artificiality of the standardized test situation, however, which makes possible first, careful study of the interactive effects of motivation and testing conditions in determining performance and, second, recommendations for specific changes in teaching and testing practices based on empirical research. In most areas of educational policy, recommendations are, of course, often far removed from data due to the complexity of the phenomenon. Educational testing policy appears to be one of the few important areas in education allowing for fairly substantial and fairly direct ties between research data and policy recommendations.

Assessment of Motivation in Testing Programs

One clear and specific recommendation from the extensive research on motivation and test performance reviewed here is the desirability of incorporating motivational indices into building, school district, state, and national achievement testing programs. This allows for assessment of relationships between motivation and test performance which would test for the presence of motivational test bias, either in general or for students of certain ages, sociocultural backgrounds, educational experiences, etc.

The collaborative project between the Illinois State Board of Education and the University of Illinois, Urbana-Champaign, referred to earlier (see Brennan & Maehr, 1979; Fyans, 1979; Hill, 1979b; Linn, 1979) demonstrates the feasibility and usefulness of such a measurement effort at the state level. In this testing program, measures of test anxiety (relaxed form of the scale) success-failure attributions, and continuing motivation were collected for some 7200 students in grades 4, 8, and 11. As noted, increasingly strong relationships were found among all three motivational variables and reading/mathematics test performance across the public school years under study for students of various sociocultural backgrounds. Ongoing research is suggesting that a 7-item form of the relaxed test anxiety scale, a 2-item defensiveness scale, and a 2-item measure of success-failure attributions which take a total of only 10-12 minutes to administer in grade 4 and less than 10 minutes in grade 11, may be a quite useful set of motivational measures (see Fyans, 1979b; Hill, 1979a). These shortened motivational measures make assessment of motivation test bias in achievement testing programs even more efficient and feasible. The need for such data is particularly critical in new minimal competency programs in which the potential for interfering motivational effects is particularly high due to the strong consequences of failure. Indeed, the state level data indicated a 33% score difference between high and low test-anxious eleventh-grade students (Fyans, 1979b) on a mathematics examination. This massive effect of test anxiety upon test performance would make a competent student (say 95th percentile) appear to be in need of Minimal Competing Remediation (test score of 62%)!

Modifications in Testing Procedures

Research reviewed earlier suggests a number of ways (achievement) testing practices can be modified to minimize or eliminate motivational dynamics which interfere with test performance and result in an underestimate of the skills a student possesses and other forms of achievement. We will comment briefly on several of the testing parameters showing greatest promise in terms of reducing motivational test bias.

There is considerable research data pointing to time pressure as an aspect of achievement/educational testing which interferes with the performance of high test-anxious students more than that of other students. Discussions with teachers and other school staff also frequently point to time limits and potential time pressure as an important source of anxiety and other negative motivational tendencies. Studies reviewed earlier suggest that the removal of time limits often results in dramatic increases in the performance of high anxious-low performing students, indicating that they know the material being tested but experience motivational or other test-taking problems under standard testing conditions.

It has been our experience testing elementary and junior high school students with a variety of achievement tests that increasing time limits between 50 and 100% allows nearly all children to finish a test comfortably. Since classroom teachers give many achievement tests, it is relatively easy and nondisruptive to have children who finish early under the longer time limits simply work at their desks on silent sustained reading or other materials. Standardized tests would have to be renormed, of course, if longer time limits are used. But these norms might be initially developed during pilot testing of new items, content areas, and the like, of the tests. Such testing would also provide useful data on the effects of the new time limits (especially if motivational measures were collected, too). The cost of renorming tests and the longer testing sessions would be justified by the more valid scores resulting from the removal of time pressure and the accompanying motivational test bias.

There is also extensive evidence that failure, especially repeated failure, further lowers the performance of children showing a pattern of negative motivation and low test performance, for example, children a year or more below norm on a standardized test. The problem is made worse by the fact that students from various grade levels take the same achievement test (4th through 6th graders, 5th through 8th graders, etc.). It seems inexcusable from what we know about motivation to expose an upper elementary grade child who is performing well below norm to a test given to students two or three grade levels above his own. This virtually guarantees almost total failure, strong negative motivational dynamics, and a test score with little meaning.

There are several possibilities for dealing with the effects of repeated failure. A simple one is to give students information about the difficulty level of the test, what grade levels of students take the test, that all students are not expected to get all of the answers correct, etc. A second strategy is to bracket the ability of the test around the student's current level of functioning. Students below norm could be given tests typically taken by younger students, and perhaps students scoring above norm could be given somewhat more advanced tests. At a minimum, tests could be given that are developed for each grade level in reading, mathematics, etc., at the elementary and junior high levels instead of being given to children in two, three, or four adjacent grade levels; this would eliminate the most serious mismatches in the student's ability level and the test difficulty level in current testing programs. Not all of the present standardized achievement tests, of course, are given to multiple grade levels.

The final testing parameter worthy of discussion here concerns testing instructions, response formats, and answer sheets used in standardized testing. Initial research results and extensive discussions with teachers and other school personnel suggest that these non-content cognitive demands of tests may be important determinants of motivational test bias. These effects may be strongest when the student first takes a standardized test, typically in the early elementary school grades, or when some new kind of demand is introduced, such as the use of computerized answer sheets about the fourth grade. The use of a full scale practice test given prior to the actual testing days is recommended for the early and middle elementary school years. Such a test would expose children to the instructions, kinds of questions and answer formats, answer sheets, and several sample problems in each area to be subsequently tested. This would avoid children becoming confused about any of these testing parameters during the test, especially the high anxious-low performing child. Such test-taking practice should be particularly helpful on tests with demanding time limits or on difficult tests, for which test pressure would be strong. A practice test also provides an opportunity for the teacher and students to discuss other aspects of testing reviewed here.

Development of New Testing Programs

It is recognized that much more data is needed on the effects of motivation on test performance, and it will take a good deal of time to develop new testing procedures and restandardize achievement tests in ways suggested here. In the meantime, one option in testing programs at the school building, district, and other levels is to develop a dual testing program in which students are tested twice, once under standard testing conditions and once under optimizing testing conditions. Order of testing, of course, would have to be examined and controlled for through some counter-balanced test-retest designs and the like. For simplicity of presentation, we will consider the case here where all children take an achievement test under standard conditions first and some or all students retake the test under optimizing conditions second (with test-retest control groups).

Students to be tested in the optimizing testing condition might be chosen on the basis of motivation-performance patterns shown during the standard testing, teacher recommendations, a low performance cut-off, or other means. Here students might be given a full practice session, more time, a less difficult test, and tested individually. The main interest would be in whether students showed marked gains in the optimizing testing program over and above the typical test-retest performance gain; such a result would provide evidence, first, that the student knew more than shown during the standardized testing, and second, that the student had a motivational or test-taking problem. Such a student should in particular benefit from teaching activities designed to help students master test-taking skills and develop positive motivation in standard testing situations. Results from the optimal testing condition should also be most useful for the teacher in terms of diagnosing what children have learned and what their educational needs are. Such a dual testing program might initially focus on the high anxious-low performing student and eventually include other students if the optimal testing procedures prove effective and useful. As suggested earlier, the least anxious-highest performing students may actually perform better under standard testing conditions, again pointing to the desirability of a dual testing program.

A dual testing program is particularly desirable for minimal competency testing programs. The very threat of failure which could lead to grade repetition, exclusion from desired courses, failure to graduate from high school, etc., should greatly enhance testing pressure and trigger strong negative motivational dynamics. Minimal competency tests which (for anxious students can) involve stringent time limits, complex instructions, fairly difficult materials, and other factors discussed here would further exacerbate interfering motivational dynamics.

Since the high anxious-low performing student of central interest in this chapter is most likely to fail minimal competency tests, it seems particularly important that these tests be developed and given in ways which minimize motivational test bias. Otherwise students will fail because they are unable to show what they know under the particular testing conditions, *not* because they don't possess the competencies being measured. Again, we make the fundamental distinction between knowing the material and being able to show that knowledge under certain artificial testing conditions.

Development of New Teaching Activities and Integrated Testing Programs

Our development of new teaching activities and classroom experiences to help children acquire test-taking skills and positive test motivation are in the early stages and just a few comments will be made here. It seems clear, on the one hand, that many of the behaviors the child brings to the achievement test situation come from classroom experiences. On the other hand, it has also been suggested that the standardized achievement test situation differs in a number of ways from classroom testing situations, especially for elementary school children. One way to teach children skills to cope with demands of standardized testing, then, is to slowly introduce these demands into the classroom (testing) activities over the elementary school years as the teacher shows the student how to deal with them. Taking time pressure, as an example, young elementary school children might always be given plenty of time to do workbook assignments that are closest to tests (assuming that the children stay on task). Time limits can be gradually introduced to classroom tests as students learn over dozens of classroom experiences that they may not be able to finish a test no matter how hard and efficiently they work. Children can be taught to persist, to work as quickly as possible without undue errors, and to derive satisfaction from doing their best (and not necessarily finishing a test and doing all of the problems). In effect, the teacher is teaching children how to cope with and function effectively under time pressure and to accept whatever performance level results from maximal, appropriate effort.

In a similar manner, teachers can gradually introduce and teach children how to deal with difficult—sometimes impossible—problems on tests, with complex instructions and answer sheets, with competition and social comparisons, and with other aspects of testing which mediate motivational test bias.

As in the ongoing longitudinal research and demonstration project described earlier, the school's basic skill/achievement testing program can be carefully coordinated to the classroom teaching activities. Highly evaluative aspects of testing conditions (time pressure, etc.) should be introduced after most students at that age level have mastered the relevant testing skills, developed positive motivation in such testing conditions, and learned to do their best when tested in these conditions. School districts can, then, select (achievement) tests to be given at different elementary school grades having testing demands which integrate well with the classroom testing/teaching programs.

Students could be ready to take standardized tests as they are now given by the end of the elementary school years, assuming that test administration will not change quickly. It would be even better to extend the transition period of preparing students for highly evaluative testing into the middle-school or junior high school level. Integration of classroom teaching activities (outside of the self-contained classroom at the junior high level) and testing programs (now across the elementary and junior high school buildings) becomes much harder, however, across these additional years.

Epilogue

Lori is a fourth grader in the same Jefferson Elementary School classroom that Mark attended seven years earlier. There has been relatively little change in the make-up of the school's student population. Lori also does about average school work, but unlike Mark, she does about as well on achievement tests as in her classroom work, usually a little above the 50th percentile.

It is September and the teacher is preparing to give this fall's school achievement test. Lori is used to being tested, having taken a number of spelling, arithmetic, and other classroom tests over the past several years. Several days ago the teacher gave a "practice test" so that everyone in Lori's class would understand how to take the test. The teacher reminded that learning to take tests and doing one's best on them was one of the things to be learned in school. Lori had practice with all of the instructions and knows the kinds of problems that will be on the test. She has learned to use the computerized answer sheet; this was something new to her and the rest of the class, and it took a while to master its use during the several short practice tests. The teacher mentioned that fifth and sixth graders take the same regular test as Lori's class will, so the test will be difficult. Lori is used to taking tests with difficult problems in school, however, and she knows that what is important is that she does all of the ones she can and not to worry if she can't do some of the problems. The practice test, in fact, included several practice problems Lori had never seen before and simply could not do, and she knows now that she should just skip over such problems and not become concerned about them.

The teacher introduces the test and explains that the class will be taking several tests over the next several days to find out what each student knows. The standard instructions developed by the testing company have some additions since Mark took the test. The teacher mentions, for example, that if the child does not know a problem, he should simply move on to the next one and that no one is expected to get all of the problems correct. The teacher indicates that there will be a time limit on each test, but this does not bother Lori. She is used to taking tests with time limits in class and knows that she may not finish a test. She also knows that what is important is to keep working hard and to work fairly quickly but not so fast as to make careless mistakes. During the test, some classmates work more quickly than Lori and some more slowly, but she does not pay a lot of attention to this.

The testing goes on for two more days, and Lori continues to work hard at each section. She does better on some subtests than others and although she does not know it, her work places her at about the 60th percentile. She has experienced some nervousness during the testing, but generally speaking has done well on the test, answering correctly most of the problems she knows how to do.

Lori also is just one hypothetical child taking part in a hypothetical teaching-school testing program and a representative achievement test in a typical fourth-grade classroom. But her case illustrates the positive gains in both self-confidence/positive motivation and achievement/other test performance that may be possible to achieve if motivational test bias can be minimized or better yet eliminated. To do so will take considerable initial effort on the part of her teacher and school district and a non-trivial amount of Lori's classroom time. But Lori's performance on achievement, aptitude, and other educational tests will play a critical role in her educational future, helping determine what she is taught in elementary school, what classes she takes in secondary school, whether she graduates, if and where she receives further education, and possibly what jobs she is offered and her success and promotions in those jobs. And in contrast to Lori, there are hundreds of thousands if not millions of children in this country from all sociocultural backgrounds with test-taking histories like Mark's who suffer the harmful consequences of motivational test bias. If the kinds of results reviewed here continue to accumulate and become stronger as research moves further into classroom teaching activities and school/general educational testing programs, then the time and effort needed to provide children like Mark with the test-taking history, classroom experiences, and confidence/testing success of Lori will be more than justified and a significant contribution to their education and lives.

REFERENCES

Atkinson, J. W. *An introduction to motivation.* Princeton, New Jersey: Van Nostrand, 1964.

Atkinson, J. W., & Feather, N. T. *A theory of achievement motivation.* New York: Wiley, 1966.

Atkinson, J. W., & Raynor, J. O. *Motivation and achievement.* Washington, D.C.: J. H. Winston, 1974.

Bodine, R. J., Hill, K. T., & McPhail, A. *A general model for report card change.* Paper presented at a conference on "Beyond Report Card Grades: Problems and Alternatives," University of Illinois, Urbana, May, 1976.

Bowers, K. S. Situationism in psychology: An analysis and a critique. *Psychological Review*, 1973, *80*, 307-336.

Brennan, M. & Maehr, M. L. *Description and theoretical overview of the Illinois Office of Education/University of Illinois at Urbana-Champaign project on motivation and achievement.* Paper presented at the annual meetings of the American Educational Research Association, San Francisco, April, 1979.

Coleman, J. S., Campbell, E. O., Hobson, C. J., McPartland, J., Mood, A. M., Weinfeld, F. D., & York, R. L. *Equality of educational opportunity.* U. S. Department of Health, Education, and Welfare. Washington, D.C.: U. S. Government Printing Office, 1966.

Crandall, V. C. Achievement behavior in young children. *Young Child*, 1967, *1*, 165-185.

Cronbach, L. J. Five decades of public controversy over mental testing. *American Psychologist*, 1975, *30*, 1-14.

Dusek, J. B., & Hill, K. T. Probability learning as a function of sex of the subject, test anxiety, and percentage of reinforcement. *Developmental Psychology*, 1970, *3*, 195-207.

Dusek, J. B., Mergler, N. L., & Kermis, M. D. Attention, encoding, and information processing in low- and high-test-anxious children. *Child Development*, 1976, *47*, 201-207.

Dweck, C. S. The role of expectations and attributions in the alleviation of learned helplessness. *Journal of Personality and Social Psychology*, 1975, *31*, 674-685.

Dweck, C. S. Learned helplessness and negative evaluation. *UCLA Educator*, 1977, *19*, 44-49.

Eaton, W. O. Profile approach to longitudinal data: Test anxiety and success-failure experiences. *Developmental Psychology*, 1979, *15*, 344-345.

Ekehammar, B. Interactionism in personality from a historical perspective. *Psychological Bulletin*, 1974, *81*, 1026-1048.

Endler, N. S., & Magnusson, D. Toward an interactionist psychology of personality. *Psychological Bulletin*, 1976, *83*, 956-974.

Feld, S. C., & Lewis, J. The assessment of achievement anxieties in children. *Achievement-Related Motives in Children*, 1969, 151-199.

Fyans, L. J., Jr. *Test anxiety, test comfort, and student achievement test performance.* Paper presented at the Educational Testing Service, Princeton, New Jersey, July, 1979. (a)

Fyans, L. J., Jr. *Sociocultural developmental differences in continuing motivation, anxiety, and achievement attributions.* Paper presented at the annual meetings of the American Psychological Association, New York, September, 1979. (b)

Gezi, K. Bilingual-bicultural education: A review of recent research. *California Journal of Educational Research*, 1974, *25*, 223-239.

Hastings, J. T. *Motivation, measurement, and the illusion of minimal competency.* Paper presented at the annual meetings of the American Educational Research Association, San Francisco, April, 1979.

Hill, K. T. Social reinforcement as a function of test anxiety and success-failure experiences. *Child Development,* 1967, *38,* 723-737.

Hill, K. T. Anxiety in the evaluative context. *Young Child,* 1972, *2,* 225-263.

Hill, K. T. Individual differences in children's response to adult presence and evaluative reactions. *Merrill-Palmer Quarterly,* 1976, *22,* 99-104, 118-123.

Hill, K. T. The relation of evaluative practices to test anxiety and achievement motivation. *UCLA Educator,* 1977, *19,* 15-21.

Hill, K. T. *Relation of motivational variables to measured achievement and implications for achievement assessment procedures.* Paper presented at the annual meetings of the American Educational Research Association, San Francisco, April, 1979. (b)

Hill, K. T. *Eliminating motivational testing error by developing optimal testing procedures and teaching test taking skills.* Paper presented at the Educational Testing Service, Princeton, New Jersey, July, 1979. (a)

Hill, K. T., & Eaton, W. O. The interaction of test anxiety and success/failure experiences in determining children's arithmetic performance. *Developmental Psychology,* 1977, *13,* 205-211.

Hill, K. T., & Sarason, S. B. The relation of test anxiety and defensiveness to test and school performance over the elementary school years: A further longitudinal study. *Monograph of the Society for Research in Child Development,* Serial No. 104, 1966, *31,* (Whole No. 2).

Holtzman, W. H., Diaz-Guerrero, R., & Swartz, J. D. *Personality development in two cultures: A cross-cultural longitudinal study of school children in Mexico and the United States.* Austin: University of Texas Press, 1975.

Houts, P. L. (Ed.). *The myth of measurability.* New York: Hart Publishing Company, 1977.

Hutt, M. L. Consecutive and adaptive testing with the revised Stanford-Binet. *Journal of Consulting Psychology,* 1947, *11,* 93-104.

Jencks, C., Smith, M., Acland, H., Bane, M. J., Cohen, D., Gintis, H., Heyne, B., & Michelson, S. *Inequality: A reassessment of the effect of family and schooling in America.* New York: Harper and Row, 1973.

Lazarus, M. Testing: Measure of achievement of test-taking ability. *New York Times,* May 1, 1977.

Lekarczyk, D. T., & Hill, K. T. Self-esteem, test anxiety, stress, and verbal learning. *Developmental Psychology,* 1969, *1,* 147-154.

Linn, R. L. *Achievement test item characteristic effects, motivational and sociocultural item test bias, and teacher judgments of items.* Paper presented at the annual meetings of the American Educational Research Association, San Francisco, April, 1979.

Maehr, M. L. Continuing motivation: An analysis of a seldom considered educational outcome. *Review of Educational Research,* 1976, *46,* 443-462.

Maehr, M. L. Turning the fun of school into the drudgery of work: The negative effects of certain grading practices on motivation. *UCLA Educator*, 1977, *19*, 10-14.

Maehr, M. L. *Theoretical overview of Illinois Office of Education/University of Illinois, Urbana-Champaign project on motivation and achievement.* Paper presented at the annual meetings of the American Psychological Association, New York, September, 1979.

Maehr, M. L., & Stallings, W. M. Freedom from external evaluation. *Child Development*, 1972, *43*, 177-185.

Masters, J. C. Social comparison in young children. *Young Child*, 1972, *2*, 320-339.

National School Boards Association. *Standardized achievement testing.* Washington, D.C.: NSBA, 1977.

Nottelmann, E. D., & Hill, K. T. Test anxiety and off-task behavior in evaluative situations. *Child Development*, 1977, *48*, 225-231.

Plass, J. A. *Optimizing children's achievement test strategies and performance: The role of time pressure, evaluation anxiety, and sex.* Unpublished master's thesis, University of Illinois, Urbana-Champaign, 1979.

Plass, J., & Hill, K. T. *Optimizing children's achievement test performance: The role of time pressure, evaluation anxiety, and sex.* Paper presented at the Annual Meetings of the Society for Research in Child Development, San Francisco, March, 1979.

Quinto, F., & McKenna, B. *Alternatives to standardized testing.* Washington, D.C.: National Education Association, 1977.

Ramirez, M., III, & Price-Williams, D. R. Achievement motivation in children of three ethnic groups. *Journal of Cross-Cultural Psychology*, 1976, *7*, 49-60.

Ruebush, B. K. Anxiety. In H. W. Stevenson, J. Kagan, & C. Spiker (Eds.), *Sixty-second yearbook of the national society for the study of education. Part I: Child psychology.* Chicago: University of Chicago Press, 1963, 460-517.

Salili, F., Maehr, M. L., Sorenson, R. L., & Fyans, L. J., Jr. A further consideration of the effects of evaluation on motivation. *American Educational Research Journal*, 1976, *13*, 85-102.

Sarason, I. G. Experimental approaches to test anxiety: Attention and the uses of information. C. D. Spielberger (Ed.), *Anxiety: Current trends in theory and research.* New York: Academic Press, *2*, 1972.

Sarason, S. B., Davidson, K. S., Lighthall, F. F., Waite, R. R., & Ruebush, B. K. *Anxiety in elementary school children: A report of research.* New York: Wiley, 1960.

Sarason, S. B., Hill, K. T., & Zimbardo, P. G. A longitudinal study of the relation of test anxiety to performance on intelligence and achievement tests. *Monographs of the Society for Research in Child Development*, Serial No. 98, 1964, *29*, (Whole No. 7).

Shea, B. M. Schooling and its antecedents: Substantive and methodological issues in the status attainment process. *Review of Educational Research*, 1976, *46*, 463-526.

Smith, C. P. *Achievement-related motives in children.* New York: Russell Sage, 1969.

Spielberger, C. D. *Anxiety: Current trends in theory and research.* New York: Academic Press, 2, 1972.

Veroff, J. Social comparison and the development of achievement motivation. In C. P. Smith (Ed.), *Achievement-related motives in children.* New York: Russell Sage, 1969, 46-101.

Weiner, B. *Theories of motivation.* Chicago: Markham Publishing Company, 1972.

Weiner, B. An attributional approach for educational psychology. In L. S. Shulman (Ed.), *Review of research in education, 4.* Istasca, Ill.: F. E. Peacock, 1977. (A publication of the American Educational Research Association.)

Weiner, B., Frieze, I., Kukla, A., Reed, L., Rest, S., & Rosenbaum, R. M. Perceiving the causes of success and failure. In E. E. Jones, D. Kanouse, H. H. Kelley, R. E. Nisbett, S. Valins, & B. Weiner (Eds.), *Attribution: Perceiving the causes of behavior,* McCaleb-Seiler, 1971.

Williams, J. P. *Individual differences in achievement test presentations and evaluation anxiety.* Unpublished doctoral dissertation, University of Illinois, Urbana-Champaign, 1976.

Wine, J. Test anxiety and direction of attention. *Psychological Bulletin,* 1971, 76, 92-104.

Wirtz, W., et al. *On further examination: Report of the advisory panel on the scholastic aptitude test score decline.* New York: College Entrance Examination Board, 1977.

Zigler, E., Abelson, W. D., & Seitz, V. Motivational factors in the performance of economically disadvantaged children on the Peabody Picture Vocabulary Test. *Child Development,* 1973, *44,* 294-303.

Zigler, E., & Butterfield, E. C. Motivational aspects of change in I.Q. Test performance of culturally deprived nursery school children. *Child Development,* 1968, *39,* 1-14.

Zigler, E. F., & Harter, S. Socialization of the mentally retarded. In D. A. Goslin (Ed.), *Handbook of socialization theory and research.* Chicago: Rand McNally, 1969.

Chapter 5

Effects of Failure: Alternative Explanations and Possible Implications[1]

MARGARET M. CLIFFORD
University of Iowa

Is failure[1] as devastating as popular opinion and humanist psychologist imply (Glasser, 1969)? Is there conclusive evidence to support the prevalent practice of minimizing the amount of failure students experience? Are the levels of success associated with the use of programmed materials, inflated grades, and mastery learning techniques ensuring optimum student motivation and cognitive development? A review of theories and research related to the effects of failure suggests that these are not naive questions and cannot be given simplistic answers. Rather, such a review leads one to conclude that educators who teach by the maxim, *"Nothing succeeds like success,"* at least sometimes may be doing students more harm than good.

There are several psychological theories which consider the effects of failure on human behavior: among these are the theories of frustration, learned helplessness, reactance, attribution, and achievement motivation. A brief description of each of these theories and a sampling of related literature will be presented. Conclusions and possible implications for future research as well as for educational practice will then be proposed.

---

*Reprinted by permission of author and publisher of *Educational Psychologist*, 1979, *14*, 44-52. Copyright (1979) by the American Psychological Association.

[1]Failure and error-making are used interchangeably in this article and simply imply an incorrect response to a task or situation.

Frustration

Frustration may be defined as *a state of increased arousal resulting from failure to obtain the type of reinforcement previously experienced under similar conditions.* Amsel (1958) argues that frustration resulting from the denial of an anticipated reinforcer has drive or energizing properties which increase the speed or intensity of subsequent behavior. Amsel and Roussel (1952) demonstrated that when rats were denied food in a goal box located halfway through a runway they advanced to the end of the goal box (where food was always received) more quickly than when they were fed at the midway point. Haner and Brown (1955) demonstrated frustration effects with children who were required to place 36 marbles into a pegboard to win a prize. The experimenter surreptitiously caused the marbles to roll out of place at various stages of task completion. Children had been forewarned that if any marbles were to fall out of place they were to push a lever to reset the board and begin the task again. Supportive of the experimenters' prediction, it was found that the closer to task completion the child was when the marbles were dislodged, the harder he pushed the reset lever.

Amsel (1972) contends that an organism exposed to partial reinforcement actually learns to perform in the absence as well as the presence of a reinforcer. Therefore, the organism is more resistent to extinction. That is, Amsel attributes persistence during extinction to the fact that stimuli associated with the reinforced trials are also associated with the nonreinforced frustration trials present in any partial reinforcement situation. Continuously reinforced subjects on the other hand do not learn the frustration-related goal-approach response and, therefore, show little persistence during extinction.

Jones, Nation, and Massad (1977) used Amsel's (1958) extension of frustration theory to explain the performance of college students who had previously experienced various levels of failure. In a pair of experiments they examined different levels of reinforcement as possible means of protecting subjects from the devastating effects of subsequent failure. In the first phase of their initial study, subjects experienced either 0%, 50%, or 100% success on a set of discrimination problems. In the second phase all subjects experienced 100% failure on a similar set of problems. In the third and final phase, performance on a mechanical problem-solving task was assessed and used as the dependent measure. Subjects experiencing 50% success in phase 1 performed significantly better than subjects in either of the other two groups. Somewhat similar results were reported for the second study for which performance on anagrams served as the dependent measure and for which two additional control groups (intended to serve as baselines of performance) were developed. One group was given soluble problems during phase 1 and phase 2, and one group was exposed to the problems used in phase 1 and 2 but not allowed to work them or receive any feedback. In addition to the superior performance of the 50% success group over that of the 0% and 100% success groups, it was reported that there was no difference between the two control groups or between either of the controls and the 50% success group.

The authors argued that 50% success provided a partial reinforcement situation in which frustration (failure to obtain success feedback) became associated with the approach response (direction of attention and problem-solving effort toward subsequent problems). This pairing of frustration and approach behavior was used to explain the resistance to failure or the relatively high performance demonstrated during phase 3 by subjects in the 50% success group as compared to those in the 100% and 0% success group. In short, subjects assigned to the 50% success group during phase 1 were thought to have been conditioned to attempt problem-solving *in spite of failure*.

A simple comparison of the three treatment groups would indeed suggest that frustration enhances performance. However, a more careful comparison of these groups against the controls suggests something less supportive of frustration theory: The 50% success group (frustration) did not perform significantly better than either of the controls used in the second study, but both the 0% and 100% success groups performed significantly poorer than these controls. Thus the results of this study might better be explained in terms of the *detrimental* effects of 0% and 100% success feedback rather than in terms of the *enhancing* effects of frustration (50% success feedback). In any event, the usefulness of frustration theory for explaining the effects of failure must be judged in relationship to alternative motivation theories.

Learned Helplessness

Cognitive psychologists studying failure within the context of learned helplessness have emphasized the detrimental effects of failure. They generally content that *excessive failure leads the subject to perceive an independence between his behavior and the consequences of that behavior*. This cognition or perception is postulated to be the cause of decreased or abnormal performance in subsequent situations where successful performance would otherwise be expected (Seligman, Maier, & Soloman, 1971). Learned helplessness theorists argue that S-R theories, which include frustration, are far less appropriate and useful for explaining the effects of failure, than are cognitive theories:

> The learned helplessness hypothesis has been stated in cognitive language, whereas most of the alternative views have been stated in S-R language. We have found it difficult to even approach the sort of phenomena that we have tried to explain with a S-R framework, and have found the cognitive theorizing to be more fruitful, and to reflect more accurately those processes that we feel to be reflected in behavior (Maier & Seligman, 1976, p. 41).

Led by Seligman and Maier (1967), learned helplessness researchers have repeatedly demonstrated the detrimental effects of failure (Overmier & Seligman, 1967; Seligman & Groves, 1970: Seligman, Maier & Geer, 1968) and have also shown that prefacing failure with high levels of success, assumed to result in perceptions of control or a dependency between one's behavior and its outcome, greatly mitigates these detrimental effects (Seligman & Maier, 1967; Seligman, Rosellini, & Kozak, 1975; Hiroto & Seligman, 1975). Such evidence has been used to support the contention that cognitions rather than the lack of reinforcements as such are major determinants of the effects of failure. Yet discrepancies within the learned helplessness literature raise questions regarding the exact nature of the relationship between failure and subsequent performance and also suggest that helplessness may not be as reliable a phenomena as originally postulated.

While much of the learned helplessness research supports a decreasing monotonic relationship between perceived lack of control and performance, not all of the research is consistent with that model. Thornton and Jacobs (1971), for example, found that introductory psychology subjects experiencing inescapable shock gained more on mental ability as measured from pretest to posttest than did subjects experiencing avoidable shock or no shock training. Roth and Bootzin (1974) published results which likewise contradicted more traditional learned helplessness studies. In their study introductory psychology subjects receiving random reinforcement (assumed to produce perceptions of uncontrollability or independence between behavior and consequences) on a concept learning task, exhibited more controlling behavior in a subsequent problem-solving situation than did subjects previously receiving contingent reinforcement or subjects receiving no pretreatment. Roth and Kubal (1975) using male and female undergraduates demonstrated both the facilitative and detrimental effects of failure; they showed that the detrimental effects became evident and increasingly pronounced only after a given amount of failure had been experienced, and that this trend occurred more readily for tasks of high as opposed to low importance. All three of these studies are at least compatible with the hypothesis that a curvilinear as opposed to a decreasing monotonic relationship exists between perceived lack of control (assumed to result from failure) and performance. Such a model implies that *moderate amounts of failure optimize performance.* A more formal presentation of this postulated curvilinear relationship, implying both the facilitative and detrimental effects of failure, has been provided within the framework of reactance theory by Wortman and Brehm (1975). Others such as Miller and Norman (1979) and Abramson et al. (1978) have suggested that learned helplessness needs to be reinterpreted in terms of attributions or causal explanations.

Reactance Theory

Brehm (1966, 1972) postulates that an individual whose behavioral freedom is threatened is motivationally aroused and attempts to restore his freedom. This *motivational response to threat-to-freedom is called reactance.* The effects of reactance include changes in the attractiveness of uncontrollable outcomes, attempts to engage in the threatened behavior, and expressions of aggressive and hostile feelings (Wortman & Brehm, 1975).

If one assumes the "freedom to succeed" is widely prized and views lack-of-control-for-outcomes as a threat to that freedom, it can be argued that reactance theory is highly applicable to failure situations. The parameters of reactance theory are expectation for freedom, strength of threat, and importance of threat-to-freedom. These parameters are similar to those associated with other motivational theories which attempt to explain the effects of failure (e.g., need achievement and learned helplessness).

Perceiving a similarity of focus between reactance theory and the learned helplessness model, Wortman and Brehm (1975) attempted to integrate them. They argued that loss of freedom—including unwanted and unexpected failure—will initially produce an increase in performance followed by a gradual and continuing decrease in performance. They further suggest that task importance will accentuate and accelerate the peak and drop of this function. That is, as an outcome increases in importance, less threat-to-freedom or failure is needed to produce an increase in performance and subsequently elicit a drop in performance. Both the facilitative and detrimental effects of threats are expected to be intensified by task importance. Wortman and Brehm argue that the devastating effects of failure, sometimes referred to as learned helplessness, are representative of only part of their model— the part associated with extended threat-to-freedom or failure. They proceed to suggest that attribution for failure may be an added factor which determines the nature and intensity of either the facilitative or detrimental effects of failure. There is increasing support for this position (Miller & Norman, 1979; Abramson et al., 1978).

Attribution Theory

Attribution theory attempts to explain feelings and behaviors in terms of *perceived cause-effect relationships* (Jones et al., 1972). The nature of the causal attribution associated with the success or failure outcome has been found to affect the evaluative judgment of that outcome (Rest et al., 1973; Lanzetta & Hannah, 1969). For example, greater punishment is administered to individuals who supposedly fail at an easy as opposed to a difficult task, for it often is assumed that such individuals lacked effort (Lanzetta & Hannah, 1969). In a series of studies in which college and high school subjects were asked to role play a teacher evaluating student performance, it was reported that evaluations were positively related to the amount of effort supposedly expended by pupils and inversely related to the level of ability these pupils were reported to have had (Weiner & Kukla, 1970).

Attributional effects have not only been observed in situations in which one individual judges the work of another, but also in situations in which subjects react to their own success and failure under different attributional conditions. For example, Wortman et al., (1976) demonstrated that undergraduate females led to attribute their own failure to lack of ability as opposed to task difficulty reported experiencing relatively more stress following failure on a problem-solving task. Furthermore, individuals encouraged to attribute their failure to task difficulty reported no more stress than subjects who experienced success during training. Contrary to prediction, performance on a perceptual judgment and concept learning task, administered after the failure and attribution manipulation, was higher for the subjects who attributed failure to incompetency and reported experiencing high stress, than it was for subjects who attributed failure to task difficulty and reported less stressfulness. Subjects given ability attribution for failure, later expressed that they were more "involved" with the task than subjects given task difficulty attributions. Thus, at least in this instance self-reports of stressfulness, assumed to result from attributing failure to inability, were also accompanied by self-reports of involvement or concern.

The authors postulated that the "stress experienced by subjects who fail to control aversive stimulation is not a function of the lack of control as such, but of the attribution of causality that they make for their failure to exert control" (Wortman et al., 1976, p. 311). It was observed that "subjects whose performance led them to feel relatively incompetent... performed better than subjects whose performance had no negative implications for their competence" (p. 312). This superior performance was not only manifested in the same situation but generalized to a different set of problems presented under different circumstances. Thus, an attribution associated with threat to perceived competence, rather than threat to control, was offered as the predominant determinant of increased performance.

This would appear to contradict findings which suggest that attributions of failure to ability lead to increased learned helplessness while attributions of failure to task difficulty produce little or no helplessness (Dweck & Reppucci, 1973; Klein et al., 1976; Tennen & Eller, 1977). Indeed, Miller and Norman (1979) have challenged Wortman et al.'s (1976) data interpretation and have argued that their subjects' awareness that the study was designed to examine the effects of noise on performance, coupled with the absence of noise during final testing, provided a situation of increased expectancy for success by those individuals who previously attributed failure to low ability. Miller and Norman also contend that there would be no such change in expectancy for individuals attributing failure to task difficulty. This line of argument lacks consistency, however. For if the absence of noise during testing could lead to increased ability expectations, why could it not also lead to decreased task difficulty expectations? On the other hand, whether or not ability attributions following failure enhance or hinder performance may in part be a matter of the certainty or uncertainty associated with one's attributions. There is in fact some evidence to support such a hypothesis.

Lee (1978), conducting a study designed to reexamine

helplessness and attributional effects simultaneously, manipulated level of failure with objective feedback and ability attributions with normative feedback. In the first phase of her study she gave college students a figure-matching task on which they received either 85% or 50% success feedback. Subjects in each condition were then told that their performance, relative to other students, was either "excellent" or "poor." In the second phase of the study all subjects experienced 90% failure on a similar figure-matching task. In the third and final phase subjects worked a third such task which provided the dependent measure. Lee failed to find a main effect for attribution or initial feedback level, but did report an interaction between these factors: Subjects receiving 85% success followed by norms which suggested they had done "excellent," tended to perform more poorly following helplessness training (90% failure) than did subjects initially receiving 85% success followed by norms which suggested that their performance was "poor." On the other hand, subjects who received 50% success feedback and were told they had done "excellent," tended to do better than those who received 50% sucess and were told that their performance was "poor." Furthermore, there was no significant difference between the 85% success-"excellent" group and the 50% success-"poor" group.

The lack of difference between the 85%-"excellent" and the 50%-"poor" group was judged to be in conflict with previously cited evidence used to support the contention that lack of control (presumably manipulated through feedback) was the major determinant of helplessness. Failure to find that the 50% success group performed superior to the 85% group provided grounds for challenging the frustration theory explanation. Failure to find the reverse effect was viewed as a challenge to the more traditional reinforcement explanation, typified in the cliche, "nothing succeeds like success." For the observed interaction between feedback and ability attributions, Lee offered a resultant discrepancy explanation which is consistent with dissonance theory (Festinger, 1957). She postulated that the degree to which cognitions are discrepant (e.g., low feedback is followed by norm information suggesting "excellence") rather than the degree to which there is perceived control, (optimized in the 85% feedback "excellent" condition) determines the degree to which there is resistance to failure and/or delays or reductions in helplessness.

An alternative way to interpret Lee's results is within the context of an inability to make trustworthy attributions. It may be that a contrast of success and failure feedback, or failure feedback for one who generally holds success expectancies, leads to an uncertainty about cause-effect relationships and prompts a response directed at resolving the uncertainty in the most favorable way. Efforts toward this end would undoubtedly be greater for an individual attributing failure to ability than for an individual attributing failure to task difficulty. Such efforts would also be more likely to occur when helplessness training is relatively brief. Thus the degree of uncertainty as well as the type of attribution may determine the effects of failure.

Achievement Motivation Theory

Achievement motivation theory and related research suggests that reactions to failure (as well as success) are predominantly determined by perceived task difficulty and either or both of two personality variables; namely, motive to succeed (M_s) and motive to avoid failure (M_{af}), the latter of which is often defined as anxiety. Since these are assumed to be inversely related, studies including either as well as both of these personality variables have been accepted as relevant to this theory. The highly symmetrical model proposed by McClelland et al. (1953) and elaborated by Atkinson (1957, 1958) implies that behavior related to achievement is an inverted "U"-shaped function of perceived probability of success for individuals with $M_s > M_{af}$ and a "U"-shaped function of perceived probability of success for individuals with $M_{af} > M_s$. Furthermore the "U"-shaped function associated with $M_{af} > M_s$ individuals represents avoidance behavior only while the inverted "U"-shaped function associated with $M_s > M_{af}$ individuals represents approach behavior only. These two functions join each other at the ends of the perceived probability of success (P_s) continuum or where $P_s=1.00$ (assured success) and where $P_s=.00$ (assured failure).

Nygard (1975), following an extensive review of the literature, suggested a revision of this model: He identified empirical evidence which supported a stronger interaction between personality type ($M_s > M_{af}$ and $M_{af} > M_s$) and perceived probability of success (P_s). Therefore, he argued the "U"-shaped and inverted "U"-shaped function typifying the behavior of $M_{af} > M_s$ and $M_s > M_{af}$ individuals, respectively, should intersect at success probabilities of about .15 and .85. This would suggest, among other things, that $M_{af} > M_s$ individuals have a greater tolerance for high degrees of failure and perform relatively better under such conditions as compared to $M_s > M_{af}$ individuals.

Compatible with this interaction model are the results of several studies: Feather (1961) demonstrated that $M_s > M_{af}$ men persist longer on an insoluble task when P_s is said to be .70 than when P_s is said to be .05, while the reverse is true for $M_{af} > M_s$ men. Weiner (1966) demonstrated that college students with low anxiety (assumed comparable to $M_s > M_{af}$) performed significantly better than students with high anxiety ($M_{af} > M_s$) on an objectively easy paired associate task to which failure norms were attached (implying high success); while low anxiety students performed worse than high anxiety students when presented with an objectively difficult paired associate task to which success norms were attached (implying moderately low success). Kukla (1974), atempting to eliminate the confounding of objective and normative task difficulty which was present in Weiner's study, designed a study in which objective difficulty was held constant. He demonstrated that high Ms college students performed better than low Ms students when they were led to believe that the task was moderately difficult and that the reverse pattern emerged when students were led to believe the task was easy.

Not all need achievement evidence, however, is supportive of this interaction between perceived task difficulty and $M_s >$ and/or M_{af} personality measures. Atkinson and Litwin (1956) for example, failed to find any evidence that male college students for whom $M_{af} > M_s$ avoided moderately difficult tasks in preference for low or high difficult tasks. Klinger (1966), following a rather extensive review of achievement motivation literature, argued that only about half of the relevant published studies cited significant relationships between measure of need achievement and performance supportive of the theory. He also contended that postdictive rather than predictive relationships are often discussed and defended and that "the pattern of hypothesis confirmation presents the theory with difficulties" (p. 291).

Researchers have attempted to improve this theory by introducing new variables. Raynor (1969, 1970), for example, suggested that perceived instrumentality (PI) or what might be described as the long-term value of success at a task is an important determinant of achievement behavior. Weiner (1967) has emphasized the need to add to the model the concept of inertial tendency (T_g), defined as that portion of motivation remaining from the preceeding trial and augmenting the next trial. Weiner has also made numerous efforts to integrate need achievement and attribution theory (1974). However, in its present form, it would appear that achievement motivation theory provides a relatively weak basis for explaining or predicting the effects of failure. Efforts need to be made to reduce the lack of parsimony inherent in the interdependency of the theory's probability and incentive terms, to improve the instruments with which the motive terms (M_s and M_{af}) are operationally defined, and to further resolve the sex-bias orientation of the theory which limits its generalizability predominantly to males.

Conclusions and Discussion

Failure has been demonstrated to have both detrimental and enhancing effects upon subsequent performance. These effects have been explained in terms of frustration, learned helplessness, reactance, attribution, and achievement motivation theory. Types of reinforcement, expectations, perceptions of control, causal explanations, and personality characteristics have been examined as determinants of the effects of failure. Yet there are many theorectical as well as practical concerns to be addressed.

Among the theoretical concerns, is the task of identifying the relative merits of the alternative explanations. Pitting one explanation against another in search of the "best" explanation may be inefficient if not useless. The simultaneous investigation of two or more theories, however, followed by attempts at theory integration, as exemplified by Lee (1978) and Wortman and Brehm (1975), may be a highly constructive way to improve our explanation, prediction, and control of the human effects of failure.

Reworking theories to explain data, rather than reworking data to defend theories is also likely to offer great payoffs. Weiner (1979), Miller and Norman (1979), and Abramson et al. (1978) exemplify this productive approach in their recent works which present theory modifications based on accumulated evidence of the nature and functions of attributions.

While data supporting the role of cognitions in general and attributions in particular in failure situations is impressive, two cautions are in order. First, we must not prematurely devalue other types of explanations. The behavior of the toddler reaching in vain for the ball under the sofa and behavior of the package-laden shopper repeatedly pressing an elevator button may be better explained by frustration within the S-R tradition than by cognitions. Individuals who are not easily aroused by discrepant cognitions, those who tend to be nonanalytical, and those who have learned to live comfortably in unpredictable environments may behave in ways that are difficult to explain in terms of attribution theory. We dare not assume that all humans at all times are as inclined to identify causal explanations for phenomena as is the college student whose occupation it is to study cause-effect relationships.

A second caution in developing cognitive explanations for responses to failure is warranted: We must incorporate into our motivational theories our knowledge of human cognitive development. If the causal explanations of young children differ in nature and sophistication from those of adults (Piaget & Inhelder, 1969), and if the causal explanations of the cognitively advanced adult differ from those of the cognitively disadvantaged adult, one might predict differences in responses to failure depending upon an individual's level of cognitive development. The dearth of research examining this *developmental* factor within the framework of motivational theories is incongruous with the mounting evidence emphasizing the determining nature of cognitions.

In addition to the theoretical concerns—which include identifying the circumstances under which motivational theories are relatively useful, integrating existing motivational theories, emphasizing the modification of motivational theories rather than an uncritical defense of them, and incorporating the factor of cognitive development into motivation theories—there are applied concerns to which we must turn our attention if the effects of failure are to be better understood and controlled within an educational setting.

Both educational philosophers and psychologists need to reassess the value and evidence that has been associated with academic success, mastery, and excellence as operationally defined by our educational system. There may be grounds for postulating that the inflated grades, mastery learning programs, and individualized instruction designed to maximize success for every student, may be creating less than optimum levels of anxiety, frustration, or reactance and may be producing situations in which attributional effects tend to be weak or nonproductive. Success which comes easy is not likely to be as highly valued as success which is difficult to achieve. With our educational practices designed to ensure success we may be conditioning students to be intolerant of performance which is less than perfect, to be conservative risk takers in learning situations, to retreat when they encounter failure, and to covet the known rather than to venture into the unknown.

Given the theories and research just reviewed it might be argued that there is a *need to identify optimum levels of failure in various learning situations,* discriminate students who are likely to benefit from failure from students less likely to do so, develop in educators and students a tolerance for failure in instructional settings, and suggest techniques through which constructive responses to failure can be reinforced. Laboratory and field studies might be designed to address questions such as the following:

(1) What are the effects of reinforcing the correction of

nondeliberate errors in addition to reinforcing initially correct responses?
(2) Does enabling and encouraging students to increase the level of risk (probability of error) associated with learning activities enhance learning?
(3) Will discriminating clearly between learning and testing activities conducted in schools encourage greater risk-taking in the former and improved performance in the latter?
(4) Does failure differentially affect students of different levels of cognitive development?
(5) What are the effects of encouraging students to associate failure with having made an "overly courageous attempt" rather than with "lack of ability or effort"?

Initially it may seem important to identify the students who benefit from error-making and those for whom it appears detrimental. Measures of competitiveness, past histories of academic success and failure, parental use of rewards and punishment, and psychological measures such as self-concept may provide a clue for such discrimination. On the other hand, perhaps a more constructive endeavor would be to modify the attitudes and behavior of the individual who strongly resists all threat of academic error-making—to reward him when he increased his intellectual risk-taking and to model for him constructive responses to failure.

Educational games can be designed which would allow students to choose risk levels and allow experimenters to manipulate reward contingencies. Game rules (e.g., use of bonus and penalty points and forfeiture of play) could be used to accentuate the value of initial success or the value of error correction or the value of high risk-taking. Games of solitaire as well as group activities might be used to investigate the effects of failure given different goals. Such games could be used in laboratory as well as field research to assess the effects of failure on learning, persistence, retention, and attitude.

Today's popular common sense notions regarding the academic, psychological, and emotional effects of success and failure need to be empirically challenged, and there are several theories which can provide exciting frameworks for doing so. Furthermore, the data resulting from this type of educational research will undoubtedly strengthen both our empirical and theoretical knowledge of the effects of failure.

REFERENCES

Abramson, Lyn Y., Seligman, Martin E.P., & Teasdale, John D. Learned helplessness in humans: Critique and reformulation. *Journal of Abnormal Psychology*, 1978, *87*, (1), 49-74.

Amsel, A. The role of frustrative nonreward in noncontinuous reward situations. *Psychological Bulletin*, 1958, *55*, 102-119.

Amsel, A. Behavioral habituations, counterconditioning, and persistence. In A. Block and W.K. Prokasy (Eds.), *Classical conditioning II*. New York: Appleton-Century-Crofts, 1972.

Amsel, A. & Roussel, J. Motivational properties of frustration: I. Effect on a running response of the addition of frustration to the motivational complex. *Journal of Experimental Psychology*, 1952, *43*, 363-368.

Atkinson, J.W. Motivational determinants of risk-taking behavior. *Psychological Review*, 1957, *64*, 359-372.

Atkinson, J.W. *Motives in fantasy, action, and society*. D. Van Nostrand Company, Inc., 1958.

Atkinson, J.W., & Litwin, G.H. Achievement motive and text anxiety conceived as motive to approach success and motive to avoid failure. *Journal of Abnormal Social Psychology*, 1956, *53*, 361-366.

Brehm, J.W. *A theory of psychological reactance*. New York: Academic Press, 1966.

Brehm, J.W. *Responses to loss of freedom: A theory of psychological reactance*. Morristown, N.J.: General Learning Press, 1972.

Dweck, C.S., & Reppucci, N.D. Learned helplessness and reinforcement responsibility in children. *Journal of Personality and Social Psychology*, 1973, *25*, 109-116.

Feather, N.T. The realtionship of persistence at a task to expectation of success and achievement-related motives. *Journal of Abnormal and Social Psychology*, 1961, *63*, 552-561.

Festinger, L. *A theory of cognitive dissonance*. Evanston, Ill.: Row & Peterson, 1957.

Glasser, William. *Schools without failure*. New York: Harper & Row, 1969.

Haner, C.F., & Brown, J.S. Clarification of the instigation to action concept in the frustration-aggression hypothesis. *Journal of Abnormal and Social Psychology*, 1955, *51*, 204-206.

Hiroto, D.S., & Seligman, M.E.P. Generality of learned helplessness in man. *Journal of Personality and Social Psychology,* 1975, *31,* 311-327.

Jones, Edward E., Davis, Keith K., & Gergen, Kenneth J. *Attribution: Perceiving the causes of behavior.* Morristown, N.J.: General Learning Press, 1972.

Jones, S.L., Nation, J.R., & Massad, P. Immunization against learned helplessness in man. *Journal of Abnormal Psychology,* 1977, *86,* 75-83.

Klein, D.C., Fencil-Morse, E., and Seligman, M.E.P. Learned helplessness, depression, and the attribution of failure. *Journal of Personality and Social Psychology,* 1976, *33,* (5), 508-516.

Klinger, E. Fantasy need achievement as a motivational construct. *Psychological Bulletin,* 1966, *66,* 291-308.

Kukla, A. Performance as a functin of resultant achievement motivation (perceived ability) and perceived difficulty. *Journal of Research in Personality,* 1974, *7,* 374-383.

Lanzetta, J.T., & Hannah, T.E. Reinforcing behavior of "naive" trainers. *Journal of Personality and Social Psychology,* 1969, *11,* 245-252.

Lee, Jongsook. Effects of absolute and normative feedback on performance following failure. Master's thesis, University of Iowa, 1978.

Maier, S.F., & Seligman, M.E.P. Learned helplessness: Theory and evidence. *Journal of Experimental Psychology: General,* 1976, *105,* 3-46.

McClelland, D., Atkinson, J.W., Clark, R.A., & Lowell, E.L. *The achievement motive.* New York: Appleton-Century-Crofts, 1953.

Miller, I.W., & Norman, W.H. Learned helplessness in humans: A review and attribution theory model. *Psychological Bulletin,* 1979, *86,* (1), 93-118.

Nygard, R. A reconsideration of the achievement-motivation theory. *European Journal of Social Psychology,* 1975, *5,* 61-92.

Overmier, J.B., & Seligman, M.E.P. Effects of inescapable shock upon subsequent escape and avoidance responding. *Journal of Comparative and Physiological Psychology,* 1967, *63,* 28-33.

Piaget, J., & Inhelder, B. *The psychology of the child.* New York: Basic Books, Inc., 1969.

Raynor, J.O. Future orientation and motivation of immediate activity: An elaboration of the theory of achievement motivation. *Psychological Review,* 1969, *76,* 606-610.

Raynor, J.O. Relationship between achievement-related motives, future orientation, and academic performance. *Journal of Personality and Social Psychology,* 1970, *15,* 28-33.

Rest, S., Nierenberg, R., Weiner, B., & Heckhausen, H. Further evidence concerning the effects of perceptions of effort on ability on achievement evaluation. *Journal of Personality and Social Psychology,* 1973, *28,* 187-191.

Roth, S., & Bootzin, R.R. The effects of experimentally induced expectancies of external control: An investigation of learned helplessness. *Journal of Personality and Social Psychology,* 1974, *29,* 253-264.

Roth, S., & Kubal, L. Effects of noncontingent reinforcement on tasks of differing importance: Facilitation and learned helplessness. *Journal of*

Personality and Social Psychology, 1975, *32,* 680-691.

Seligman, M.E.P., & Groves, D. Nontransient learned helplessness. *Psychonomic Science,* 1970, *19,* (3), 191-192.

Seligman, M.E.P., Maier, S.F., & Geer, J.H. Alleviation of learned helplessness in the dog. *Journal of Abnormal Psychology,* 1968, *73,* 256-262.

Seligman, M.E.P., Maier, S.F., & Solomon, R.L. Unpredictable and uncontrollable aversive events. In F.R. Bush (Ed.), *Adversive conditioning and learning.* New York: Academic Press, 1971.

Seligman, M.E.P., & Maier, S.F. Failure to escape traumatic shock. *Journal of Experimental Psychology,* 1967, *74,* 1-9.

Seligman, M.E.P., Rosellini, R.A., & Kozak, M.J. Learned helplessness in the rat: Time course, immunization, and reversibility. *Journal of Comparative and Physiological Psychology,* 1975, *88,* 542-547.

Tennen, H., & Eller, S.J. Attributional components of learned helplessness and facilitation. *Journal of Personality and Social Psychology,* 1977, *35,* (4), 265-271.

Thornton, J.W., & Jacobs, P.D. Learned helplessness in human subjects. *Journal of Experimental Psychology,* 1971, *87,* 367-372.

Weiner, B. The role of success and failure in the learning of easy and complex tasks. *Journal of Personality and Social Psychology,* 1966, *3,* 339-344.

Weiner, B. Implications of the current theory of achievement motivation for research and performance in the classroom. *Psychology in the Schools,* 1967, *4,* 164-171.

Weiner, B. *Achievement motivation and attribution theory.* Morristown, N.J.: General Learning Press, 1974.

Weiner, B. A theory of motivation for some classroom experiences. *Journal of Educational Psychology,* 1979, *71,* (1), 3-25.

Weiner, B., & Kukla, A. An attributional analysis of achievement motivation. *Journal of Personality and Social Psychology,* 1970, *15,* 1-20.

Wortman, C.B., & Brehm, J.W. Responses to uncontrollable outcomes: An integration of reactance theory and the learned helplessness model. In L. Berkowitz (Ed.), *Advances in experimental social psychology,* (Vol. 8). New York: Academic Press, 1975.

Wortman, C.B., Panciera, L., Shusterman, L., & Hibschuser, J. Attributions of causality and reactions to uncontrollable outcomes. *Journal of Experimental Social Psychology,* 1976, *12,* 301-316.

Chapter 6

Motivation and Educational Productivity: Theories, Results, and Implications

HERBERT J. WALBERG
MARGARET UGUROGLU
University of Illinois at Chicago Circle and at the Medical Center

A science and profession of education require a general theory of learning productivity that is empirically tested and proven in educational practice. Theories can be ranked not only on their scientific parsimony, comprehensiveness, and explicitness but also on how well they fit the facts and are useful in the practice of education. By such criteria, education is neither a science nor a profession.

It seems, however, that educational research is reaching new levels of scientific maturity. Walberg, Schiller, and Haertel (1979), in an analysis of reviews of educational research in the last decade, show that six factors predict with law-like regularity, cognitive, affective, and behavioral learning outcomes and gains: *student motivation and ability; intensity* and *amount of instruction;* the *social-psychological climate of the classroom group;* and the *parent stimulation of the child in the home environment*. A series of explicit, operational theories of learning productivity including these factors may be specified for testing. This chapter discusses a series of such general educational-productivity theories, the role of motivation in these theories, and selected empirical research that illustrates linkages of motivation to other constructs within a generalized causal framework. The chapter concludes with a number of research and practical implications.

Before turning to theories, it may be helpful to: distinguish three areas of educational inquiry, explain our emphasis on productivity, and discuss the issue of causality in educational research. The first area of educational inquiry is normative and concerns human values, chiefly the admissibility, morality, and ethics of educational means as well as the priority of goals. The second area of inquiry is scientific and concerns the measurement (or observation) and causal relations of means and goals. (The second area overlaps the first to the extent that social surveys of expert or public opinion on normative values are scientific.) The third area of educational inquiry is prescriptive: given the educational values and causal facts pertinent to admissible means, the most productive program to obtain the goals may be specified. Given the dollar or other resource cost of the means or even the amount of teacher and student time required for them, the most cost-efficient program may be specified.

In an era of increased concern about the productivity of public institutions, educators must face normative, descriptive, and prescriptive problems and attempt to give a reasoned account of the decision-making process of allocating scarce public resources and efforts to the accomplishment of valued goals. Education is the largest industry in the U.S. It consumes approximately a tenth of the gross national product and involves nearly all the people for a substantial period of their lives and perhaps a quarter to a third at any given time. Although questions of productivity may sound simplistic and inhumane in an era where the fruits of industrial civilization are increasingly questioned, even small increases in educational productivity can lead not only to lower costs but considerably greater learning and savings of time—an irreplaceable value in life.

Productivity and Causality

In the 1970s, moreover, public interest in educational productivity and cost-efficienty rose. *Time Magazine*, on November 14, 1977 reported that during the period 1962 through 1976 costs per student in the public schools increased from $400 to an inflation-adjusted $750. But the average scores on the College Entrance Examination and the Scholastic Aptitude Test (SAT) for national samples declined from 470 to 430 on the Verbal section and from 500 to 470 on the

Mathematics section. *Newsweek* on November, 7 1977 also noted the SAT decline and reported for the period 1972 through 1977 that total professional staff in all United States public schools rose 8%, the cost per student rose 21% (with inflation correction), while the number of students and the number of schools each fell 4%. Citing these facts is not meant to imply that the SAT scores should be considered the only outcome of interest or that costs per student are the best indication of efficiency. The point is, however, that raising questions about productivity is timely and in the public interest.

To determine educational policy requires not only the valuing of certain ends but the assumption that the means employed indeed cause the ends in question. Simply trying one means and then another unsystematically without assessing their effects, as is usual in education, cannot efficiently accumulate causal and practical knowledge. Experiments with random assignments to treatments are ideally suited to probe or assess causality; but Gilbert, Light, and Mosteller (1976) in a recent review were only able to find what they considered to be two well-executed field experiments in education. Although the stringency of their criteria might be relaxed to include more experiments, it seems that experiments are difficult to implement in natural settings of education and the laboratory experiments are difficult to generalize. Two alternatives, quasi-experiments and path analysis, also present difficulties in elucidating causation. Quasi-experiments in practice usually only control for one or two rival causes rather than a reasonably complete list of the known correlates of educational achievement; and path analysis and structural equations as employed in educational, psychological, and sociological research usually beg causal questions by assuming rather than probing their answers. Thus educational research workers who wish to base policy on creditable causality confront the dilemma of choosing among experiments that are difficult to implement or generalize, and quasi-experiments and path assumptions that are difficult to defend. A promising way to resolve this dilemma, it is argued below, is to include in educational theory as well as in both experimental and nonexperimental research the chief known correlates of educational achievement.

Theories of Educational Productivity

This section draws on educational, psychological, and sociological research on six learning correlates and integrates the findings within econometric production theory. The discussion starts with the simplest one-factor models and adds additional degrees of complexity as indicated in Table 1.

Reviews of research (Walberg, Schiller, and Haertel, 1979) indicate that six production factors are consistently related to educational outcomes, and the approximate amounts of partly overlapping variance in cognitive learning they predict are:

Motivation, 11%
Ability, 60%;

Table 1

Theories of Educational Productivity

Theory	Equation	Factors	Coefficients	Figure
1. One-Factor	$L = f(I); L = a + bI$	1	2	1a
2. Two-Factor	$L = a + bA + cI$	2	3	2a
3. Interactive	$L = a + bA + cI + dAI$	2	4	1b
4. Six-Factor	$L = a + bM + cA + dT + cI + fC + gH$	6	7	3a
5. Diminishing Returns	$L = aM^b A^c T^d I^e C^f H^k$ where b, c, d, e, f, g are less than 1	6	7	1c, 2b
6. Causal Multiplicity	$h(L,M,A,T,I,C,H) = f(L,M,A,T,I,C,H)_{t-1, t-2}...$	7	?	3b, 4
7. Multiple	$i(L^1, L^2,...) = f(M,A,T,I,C,H)$	?	?	5

Note: The upper-case symbols are the initial letters of the capitalized factors: Learning, Motivation, Ability, Time and Intensity of instruction, Class and social-psychological environment, and intellectual stimulation of the child in the Home; the lower-case symbols are empirical constants except t which indicates a numbered point in time.

Time or quantity of instruction, 20%
Intensity or quality of instruction, 20%;
the social-psychological *Climate* of the classroom group, 60%; and
the educational stimulation of a child by the parents in the *Home*, 20%.

Affective and behavioral learning are undoubtedly worthy goals, but the corpus of research on these outcomes is insufficient to yield even crude variance estimates.

Even though school learning must involve the six factors, much educational research concerns only single, possibly randomly-manipulated independent variables. In a typical experiment or quasi-experiment contrasting an old and new instruction, for example, the first regression equation in Table 1 estimates (a) the mean posttest score of the group under the old method of instruction; and (b) the superiority (or inferiority) on the posttest of the new method of instruction in contrast to the old. If the quality or intensity of instruction is a continuous variable then the constant (a) estimates the level of learning when intensity is at zero, and (b) estimates the additional amount of learning with each one-unit increase in intensity; Figure 1a illustrates increasing, instructional intensity linearly associated with increasing amounts of learning.

The next most common implied theory of educational production found in research literature contains two prediction factors; typically ability is the covariate and interest centers on whether instruction or some other production factor produces gains in learning with ability controlled. The second equation in Table 1 assumes that the production factors are additive and can trade-off or substitute linearly for one another as shown in Figure 2a. Thus, for example, given amounts of time can substitute for lesser amounts of ability or motivation.

The next complication involves interactions among the production factors. Figure 1b illustrates the hypothesis that a production factor has a greater effect at one level of another production factor than it does at another level; for example, additional increments in intensity of instruction may benefit students with one level of motivation or ability (x) more than

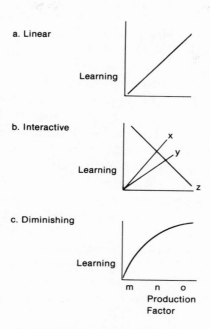

Figure 1. Linear, interactive, and diminishing returns

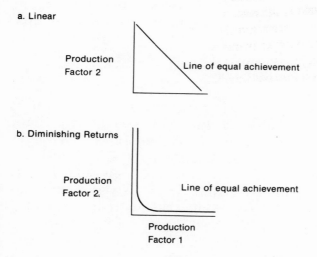

Figure 2. Linear and diminishing substitution.

those at another level (y). The same increments in instructional intensity might actually produce decrements in learning at level z of motivation or ability. These two types of hypothesized but seldom-replicated complications are known respectively as ordinal and disordinal interactions.

It is dangerous to infer causality from one-factor, two factor, and interactive theories of learning shown in Table 1 unless the causal directions are known with certainty or students are assigned at random to levels of the factors, since omitted variables such as motivation, or home or class environments may cause the production factors which in turn cause learning. An enthusiastic class, for example, may stimulate the teacher and cause students to spend more time on their lessons and homework. One way to begin to solve this problem is to estimate a more completely specified equation, that is, to include all six known production factors in theory as well as in data sets. The six-factor equation, however, requires seven coefficients; and including first order interactions requires 15 more coefficients for a total of 22. Allowing for simple curvature in the relations of learning to the production factors, moreover, requires six additional coefficients for squared variables. Such an equation would be unparisimonious and unlikely to be replicated because of the large correlations ("colinearity") among the main, multiplicative, and squared variables.

Diminishing Returns

A "Cobb-Douglas" or "log-linear" equation which includes interactions and curvature of a specific type that is found useful in econometric production studies (Theil, 1971) simply requires taking the *logarithms* of all variables before calculating an ordinary least-squares regression. The equation is parisimonious in requiring only seven coefficients and yet seems more plausible and potentially useful than the linear and additive or the usual interactive equations.

First, it seems improbable as implied by the linear model that increasing any production factor will bring about the same linear returns to learning indefinitely as implied by

Figure 1. Increasing one factor while holding all the others constant in industrial or agricultural productivity research usually leads to diminishing returns (Figure 1c). For example, increasing capital or labor, or the number of horses, plows, or farmers, or the amounts of land or seed, while holding the remaining factors constant, eventually leads to a smaller and smaller increase in yield with each additional increment in the factor (Figure 1c). The equation also implies that (Figure 2b) only prodigious amounts of a factor can trade-off or compensate for minimal levels of another factor.

The returns, at least in theory, could eventually become not only miniscule but negative which implies downward slope (not shown). For example, the addition of too many farmers or too much seed for a fixed quantity of land would eventually be counter-productive. The analogy in psychology is the Yerkes-Dodson law in which motivation increased beyond a certain point becomes debilitating anxiety which lowers performance. In industry and agriculture, common sense generally prevails and cost-conscious managers attempt to raise factors that will raise output the most at the least cost, for example, those at point m in contrast to n in Figure 1c and to avoid raising any factor beyond the point of negative returns (point o).

Second, the diminishing-returns production function is multiplicative, which implies that any factor at its zero point will result in zero learning since any number multiplied by zero equals zero. Although education and psychology lack ratio measures (with true zero points) of many important aspects of human nature and social environments, the concept is a useful one. For example, it implies that, even though motivation accounts on average for only about 11% of the variance in school achievement, or about a fifth of that accounted for by ability, a child completely lacking in motivation to learn cannot learn at all; or more realistically, given measurement limitations, *a very poorly-motivated child can learn little even if all the other factors are at moderate or high levels.* Similarly, children given no time or no instruction, or presented with content they already know or that is completely beyond them, cannot learn. Such plausible examples suggest that all six factors including motivation should be included in future research on learning productivity to investigate unusual, uneven and diagnostically-valuable profiles of students on the factors as well as to obtain more precise estimates of the weights of the factors in concert.

Third, the log-linear model is readily interpretable. Weights b through g indicate the percent increment (or "return") in learning associated with a 1% increase in each respective factor. As estimate of d of .75, for example, implies that increasing instructional time by 10 percent is associated with an increase in learning of 7.5 percent. The understanding of associated percentage changes and forecasts by researchers and much of the lay public give the log-linear model appeal. Because the scales of the variables in the first four equations are uncalibrated, the results of research employing them are difficult to interpret to education practitioners and policy makers.

Multiplicity

Despite the several appealing features of the fifth equation, additional complexities require probing and estimation. The first five equations assume one-way causality in that each factor is assumed to be a sole or joint determinant of learning. As illustrated in Figures 3 and 4, however, the flow of causality may be reversed, reciprocal and mutually reinforcing, or mediated. "To them that hath shall be given;" students who are learning more may become further motivated. Those who get off to a bad start in school or perhaps in a new subject may become less motivated, learn at a lower rate, and fall increasingly behind in both learning and motivation. The flow of causality may also entail instructional quality and quantity and other factors in highly complex ways. In addition, the stimulating quality of the home environment may operate through, or be mediated by, motivation and ability. Without random intervention in such a complex system, there is little hope of sorting the preponderant causal flows unless econometric analyses (Theil, 1971) are applied to time-series measurements on these variables.

Figure 5 and the last equation in Table 1 suggest that learning output is by no means unitary. Production economics

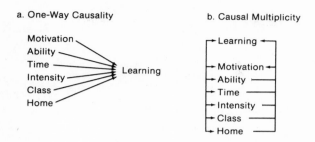

Figure 3. Two types of causal directionality.

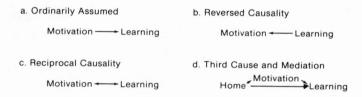

Figure 4. Causal directions among several variables.

offers caution in trying to maximize only one output, and similar caution is necessary in education. The law of increasing relative costs in economics states that to produce equal additional amounts of one good, other things being equal, larger and larger amounts of other goods must be sacrificed; guns and butter in a national economy are the usual textbook examples but the principle applies generally to combinations of goods in smaller public and private sub-sectors of society. A student, by analogy, may be moderately good at academics and athletics (point a in Figure 5); but to become excellent in athletics he or she would become less than mediocre at academics (point b). First-rank performance in national competition in swimming, chess, ballet, and in many other competitive activities may require very high levels on all the factors productive of the performance at the price of minimal or zero progress on other worthy performances and goals.

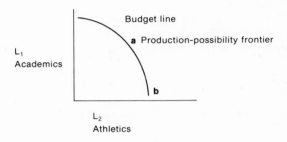

Figure 5. Learning trade-offs.

The compensations or trade-offs between possible accomplishments and the difficult choice of the degree of all-roundedness weighed against first-rank specialization must be considered not only in psychological but also economic terms. With a given budget of time and other factors and equal ability of several goals, (Figure 5), for example, it pays to be a generalist. Modern societies, on the other hand, hold out immense rewards at least over the short term to youth such as Mark Spitz, Bobbie Fisher, and Dorothy Hamil who sacrifice much to be first in national and world rankings. Such difficult choices are close to the core of self-hood and individuality and are not for economists and psychologists to make for educators and students; but we can more fully explicate the causal structure or productive performance in a variety of endeavors as well as the benefits and costs broadly considered in such terms as personal satisfaction, contribution to society, time, quality of instruction required, and career opportunities.

In summary, general theories of educational productivity suggest a wide array of causal possibilities that deserve investigation. To meet the standards of normative, descriptive, and prescriptive educational inquiry suggested in the introductory section will require a considerable amount of research involving all six production factors. Because of the specialized interests of educational researchers, they have almost always investigated the relations of one or two factors to learning rather than a comprehensive set. Virtually none of the past research, moreover, has been analyzed in ways that probe the full possible multiplicity of independent and dependent variables and causal directions and that take into account the relative trade-offs of both inputs and outputs.

Not only do researchers investigating a particular factor such as instruction tend to ignore research on other factors such as motivation and the social environment of the class, but they also seem to ignore work within their own special area. Partly overlapping theories within each area appear to be unrelated and unsynthesized, and empirical research often ignores theories or is only vaguely or implicitly related to them. Educational research abounds with one-shot speculation, one-shot theories that are unrelated to prior related conceptions, one-shot hypotheses, and one-shot measurement instruments. Although freedom of inquiry is an undisputed value, continuation of the present course is unlikely to lead to theoretical consolidation, theoretically-relevant empirical research, and replication—the essense of science. To make education more productive, moreover, requires a list of agreed upon theoretical sub-constructs within each factor, valid measurements of each, a specification of their causal relations to one another and to sub-constructs of other factors, and proven leverage points within the causal system for producing the maximum returns to learning.

Theories of Motivation

Theories of motivation, although they have a long, distinguished and a potentially important future in educational-productivity research, suffer from a number of problems just discussed. Table 2, for example, shows a chronological list of major theories of motivation beginning with the James-Lange theory of emotion. From James through the first third of the twentieth century, psychologists tended to emphasize the association of behavior, reinforcements, and ideas with emotions and drives; since Tolman and Lewin, however, cognition increasingly became the dominant element in general psychology and even in motivation theory that had been more closely

Table 2
Motivation Theories

Author	Approximate Date of Theory	Classification	Theory	Major Constructs
James	1884	Associative	James-Lange Theory of Emotion	Thoughts/feelings follow behavior.
Lange	1885	Associative	James-Lange Theory of Emotion	Thoughts/feelings follow behavior.
Freud	1900	Psychoanalytic	Instinctual Drives	Approach avoidance conflict; subconscious drives.
Thorndike	1913	Associative	Law of Effect	Stimulus—Response bond; satisfiers—annoyers.
Hull	1930	Associative	Drive Theory	Stimulus—Response; drives times habits equal behavior.
Tolman	1932	Cognitive	Drive Stimulation	Needs; expectancies; goals; incentives
Lewin	1936	Cognitive	Field Theory	Cognitions as causal factors; levels of aspiration; choices toward goals to satisfy needs.
McClelland	1953	Cognitive	Achievement Motivation	Need for achievement; cues; affect precedes behavior.
Skinner	1953	Behavioral	Operant Conditioning	Stimulus—Response; emotions/cognitions purged from theory.
Rotter	1954	Cognitive	Social Learning Theory	Expectancy; locus of control.
Spence	1956	Behavioral	Quantitative Stimulus—Response	Drives; habits; incentives.

Author	Approximate Date of Theory	Classification	Theory	Major Constructs
Atkinson	1957	Cognitive	Achievement Motivation	Need for achievement; expectancies; incentives; approach success/avoid failure.
Heider	1958	Cognitive	Locus of Causality	Perceptions; attributions.
Berlyne	1960	Cognitive		Curiosity; novelty; arousal potential.
Crandall	1968	Cognitive		Expectancies; competence.
deCharms	1968	Humanistic	Origin and Pawn	Personal causation (relates to locus of control); goal setting.
Freire	1968	Humanistic		Self reflection.
Maslow	1970	Humanistic	Self Actualization	Self actualization, aesthetics, knowledge, esteem, belonging, safety, physiological needs.
Kagan	1972	Cognitive		Uncertainty reduction; competence; goals; imitation of models; intrinsic motivation.
Weiner	1972	Cognitive	Attribution Theory	Internal/External causation of ability, task, effort and luck and stability of each.
Cattell	1975	Affective	Dynamic Theory	Self sentiment and super ego (most important from sixteen factors).
Deci	1975	Cognitive		Perceptions and cognitions; intrinsic motivation; competence and self determination.
Maehr	1975	Cognitive	Continuing motivation	Perceived selfhood (identity, judged competence, self as initiator), social and cultural determinants.
Bloom	1976	Cognitive	Affective Entry Characteristics	Interests; attitudes; self views.
Lefcourt	1976	Cognitive		Locus of control.

allied with emotions and affect. During the last half-century, cognition wrested first place from behavior, while emotion remained in a distant third place, in psychology's classic triumvirate of thinking, feeling, and acting.

Bolles and Wittrock noted these trends in the two lead articles in the 1978 *Educational Psychologist*. Wittrock (1978) repeats the remark that "psychology lost its soul a long time ago, destroyed its mind about at the turn of the twentieth century, and is now having trouble with behavior" (p. 14). Our chronology and account but not the main point of the substellar position of motivation differ from those of Wittrock and Bolles. Wittrock argues that it is more useful to study "how teaching style influences student attention, motivation, and understanding, which in turn influence behavior" (p. 14) than it is to study how teaching directly influences student behavior because cognitive models emphasize the active and constructive role of learners.

Similarly, Bolles (1978) asks in the title of his article: "Whatever Happened to Motivation?" His answer is that motivational concepts were introduced to correct the deficiencies of mechanistic, passive views of the organism that once prevailed. Modern cognitive explanations of the goal-seeking organism, however, make it possible, he argues, to deal with behavioral and motivational phenomena without invoking motivational concepts.

McKeachie (1976), in his American Psychological Association Presidential Address, sees psychology converging on an all-American "cognitive behaviorism." But in a more recent APA Presidential Address, Smith (1978) expresses doubt in such assessments: "Even liberalized cognitive behaviorism falls short of doing justice to the vast domain that psychology has staked out for itself" (p. 1053). He argues that selfhood is the criterion of the human condition and examines this motivational concept in evolutionary, cross-cultural, and developmental perspectives. He further argues that psychology must be a historical and cultural as well as a biological science and must draw on both the causal and interpretive traditions to increase our individual and collective capacity for self-direction.

We view these contentions as valuable but academic. Even if authorities conclude that behaviorism or cognition is the first cause or *sine qua non* of psychology, our language, culture, tradition, and everyday lives show most of us that motivation and affect are keys to understanding ourselves and others. The man on the street, the teacher in the classroom, and parents in the home will insist that neither feelings nor thoughts and acts can be neglected.

Neither can we accept the case that motivation theories have little to offer educators; to the contrary, such theories offer an embarrassment of riches, and the general educational psychologist is confronted with the task of sorting out their varying implications. All the theories in Table 2 have a degree of plausibility at least in some respect; all have interesting constructs, some of which may be critical to student productivity. The important tasks in making such theories relevant to problems of the educator are: systematically consolidating the constructs across theories, ascertaining the causal relations to one another and to constructs within other domains such as instruction and the home environment, and identifying the leverage points within the causal framework that will channel motivational energies in directions desired by educators, parents, and students. It is not within the purpose and scope of this chapter to accomplish these tasks systematically and completely, but the reader may be referred to the other chapters in this book for useful approaches and to several examples suggested by recent educational research discussed in the next section of this chapter.

Recent Empirical Research

Another starting point for advancing research on motivation aimed at improving educational productivity is current psychological research on motivation and learning in classroom settings. This section discusses a recent quantitative synthesis of the correlational studies of motivation and academic learning, another quantitative review of the effects of open education on motivation and other affective traits, and a recent study of the relation of home environment and motivational development.

Motivation and Learning

Uguroglu and Walberg (1979) recently analyzed 232 correlations of motivation and academic learning reported in 40 studies with a combined sample size of approximately 637,000 students in first grade through the senior year of high school. Even a cursory examination of these studies reveals that empirical research lacks coordination with theories discussed in the last section. Most of the correlations involved general and academic self-concept, locus of control, and achievement motivation; but the measurements and their use in research do not clearly reveal explicit operationalization of constructs from motivation theories. The analyses, moreover, are restricted to Pearson product-moment correlations and other correlational indexes and do not probe the possibilities of reverse-cause, mediation, feedback, and other complex causal mechanisms discussed in earlier sections of this chapter.

Ninety-eight percent of the 232 correlations between motivation and academic achievement are positive; and the mean correlation across all samples is .338. These figures indicate that motivation is a highly consistent positive correlate of achievement but that it is associated with only about 11% of the variance in achievement on average. Motivation may be a much more potent determinant of learning than the variance estimate suggests for several reasons. First, as just mentioned, few of the instruments are strongly grounded in theory. Second, the analyses of the determinants of the magnitude of the correlation show that, with other things such as the age and sex of the samples, technical features of the motivation measure, and the area of academic achievement, held constant, measures of different motivational constructs show no significant difference in their correlation with achievement. The absence of differences in this regard means that the empirical research analyzed is insufficiently valid to help adjudicate even the gross comparative creditabilities of various theories. Third, if multiplicative laws of learning apply, then motivation (as well as each of the other production factors) is a necessary condition for learning, even though on average it may be associated with a relatively small percentage of the variance in learning.

Fourth, the typical regression-adjusted estimate of the correlation between the motivation and learning measures rises from about .24 in first grade to about .35 in twelfth grade. Although adolescents may simply be more self-insightful, it is also possible that the association strengthens with age because of reciprocal or mutually-reinforcing causality operating between motivation and learning (possibly involving in addition other production factors) such as rich-getting-richer phenomenon.

Fifth, motivation, self concept, and locus of control are poorly measured in much of the recent research; and the quality of the measures are most often gaged by a less than useful standard—the internal-consistency reliability. Although many investigators do not report the technical features of their instruments in the research analyzed, those that do most often report internal consistency, a measure of the extent to which the test measures only a single factor. Since learning and motivation itself are undoubtedly multi-dimensional, bivariate analyses of uni-factorial tests cannot reveal the full association between the constructs.

Home Environment and Motivation

Dolan (1978) recently reported moderate correlations between indexes of parent stimulation of children in their homes and measures of motivation. More than 140 correlations from 17 studies show that home environment is strongly linked with cognitive ability and achievement. Dolan's study opens the possibility that the home influence may also operate through motivation to indirectly affect learning. Since the child is influenced by parents more hours per day and for more years in his life than by teachers, the home may provide a strong leverage point for school-based programs to increase motivation, ability, and learning of many kinds.

Causal circularity and other kinds of multiplicity, of course, require consideration. Children who do well in school may bring home academic enthusiasm and glowing report cards which in turn cause parents to buy them books, and discuss advanced topics with them to stimulate them still further in a variety of ways. Two kinds of research are in order. The first would detect the preponderant causal flows among learning, motivation, home environment, and other factors. The second would ask, in view of the expense and delicate questions of confidentiality of home interviews, if motivation, ability, and the social-psychological climate of the class completely mediate home environment influence. If so, then it would be feasible to omit home environment from non-experimental research on instructional productivity under certain circumstances and still have a reasonably good chance of drawing valid causal inferences.

Open Education and Motivation

Motivation and affect are not merely a means to the ends of cognitive and behavioral learning. Educators undoubtedly try to accomplish many worthy affective goals that are presently unmeasurable or infrequently measured. But even within the domain of currently feasible measurement, motivation, affect, and related concepts have important roles to play as dependent variables or ends in their own right, as shown by two recent reviews of the effects of open education (Horowitz, 1979; Peterson, 1979: summarized by Walberg, Schiller, and Haertel, 1979). In a sample of 202 studies of open education contrasted to conventional instruction, 55% of the significant studies showed open programs superior on academic achievement. On affective and motivational outcomes, however, the results are far more striking: on self-concept, open education was superior in 88% of the significant studies; attitude toward school, 92%; independence, 95%; and curiosity and creativity, 100%. Thus, motivation and other such affective measures help make evaluation research on innovative educational programs more revealing and useful.

Conclusions

Motivation is one of several keys to effective learning. It is consistently positively correlated with measures of educational achievement; and the strength of its relation to learning is currently under-estimated for historical, technical, and theoretical reasons. On theoretical and empirical grounds, a case can be made that motivation may serve as a cause, mediator, and effect of the learning process. If school learning is a function of several multiplicative production factors each yielding diminishing returns, it appears that motivation is a necessary condition for learning and that increasing the other factors, such as the quality and amount of instruction, at great cost will be relatively fruitless if motivation remains at low levels. Psychology, once thought of as an exclusively behavioral science, has rediscovered the mind and continues to neglect the soul. But educators and parents in their wisdom know the importance of motivation and affect and look to psychologists and other researchers for ways that motivational concepts can help to increase educational productivity. Increased theoretical rigor, greater coordination of empirical research with theoretical models, better instrumentation, and econometric techniques that analyze causal multiplicity can assist in the accomplishment of this pressing expectation.

REFERENCES

Bandura, A. *Social learning theory.* Englewood Cliffs, N.J.: Prentice-Hall, 1977.

Bloom, B. S. *Human characteristics and school learning.* New York: McGraw-Hill, 1976.

Bolles, R. C. Whatever happened to motivation? *Educational Psychologist*, 1978, *13*, 1-13.

Brewster Smith, M. Perspective on selfhood. *American Psychologist*, 1978, *33*, 1053-1063.

Cattell, R. & Child, D. *Motivation and dynamic structure.* New York: Wiley, 1975.

Deci, E. *Intrinsic motivation.* New York: Plenum, 1975.

Dolan, L. The affective consequences of home support, instructional quality, and achievement. *Urban Education*, 1978, *13*, 323-344.

Gilbert, J. P., Light, R. J., & Mosteller, E. Assessing social innovations.

In A. R. Lumsdaine & C. A. Bennett (Eds.), *Evaluation and experiment.* New York: Academic Press, 1976.

Horowitz, R. Open education effects. In H. J. Walberg (Ed.), *Learning environments and effects: Evaluation, research, and policy.* Berkeley: McCutchan, 1979.

Johnson, D. J. Affective outcomes. In H. J. Walberg (Ed.), *Evaluating educational performance.* Berkeley: McCutchan, 1974.

McKeachie, W. J. Psychology in America's bicentennial years. *American Psychologist,* 1976, *31,* 819-833.

Maehr, M. L. Sociocultural origins of achievement motivation. In D. Bar-Tal, & L. Saxe (Eds.) *Social psychology of education: Theory and research.* New York: Hemisphere, 1978

Peterson, P. L. Direct instruction reconsidered. In P. L. Peterson & H. J. Walberg (Eds.), *Research on teaching: Concepts, findings, and implications.* Berkeley: McCutchan, 1979.

Shavelson, R. J., Hubner, J. J., & Stanton, G. C. Self-concept-validation on construct interpretations. *Review of Educational Research,* 1976, *46,* 407-441.

Smith, M.B. Perspectives on Self-Hood. *American Psychologist,* December, 1978, *33,* 12, p. 1053-1063.

Theil, H. *Principles of Econometrics.* New York: John Wiley & Sons. 1971.

Turner, E. V., Helper, M. M., & Drisha, S. D. Predictors of clinical performance. *Journal of Medical Education,* 1974, *49,* 338-342.

Uguroglu, M., & Walberg, H. J. Motivation and achievement: A quantitative synthesis, *American Educational Research Journal,* 1979, *16,* 375-389.

Weiss, J. The identification and measurement of home environmental factors related to achievement motivation and self-esteem. In Kevin Marjoribanks (Ed.), *Environments for learning,* Slough, Berks., England: National Foundation Educational Research, 1974, pp. 140-149.

Walberg, H. J. A psychological theory of educational productivity. Chicago: manuscript, 1979.

Walberg, H. J., Schiller, D., & Haertel, G. D. The quiet revolution in educational research. *Phi Delta Kappan,* November, 1979, *61,* 3, 179-183.

Wittrock, M. C. The cognitive movement in instruction. *Educational Psychologist,* 1978, *13,* 14-29.

Chapter 7

A Model of Direct and Relational Achieving Styles*

JEAN LIPMAN-BLUMEN
Center for Advanced Study in the Behavioral Sciences

HAROLD J. LEAVITT
Stanford Graduate School of Business

KERRY J. PATTERSON
Stanford Graduate School of Business

ROBERT J. BIES
Stanford Graduate School of Business

ALICE HANDLEY-ISAKSEN
Stanford Graduate School of Business

The extensive research on the achievement concept has centered largely on the underlying need for achievement and its attributes: the genesis and nature of the achievement motive, as well as the targets and outcomes of achievement drives. The socioeconomic, cultural, racial, and gender correlates of need for achievement also have been investigated intensely.

By contrast, the present research attends to *styles* of achieving, a relatively neglected aspect of achievement. Styles of achieving are *characteristic ways in which individuals approach achievement situations*. While the traditional literature on achievement motivation carries major and immediate relevance for styles of achieving, other research veins also are clearly germane. These include the literatures on social learning, leadership styles, small groups, personality traits and "syndromes," and sex roles, all of which relate to our

---

*We are indebted to other members of our ongoing research group: Inez Brunner, Joan Kofodimos and Eddie Reynolds. We also wish to express our thanks to Drs. Helen Farmer and Leslie Fyans who helped in the data collection and analysis in an earlier phase of this research.

conception of styles of achieving. After a brief review of those literatures, we shall present (1) the historical development of the achieving styles model; (2) etiological assumptions of the model; (3) other underlying assumptions; (4) characteristics and components of the current model; and (5) differences from other models. Following that, we shall (6) describe an instrument for measuring achieving styles.

Previous Research

Achievement Motivation

Drawing on Murray's (1938) complex taxonomy of needs, McClelland, Atkinson, Clark, and Lowell (1953) began to explicate the need for achievement (N-ACH), which McClelland later studied within an entrepreneurial context (1961).

In their 1953 formulation, McClelland et al. suggested that

> All motives are learned . . . they develop out of repeated affective experiences connected with certain types of situations and types of behavior. In the case of achievement motivation, the situations should involve 'standards of excellence' . . . and the behavior should involve 'competition' with those standards of excellence or attempts to meet them which, if unsuccessful, produce positive affect or, if unsuccessful, negative affect. (McClelland et al., 1953, pp 275.)

This work of McClelland et al. presented an "affective arousal" model of motivation which distinguished it from earlier behavioristic models (Hull, 1943; Miller and Dollard, 1941). This newer approach to motivation proposed a learned need for achievement (N-ACH), which is treated as a drive activated by selected environmental cues to engage in need-fulfilling (drive-reducing) behavior. Standards of excellence, personal responsibility, "anticipated explicit knowledge of results" defining success or failure, and uncertainty associated with outcomes are other critical components of the need for achievement (Atkinson, 1968). The arousal of N-ACH activates behavior to reduce the drive-activated tension. However, the ways people go about achieving remain ambiguous in this model.

Variants of need theory have been central to several other conceptions of motivation. In addition to McClelland's N-ACH formulation, several other models trace their roots to Murray's generative conceptualization: Maslow's (1954) growth-oriented hierarchy of needs, White's (1959) concept of personal efficacy, Horner's (1968) motive to avoid success, and de Charms' (1968) need for personal control.

In recent years, however, an increasing number of critiques of traditional need theories have appeared (e.g., Salancik and Pfeffer, 1977), generally moving toward more cognitive formulations. For example, some approaches have emphasized that expected outcomes learned through previous direct experience (Atkinson and Raynor, 1974; Bandura, 1977; Vroom, 1964) or through social comparisons (Abramson, Seligman, and Teasdale, 1978; Bandura, 1977) play an important role in the initiation of behavior. These theorists argue that the expectations that one *can* perform a certain behavior and/or achieve a desired end play an important role in determining the level of achievement-oriented behavior. Within this more cognitive framework, other researchers have stressed the importance of attributions of success and failure which differentially influence subsequent achievement efforts (Weiner, 1974), including learned helplessness (Seligman, 1975; Dweck, 1973, 1975, 1976). Deci's (1975) work on intrinsic/extrinsic motivation falls within this attribution framework.

Traits as Motives and Social Learning Theory

Considerable ambiguity exists at the interface between the concepts of "traits" and "motives." In fact, Mischel (1968) calls into question the very existence of internal motivations or drives, since most personality "traits" have not proven to be the broad response dispositions conceptualized by earlier theorists (Allport, 1937, 1966; Murray, 1938).

In his social learning theory, Bandura (1977) offers a useful framework for understanding why certain behaviors are selected over others, without recourse to the concepts of motives or traits. He argues, with supportive research findings, that people learn to expect that certain responses will result in specific outcomes and that these *expectations serve as guides for action.* The information crystallized into expectations emanates from several sources, including direct experience and vicarous learning. That is, given a certain level of ability, expectations about one's capacity to master a task are learned from one's own behavior or observations of others in relevant contexts. Given that people repeatedly confront similar task situations, "characteristic" ways of approaching goals, thus, may emerge. We shall call these characteristic ways of approaching goals "styles of achieving."

Leadership Styles, Personal Styles, and Small Group Research

In social psychology, established traditions of research have developed around the behavior of leaders and members of small groups. At least three relevant research strands may be distinguished: (1) leadership studies, (2) personal styles research, and (3) small group research.

Leadership Studies. In contrast to the traditional research on achievement, much of the leadership has attended to *styles* of leadership behavior, as well as underlying traits and motives. As early as the 1930s, Kurt Lewin and his associates were exploring systematically the impact of leadership styles on group dynamics. Lewin, Lippit, and White (1939) compared democratic with authoritarian and laissez-faire leadership styles. They reported that democratic group leadership was more likely to promote satisfaction and productivity among group members than were authoritarian or laissez-faire styles.

Later researchers repeatedly identified an important dichotomy in leadership behavior: task-orientation and people-orientation (Blake and Mouton, 1969; Fiedler, 1978; Fleishman, 1971; Likert, 1961; Stogdill and Coons, 1957; Vroom and Yetton, 1973). There is considerable agreement among students of leadership styles that some combination of task- and people-orientation by a leader effectively promotes attainment of group goals.

Personal Styles. Concern with leadership styles is closely linked to Adorno et al.'s (1950) research on authoritarian personality, as well as Christie and Geis' (1970) work on Machiavellian behavior. The authoritarian "syndrome" includes concern about acquiescence to authority and the readiness to use authority as an achievement mechanism. The concept "authoritarian personality" thus describes one set of characteristic ways for approaching both relationships and tasks.

In attempting to measure an individual's general strategy for dealing with others, Christie and Geis (1970) isolated at least three key characteristics of Machiavellian "personalities": (1) manipulative interpersonal strategies; (2) a negative view of human nature; and (3) an abstract and impersonal morality. Thus, individuals who score high on the Mach scale have a zest for competition and success, are skilled in such activities, and are likely to be successful. "Hi-Mach" types tend to initiate and control social interaction using either defensive and/or offensive manipulative tactics, as the situation requires. Interpersonal contact is essential to the success of the "Hi-Mach's" manipulative strategies. Without the physical presence of others or the opportunity to communicate directly, "Hi-Machs" do not show much Machiavellianism. These currents of instrumentality, competition, and power are germane to the achieving styles model we shall present below.

Small Group Research. Notions of leadership styles, as well as people- and task-orientation, also appear in Bales and Slater's (1955) earlier research on small group processes. From their observations of the "natural history" of small group dynamics, Bales and Slater demonstrate that members of small groups often perform differentiated functions which contribute to the accomplishment of the common task.

Bales and Slater observed two recurring classes of group functions—instrumental and expressive—and the emergence of instrumental (or task) leaders and socio-emotional (expressive) leaders to fulfill them. The instrumental-expressive dichotomy is, of course, analogous to the task- and people-orientations distinction discussed earlier. Of special interest here is Bales and Slater's argument that all small groups must carry out those two key functions if the group is to complete its chosen task. First, the group must attend to the task itself, and second, the group must deal with the emotions and interpersonal relations that arise in the process of trying to accomplish the goal. Insights from small group research have been applied inappropriately to differences in male and female behavior (Parsons, 1954); however, a resurgence of the feminist movement has led to a reevaluation of this application (Lipman-Blumen and Tickamyer 1975).

Sex Roles

Before the revitalization of the feminist movement, most research on achievement, leadership, and small groups concentrated heavily on male subjects. Female achievement received only scant attention (Angelini, 1955; French and Lesser, 1958; Lesser, Kravitz, & Packard, 1963). It took Horner's (1968) influential research on the motive to avoid success to launch a decade of intensive research efforts (Tresemer, 1976; Alper, 1973, 1974; Mednick and Puryear, 1975, 1976; Puryear and Mednick, 1974; Levine and Crumrine, 1975; Hoffman, 1972; Fleming, 1977) to confirm, deny or reinterpret women's achievement behavior.

The *vicarious achievement ethic* (Blumen, 1970; Lipman-Blumen, 1972, 1972) was described as one potent force in channeling women into traditional "feminine" occupational roles. Other researchers drew attention to the ways in which men's occupational roles and institutions consumed the achievement efforts of their wives, even while denying them equal recognition and reward. Coser and Rokoff's (1971) "greedy institutions" and Papanek's (1973) "two-person careers" described the special hazards confronting women's independent achievement efforts.

The advent of a burgeoning sex role literature sparked renewed interest in the origins, nature, and outcomes of differential socialization experiences of women and men. (For reviews of the sex role literature, see Hochschild, 1973; Lipman-Blumen and Tickamyer, 1975; Mednick and Weissner, 1975). Efforts to solve the conundrum of sex role stereotypes (Bernard, 1968; Bem, 1974; Spence and Helmreich, 1978) brought the notion of sex-linked characteristics (e.g., aggression and passivity) under serious scrutiny. Concern with sex role stereotypes also focussed analysis on sex-typed behaviors (Maccoby and Jacklin, 1974) and occupations (Epstein, 1970; Blaxall and Reagan, 1976). Efforts to sort out sex-linked personality differences have brought into sharper focus the alleged "person orientation" of women and "task orientation" of men.

The Direct/Relational Model

Historical Development

The genesis of the current model can be traced to an earlier interest in "mode of achievement satisfaction" (Blumen, 1970; Lipman-Blumen, 1972). That research related achievement modes to sex role ideology and educational aspirations among highly-educated married women. There, mode of achievement satisfaction was conceptualized as an active to passive continuum. At one extreme, the individual satisfied her achievement needs directly and completely by her own efforts, while, at the other end, the individual's achievement needs were satisfied essentially by identification with and pride in her husband's achievements. Mode of achievement satisfaction was found to be a powerful intervening variable in the relationship between sex-role ideology and educational aspirations. In short, women with a traditional sex-role ideology tended to meet their achievement needs through their husband's accomplishments. They thereby seemed to short-circuit their own educational and occupational needs.

Examination of the early Horner data (1968) suggests that the context of vicarious achievement could be used to account for some of her early findings. That is, certain types of rejection of direct achievement, which Horner coded as a motive to avoid success, could be partialled into several more refined categories, including vicarious achievement. For instance, stories in which the now-famous medical student, Ann, renounced her student role to marry a classmate and rejoice in his success might be viewed, not as a motive to avoid success, but as an example of vicarious achievement behavior. Both personal and occupational roles, we reasoned, could be understood within this framework. Thus, renunciation of the physician role in favor of nursing similarly could be treated as a movement toward a vicarious achievement role.

Since most of the early Horner research used story cues about a *single* individual, we hypothesized that the very form of the instrument introduced a bias against the expression of vicarious achievement. Thus, in our earliest joint efforts (Lipman-Blumen & Leavitt, 1976: Leavitt, Lipman-Blumen, Schaefer, and Harris, 1977), we devised a set of story cues fashioned after Horner, but with an important difference: some cues featured a *single* character confronting an achievement situation and others had *two* characters, one an achiever, the other an observer.

Since our original interest was in sex differences in achievement orientation, we varied the sex of the characters, sometimes presenting two characters of the same sex, and other times using characters of both sexes. The sex of the observer and the achiever were varied as well. In addition, we varied the nature of the relationship between the characters from close same- and cross-sex relationships to relatively distant same-sex and cross-sex relationships. Thus, we developed stories about fiancé(e)s, parents and children, best friends, and classmates, in which we varied the gender of the pairs of characters.

The open-ended story cues required a response describing possible outcomes of the story. Using data from female and male university freshmen, we undertook a content analysis of the essay-type material. These early led us to hypothesize several subtypes of vicarious, as well as direct, achieving behaviors, which we then called altrusitic vicarious, contributing vicarious, and instrumental vicarious,[1] and intrinsic direct, competitive direct, and instrumental direct.

Using these categories we developed one- and two-person stories, this time with forced-choice endings, which respondents were asked to rank order for likelihood of occurrence. Each outcome was written to represent a particular subtype of vicarious or direct achievement orientation. At that time, we also included two other categories with linked forced-choice responses: fear of success and fear of failure.[2]

Some early findings revealed that sex of the achiever in the

---

[1]Later, these terms were changed to vicarious relational, contributory relational, and instrumental relational. See p. 144 for explanation of terminology change.

[2]These categories subsequently were dropped to refine the model and simplify the instrument.

story cue, as well as the relationship between the characters, was related to perceived achieving orientations. This finding was consistent among male and female subjects. Briefly, in two-person, cross-sex stories, when the achieving character was male, both male and female respondents attributed considerable vicarious achievement to the observing female character. Thus, if Mary and Jim were classsmates, and Jim was the top student in their class, Mary's achievement needs were very likely to be seen as satisfied through her relationship with Jim. Both women and men respondents regarded Mary as vicariously satisfied; however, male subjects attributed more vicarious achievement satisfaction to Mary than did female subjects.

The closer the relationship between Mary and Jim (i.e., fiancé(e)s or parent/child, rather than classmate), the more vicarious achievement satisfaction was attributed to the female figure observing the successful male. Thus, both male and female subjects seemed to be saying that when a woman engages in a relationship with a successful man, she derives considerable vicarious satisfaction from the man's success. Closer, more intimate relationships were seen as producing higher levels of vicarious satisfaction in women. But even more distant relationships between a successful male and an observing female were seen as providing the female with a significant level of vicarious achievement satisfaction.

When the positions of observer and achiever were reversed (i.e., with the male figure observing a successful female), somewhat different responses were recorded. Male subjects reported that Jim was experiencing vicarious achievement through Mary's success; however, female subjects were more likely to see Jim as competitive and instrumental vis-a-vis Mary.

Comparison of one- and two-character stories revealed that one-character stories evoked both direct and vicarious achievement interpretations. These findings tended to confirm our earlier hypothesis that vicarious achievement had been obscured artifactually by the design of earlier studies focussing on single-character cues.

In the course of the various analyses of that early achieving styles inventory,[3] as well as the later self-descriptive instrument, several new variants of the achieving styles model emerged. The data suggested at least three new categories: *collaborative relational, reliant relational,* and *power direct.* Our analysis suggested that relationality, rather than vicariousness per se, was a broader and more useful conceptualization of the domain. Vicariousness then was seen as one subcategory within the relational domain. As a result, the labels of the remaining styles were changed as previously indicated (See footnote #1).

Working interactively between theory and data, we conceptualized the current model, comprised of two major domains: relational and direct achieving styles. Within the *relational domain,* five subcategories may be identified: vicarious relational, contributory relational, collaborative relational, reliant relational, and instrumental relational. The *direct domain* include four subcategories: intrinsic direct, competitive direct, power direct, and instrumental direct.

The Etiology of Achieving Styles

Our speculations about the etiology of achieving styles flow from a variant of a now-conventional set of views regarding how needs and means of satisfying those needs are learned (Freud, 1938; Maslow, 1954; McClelland et al., 1953; Murray, 1938). Our approach departs from these other perspectives in several key ways. First, we categorize needs into three major types: *physical* needs, assumed to be largely innate and generally operant from birth; and two classes of primarily learned needs which we (and others) call *social* and *egoistic.* Our model assumes that the key initial learning of social and egoistic needs stems from differential experiences of success and failure in the satisfaction of physical needs.[4] The important

---

[3]In our work with the earlier inventory, we were helped greatly in data collection and computer analysis by Dr. Helen Farmer and Dr. Leslie Fyans of the University of Illinois at Urbana.

[4]This is not to deny the notion proposed by J.W.M. Whiting (as described by Sears, et al., 1953, pp. 180-181) that actions which are occasionally punished and occasionally rewarded may develop into a "secondary motivational system." Thus, the means for satisfying needs may become needs themselves.

learning mechanism is assumed to be *dependency*, initially of the infant on parents and others in his/her environment, later on any other relevant person(s).

The line of argument runs as follows: Humans are caught from the outset in dependent relationships. Different individuals experience varying degrees of difficulty in satisfying their physical needs in those dependent settings. Some people find their dependent relationships satisfactory in that they can fulfill most of their physical needs through them most of the time. Others, for various reasons, find that their available repertoire of behaviors yields only long delays, irregularities and frustrations.

Individuals whose early experience is toward the successful end of the continuum are presumed, in this model, to generalize two related, but conceptually separable, perspectives. First, they will learn to value as ends in themselves these success-providing relationships. Such individuals will develop relatively strong social needs—needs for affiliation, for membership, for nurturance and the like. Second, they also will learn a generalized preference for the use of their now-proven relational means in their search for need satisfaction of all kinds. That is, they will prefer tools they know how to use successfully (even when they are not entirely appropriate)—in this case, behavioral tools for developing and maintaining relationships with others.

At the other extreme, those who, during periods of dependency, encounter difficulty in fulfilling their physical needs will learn different patterns of needs and will search for other more rewarding behavioral means. They will develop stronger needs for independence, autonomy and achievement—the egoistic needs; and they also will learn to *prefer* more independent and direct means to need satisfaction in general, because these, rather than relational, means are the ones likely to have been positively reinforced. That is, they will prefer acting directly on the world to acting indirectly on it through intervening relationships with other people. The process, of course, becomes highly differentiated as individuals experience wide varieties of more or less dependent roles.

A given means used in an effort to get what one wants may

not get one *everything* one wants, but may work tolerably well and then will continue to be used. The important relevant consequence of such satisfying (Simon, 1957) behavior is an attenuated search for better alternatives and a tendency to fix more or less firmly on those means which work reasonably well reasonably frequently. That process provides, we believe, the genesis of achieving styles.

There is plenty of room, of course, in this type of model for wide diversity both in the salience of particular needs and in the preferred means for trying to meet them. But the important assumption is that people who have developed strong social motives also are more likely to have developed a preference (and an accompanying skill) for relational means, while people who have learned stronger patterns of egoistic motives are likely to have learned a preference for direct means.

This model assumes, moreover, that while any individual's need patterns are diverse (e.g., individual with strong egoistic needs will also show *some* level of social needs, and so on), the preferred means for satisfying all needs are widely generalized and are *not* need-specific. Thus, we would expect strongly socially-motivated individuals to try to satisfy even their egoistic needs by relational means and strongly egoistically-motivated individuals to try to meet even their social needs by direct means.

This "means crossover" concept implies that some persons will use "inappropriate" relational means to satisfy egoistic needs—like needs for power. It also suggests a somewhat more curious possibility: that some individuals will try to satisfy social needs, like affiliation, by direct, rather than relational, means.

Examples of such crossovers are people who try to buy love or to coerce others into an affiliative relationship—phenomena which do not seem terribly uncommon. Or, on the other side, one might encounter individuals attaching themselves to powerful others and drawing egoistic sustenance from the accomplishment of those others.

Other Underlying Assumptions of the Achieving Styles Model

Before presenting a detailed description of the model, we

shall explicate briefly the other key assumptions underlying it. First, while motives per se are not a central focus of this paper, our framework assumes that where motives do exist, they take the form of needs or underlying tensions that must be satisfied, at least temporarily, before the tension rebuilds and claims further behavioral effort. Despite the inevitable ambiguities in conventional definitions of motives and needs (Salancik and Pfeffer, 1977), we shall treat them as synonymous.

Second, the focus of this research—achieving styles or orientations—is upon means of approaching or accomplishing goals, regardless of the substantive character of those goals or the relative strength of the underlying motives. In addition, in behavioral chains, we recognize the possibility that ends at one stage may become means at another, and vice versa. For example, someone who attains goals through a competitive style (means), ultimately may begin to develop a need for competition as an end in itself. So where one simply might begin by approaching all goals in a competitive way, eventually the achievement may become winning over or besting another (i.e., successfully competing), regardless of the substantive nature of the accomplishment.

Third, the model assumes that individuals develop characteristic styles for achieving goals, regardless of the specific nature of those goals. Thus, one may approach cognitive, emotional, and sexual goals through similar achieving styles.[5] These characteristic styles of achieving presumably are preferred above all others and are selected with reasonably predictable regularity.

Fourth, the model further assumes that individuals frequently select roles which are consonant with their preferred achieving styles and that occupancy of such roles then reinforces the styles.

Fifth, we also assume certain roles create "structural propensities" for certain achieving styles. Thus, the parent role

---

[5] For a discussion of the application of an earlier version of the model to sexual behavior, see Lipman-Blumen, J., and Leavitt, H. J., "Sexual behavior as an expression of achievement orientation," in Herant A. Katchadourian (ed.), *Human sexuality*. Berkeley, Calif.: University of California Press, 1979.

involves societal expectations of contributory and vicarious relational achieving styles.

Sixth, achieving styles have several dimensions, including flexibility, range, and intensity, which affect their particular use by any given individual.

Seventh, we assume that environmental factors (i.e., the "situation") may strongly influence the choice of achieving styles, particularly for those individuals who (1) have access to a wide range of achieving styles, and (2) have the flexibility and cue sensitivity necessary to allow movement among these styles when indicated. But other individuals without sufficient range, cue sensitivity, or flexibility consistently may select certain styles, even when those styles are dysfunctional.

Eighth, in addition to preferring and selecting predictable achieving styles, individuals with limited range or flexibility will seek to enact or create or redefine situations to make them responsive to their preferred styles.

A ninth assumption holds that *neither the two major achieving styles' domains, nor the subcategories within or across them, are necessarily mutually exclusive or independent.* Pragmatically, however, certain combinations of styles are more probable than others, and contiguous styles are more likely to be selected together than styles farther apart on the continuum.

Finally, the model assumes that achieving styles are learned relatively early in life, as we have described elsewhere in this paper; however, we also assume that this early learning is not immutable and that resocialization possibilities exist.

Overview of the Model: Direct and Relational Achieving Styles

The model proposes two major domains of achieving styles: direct and relational. A *direct achieving style* is used by individuals who confront the achievement task directly, using their own efforts of mind and body to accomplish their goal. Individuals who use direct styles of achieving act as agents on their own behalf, encountering the task without recourse to intermediaries.

Relational achievers,[6] in contrast, seek success through the medium of relationships. Individuals who utilize relational achieving styles establish, contribute to, depend on, or manipulate relationships to get what they want.

Direct Achieving Styles. Within each of these major achieving styles' domains, we can distinguish several subtypes. Within the *direct domain*, we have specified four subcategories: *intrinsic direct, competitive direct, power direct,* and *instrumental direct.*

The *intrinsic direct* style is characterized by a propensity to select, initiate, and/or seek out activities which permit direct confrontation with one's environment. Individuals who characteristically select an intrinsic direct approach to success are primarily task-oriented. They tend to experience the thrill of accomplishment for accomplishment's sake by pitting the self against the task, rather than against others. Intrinsic direct types' absorption in the task often makes them oblivious to the world. Whether working in a group or alone, individuals who use an intrinsic direct strategy maintain a primary orientation toward the task and tend to evaluate their own performance against standards of excellence which are largely independent of comparison with others.

A prototypical case of the intrinsic direct style was shown recently in a television interview. The interviewee, after retirement, had decided to build singlehandedly an east-west highway across his home state. Armed only with a dilapidated tractor and dogged determination, he had cut a nine-mile swath across land he personally had purchased. The novice road builder's explanation of this unusual undertaking: "Sometimes I just come out here and look at what I've done—that alone keeps me going."

The *competitive direct* style is characterized by a tendency to select activities which permit evaluation of one's own performance against that of others. Outperforming others or winning

---

[6]Strictly speaking, the model deals with styles of achieving, rather than types of achievers; however, for expositional purposes, we shall take several semantic liberties. Thus, we shall use terms such as "relational (or other type) achiever," "relational types," etc., to avoid more cumbersome (and technically more accurate) descriptors.

over competitors is central to this style. Whether working in a group or alone, competitive direct "types" tend to construe the situation in competitive terms. Even in situations that less competitive persons view as noncompetitive, "competitive direct" individuals often evaluate their performance in comparison with the accomplishments of others they define as relevant.

Individuals with competitive direct styles experience zest and excitement in competition and perceive competition as a spur to accomplishment. The excitement of success is diminished for the competitive direct achiever if s/he is not able to strive against competitors. This type of achiever experiences "the thrill of victory and the agony of defeat." In contrast to intrinsic direct achievers (whose hallmark is absorption in the task itself), competitive direct achievers require the additional incentive of competition against some contender (who may be known or anonymous).

The *power direct* style of achieving ordinarily involves a proclivity to select, initiate, or seek out contexts which permit control and/or organization of individuals, things, or situations as a means to task accomplishment. The power direct style includes domination and use of personal control to attain success. Individuals who select power direct means attempt to exercise close control over all factors impinging on task accomplishment. For them, almost all tasks require organization and control. They usually proceed to take charge of getting things done and often prefer leadership roles.

Although some individuals use power tactics to enhance their own positions, such an orientation is not necessarily inherent in a power direct style. In fact, power direct strategies may be used by people only minimally concerned with enhancing their own status, but extremely aware of the importance of organizational skills for implementation. Thus, many executives typically employ power direct strategies to get things done without concern for self-aggrandizement, status, prestige, or personal power (cf. McClelland and Burnham, 1976).

The *instrumental direct* achieving style is characterized by using success as an instrument for further successes. In contrast to the intrinsic direct approach (which pits an individual directly against a task), the instrumental direct style calls for the intermediate use of one achievement as an avenue to another.

Instrumental direct "types" tend to evaluate achievements for their usefulness in leading to other accomplishments and to use their accomplishments as currency for purchasing additional successes. An example: one athlete described her original interest in athletic competitions in terms of possibilities for international travel. She noted that only after beginning to break her own previous record did she become "turned on" to the athletic activity *sui generis*.

Relational Achieving Styles. Relational achieving styles, which utilize relationships as the media of achievement, encompass five subcategories: *vicarious relational, contributory relational, collaborative relational, reliant relational*, and *instrumental relational*.

The *vicarious relational* style is characterized by a tendency to achieve indirectly through identification with one or more direct achievers, or, in some instances, even with an institution. The identification that vicarious relational individuals feel for the selected achiever is so strong that they accept the accomplishments of the other as their own.

Individuals who select vicarious relational achieving styles tend to accept the "other's" achievement goals rather than to select their own. For example, the vicarious relational spouse of a direct achiever feels successful when the direct achiever succeeds even at something of no intrinsic importance to the spouse. Vicarious relational individuals treat as their own both the successes and failures of the achiever with whom they identify.

Vicarious relational achievers may satisfy their achievement needs either through a close personal relationship or simply through identifying with an achiever worshipped from afar. Sports fans who feel elated when their hero hits a home run are using vicarious relational means. In a similar (though not identical way), parents who feel successful when their child becomes famous also are using vicarious means.

The *contributory relational* style is characterized by the

tendency to achieve through contributing to another's success. Facilitating or contributing to another's performance is a recognizable feature of this style. Contributory relational "types" identify with the direct achiever, while accepting both the goals and means defined by the direct achiever.

Contributing to the direct achiever's success constitutes the contributory relational person's accomplishment. Although there is still strong identification with the direct achiever, the contributory relational achiever distinguishes the direct achiever's accomplishments from his/her own. Unlike vicarious relational "types" (who achieve simply through identification) contributory achievers take an active, but supportive, role in the accomplishment. As the label suggests, contributory relational achievers contribute to another's success without usurping the other's role or perceiving the task as principally their own.

For example, a boxer's seconds behave in a contributory relational way when they feverishly work on him between rounds. They stop his bleeding, feed him liquids, rub him down, shout encouragement, indeed do just about everything except enter the ring to confront his opponent. When the bell sounds, however, it is the boxer himself, the *direct* achiever, who steps into the ring. Nevertheless, when the match is won, the "champ's" sense of victory, of accomplishment, is shared with his seconds.

The third relational style—*collaborative relational*—is characterized by collaboration among two or more peers. Ideally, all participants perceive the achievement as their joint accomplishment. The synergistic effects of the team effort are a spur to accomplishments in this relational mode. The relationality of the style inheres in the collaborator's preference for a social context for task accomplishment.

The collaborative type usually selects collaborators on the basis of competence and potential contribution to the goal. However, secondary gains emanate from the camaraderie and support of the group. Faced with a task, collaborative relational achievers seek, even bring together, a set of coworkers with whom to share responsibilities and the rewards of victory. In contrast to the first two relational achieving styles (in which goals are selected by the direct achiever), collaborative relational types prefer that all group members share in task definition and implementation.

The prototypical milieu for the collaborative relational style is the athletic team. Here, while victory against an opposing team is the common task, each team member plays a distinctive, but mutually valued, role.

The functional relationships that are a necessary feature of the team's operation may or may not evolve into personal relationships. But smooth interaction among team members is valued, and interpersonal difficulties among the players are thought to impede success.

The *reliant relational* achieving style's most salient feature is the tendency to seek situations in which other individuals (or institutions) carry out the tasks defined by the reliant achiever. Practitioners of the reliant style set their own goals; however, they expect those with whom they have established dependent relationships to take responsibility for fulfilling these goals. Reliant relational achievers tend to perceive themselves as requiring help and support to meet their own goals. Relationships provide the medium in which the seeds of that help and support may grow.

Although reliant relational achievers establish relationships with individuals or institutions on whom they can rely, they do not develop the strong identification with the direct achiever that leads to accepting the means and goals of others as their own. Unlike vicarious and contributory achievers (who identify with selected others), reliant types establish relationships in which others identify with them.

The reliant relational type is concerned with self and self-defined goals, rather than the success of the other. Despite this concern with their own goals, reliant achievers tend to leave the *choice* of means to the direct achiever. The direct achieving member of the relationship, nevertheless, is valued, even clung to, as the source of effective goal attainment. This reliance upon the achievement skills of another, coupled with a sense of needfulness on the reliant achiever's part, distinguishes this achieving style from the instrumental relational mode. The reliant relational style is exemplified in the behavior of people who feel unable to cook and look to their spouses to provide meals.

Note that this style may encompass a widely diverse set of people. At one extreme are executives who typically delegate responsibility for decisions, allowing broad temporal and methodological discretion to their subordinates. At the other extreme, dependent people who feel incapable of accomplishing their goals and cling to others for this purpose also fall within this set. These represent two extremes of the reliant relational style, but they share the characteristic behavior of getting others to help them reach their own goals.

The *instrumental relational* style is a close cousin both to reliant relational and to instrumental direct modes. It is characterized by a propensity to use relationships as a means to achieving one's own goals. The instrumental relational style, however, differs from the reliant relational mode in important respects. The instrumental relational achiever specifies not only the task, but also the means by which others will accomplish the instrumental relational's ends. Moreover, instrumental relational types have confidence in their ability not only to define the goals and the means to success, but also to manipulate others toward the desired ends.

Another distinctive feature of the instrumental relational style is its covert manipulative nature. This covert quality distinguishes the instrumental relational type from the reliant relational achiever who more explicitly uses relationships. In addition, the instrumental relational achiever does not identify with the "other" nor attempt to evoke identification from the person perceived as the "instrument." The relationship is viewed by the instrumental relational type as an intervening step toward other accomplishments, and all relationships are evaluated in terms of their usefulness in gaining other accomplishments.

The instrumental relational achiever is the mirror image of the instrumental direct type. Here, relationships, not achievements, are used as leverage to success. The prototypical example is the lobbyist who establishes relationships to influence opinions and votes.

Dimensions of Achieving Styles

We have outlined nine distinct styles of achieving, all within the direct and relational domains. Each of these styles encompasses several other dimensions that are worthy of note: (1) the orientations of each style, (2) flexibility, (3) range, and (4) intensity.

Orientations to Self, Other, and Task

As Figure 1 suggests, degrees of self-, other-, and task-orientation vary (cf., Bass and Dunteman, 1963) as we move around the diagram. Among relational styles, orientation toward others is greatest in the vicarious and contributing relational modes, while self-orientation begins to emerge in the collaborative and reliant modes. The instrumental relational and the instrumental direct modes convey the strongest self-orientation.

In the direct domain, task-orientation is greatest in the

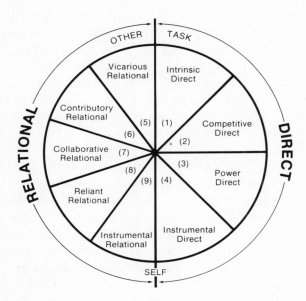

Figure 1. A model of achieving styles.

intrinsic mode and shades into self-orientation in the competitive, power, and instrumental segments. Generally, the modes in the top half of the diagram (particularly vicarious relational, contributory relational, and intrinsic direct) share a non-self orientation, focusing more on others or tasks. The instrumental segments and their adjacent modes are less oriented toward task or others and more focused on the self.

Range

For heuristic reasons, the foregoing discussion described achieving styles as if they were virtually independent and mutually exclusive. Empirically, that is rarely the case. *Most people use more than one style of achieving, employing them combinatorially or sequentially.* Access to and ability to use multiple modes are described as the range of an individual's achieving style repertoire. The greater the number of modes demonstrably accessible to an individual, the wider his/her range.

The issue of access to a mode is a serious one, particularly if we are concerned with people's needs to broaden their ranges. Access to a style of achieving merely means that an individual is capable of using that style. It offers no guarantee that in any specific instance the achiever will select that particular mode (unless, of course, it is the only mode to which one has access). While relative propensity to select one mode over another is linked to the question of intensity (which we shall discuss below), it is also linked to the "natural" ordering of the nine modes.

Indeed, the ordering of modes in the model (Figure 1) reflects the authors' view that the categories may be best thought of as key points along a continuum. Certain combinations are more probable than others. As noted earlier, adjacent modes therefore are more likely to be accessible to a given individual than are modes more distant from one another on the diagram. Several empirical findings—drawn from correlational, factor and canonical analyses—lend tentative support to this view. But that does not foreclose the possibility that any specific individual will have access to noncontiguous modes.

Socialization within different cultures may emphasize or preclude particular styles of achieving. By endorsing certain values and encouraging behavior that reflects them, entire societies may develop what Inkeles and Levinson (1969) have called "national character" structures. The nature of achievement and acceptable modes of achieving may be defined differently across disparate cultures. Thus, among the Tarahumaras of Mexico,

> Competition is dispensable. Among Mexico's Tarahumaras racing is participatory—men, women, and children, old and young, race together. It doesn't matter who finishes first—to arrive last is as honorable as to arrive first. The whole race is of interest to all spectators who measure performance against ability. What counts is that everyone who participates does his best. (Schechner, 1977, p. 223).

This noncompetitive orientation stands in sharp contrast to the American cultural value of competition, reflected early on in Benjamin Franklin's observation, " 'tis a laudable ambition, that aims at being better than his neighbor" (Franklin, 1936, p. 175). Three centuries later, the American competitive tradition remains alive and well, as former President Gerald Ford's (1974) remarks suggest:

> [We] have been asked to swallow a lot of home-cooked psychology in recent years that winning isn't all that important anymore . . . I don't buy that for a minute . . . it isn't enough just to compete. Winning is very important . . . Broadly speaking, outside of a national character and an educated society, there are few things more important to a country's growth and well-being than competitive athletics.

Cross-cultural and subcultural differences in achievement values have been the foci of other research (Salili et al, 1976; Lipman-Blumen & Leavitt, 1978). For example, McClelland (1961) explored the cultural roots and expression of the achievement motive, as well as its relation to economic development in several societies. And Miller (1973) reported natural experiments in which the high levels of cooperation among Blackfoot Indian children diminished after school integration. Thus, the range of achieving styles acceptable and, therefore, available to individuals may be influenced by cultural, as well as individual and personal, factors.

Flexibility

While range reflects the total set of achieving styles available to an individual, *flexibility* refers to the ease with which one can move within this range. Flexibility depends in part on range; that is, one obviously needs access to more than one mode in order to move in a flexible manner. Beyond this, flexibility refers to the capability to respond easily to different situations with different, presumably situationally appropriate, achieving styles. Underlying flexibility are two distinguishable factors: the capacity to recognize differential cues in the environment; and the ability to respond appropriately by selecting the relevant achieving mode(s). While it is clear that flexibility may be a learned response, it also may be inhibited through the learning mechanisms. Moreover, roles which set strict behavioral parameters may inhibit (or facilitate) flexibility.

Flexibility[7]—the capacity to move readily from one achieving mode to another—often occurs rapidly and naturally. The bench chemist, for example, may work in an intrinsic direct manner in the laboratory, but achieve in a vicarious relational mode with family members. On the other hand, flexibility may be a condition that develops over the life or career cycle. The world-class shotputter mentioned earlier described the changing uses and meanings of her athletic activities at sequential career points. The reader will recall that she initially used her athletic skills as an instrument for world travel. Later, as her skill improved, she began to experience an intrinsic satisfaction, a sense of elated achievement, in throwing for its own sake and in besting her previous athletic record.

Intensity

Intensity refers to the strength of an individual's propensity

---

[7]Flexibility raises the questions of the interface between styles of achieving, achievement motivation per se, and intention; however, it is beyond the scope of this paper systematically to probe the conceptual linkages among these concepts.

to select a particular mode of achieving over all other modes. Intensity in achieving styles varies within and between individuals, as well as among collectivities. Thus, a single individual may display varying degrees of intensity even among those achieving styles to which s/he has access. And different individuals with access to the same range of achieving modes may display differential strength or intensity.

Preliminary data suggest that, in addition to intra- and inter-individual differences, intra-group intensity variations also exist. While our own data indicate possible gender differences in intensity (as well as range), it is also likely that systematic variations would characterize other collectivities that differ on cultural, occupational, religious, socioeconomic, and comparable dimensions.

There are clear theoretical links among intensity, range, and the enacting process. *Enactment* here means the propensity of an individual to shape a situation to conform to a favored achieving style. Many examples of enacting may be found in all achieving style modes. One example: the behavior of high intensity competitive direct achievers who rearrange even non-competitive situations to permit competitive direct strategies. And individuals with a severely limited range of accessible styles, may, by default, deal with the world by reshaping it to fit their narrow repertoire.

A number of intriguing hypotheses could be advanced concerning the connections among intensity, range, and enactment. For the present, we only wish to suggest that these theorectical links are worth further exploration.

Differences from Other Models

The earlier review of the literature alluded to certain commonalities and disparities among previous models. Here it might be useful to touch briefly upon the key differences between other research efforts and the model just presented.

Much of the previous research on achievement has concentrated predominantly on discrete individuals fulfilling their achievement needs by acting directly on their environments. That research emphasis on individual achievement behavior—

conceptually and artifactually isolated—obscured the visibility of relational achievement behavior. And the concept of relational or vicarious achievement orientation introduces an additional perspective on the achievement paradigm.

As we have noted elsewhere, the emphasis in previous research has been upon motives, needs, and traits. The present research allows greater emphasis on *styles, roles,* and *structure.*

The leadership literature focuses in large part on the impact of the leader's behavior on group goal attainment. It pays rather scant attention to the leader's personal achievement (unless one assumes that the leader's achievement is entirely described by his/her effectiveness in moving the group toward group goals). Leadership studies tend to be cast in normative terms, suggesting that certain types of leadership behavior are "better than" others and are thus preferable for goal attainment. The present model is non-normative, attempting only to call attention to situational appropriateness of achieving styles.

Small group research, rooted in a structural/functional tradition, examines group dynamics. That perspective specifies the necessary functions that must be performed for group survival, including goal attainment. Within that framework, goal attainment is seen as a structural phenomenon. It is a recurring characteristic of groups which evolve roles to address certain required functions. These roles, in turn, are filled by individuals who, by personal predisposition or situational opportunity, act to help implement group goals. The present research temporarily sets aside questions of group dynamics and focuses on individuals' learned dispositions to achieve goals (whether their own, another's, or the group's) in characteristic ways.

The present research is close to some models of leadership and group behavior in its concern for the twin domains of "task" and "people." Our model's interest in people is, however, rather different from the people interest of those other approaches. Our focus is on achieving *through relationships* rather than upon concern for people per se. The instrument we shall describe in the next section was developed with these differentiating characteristics in mind.

An Instrument for Measuring Achieving Styles

Over the past few years, we have been developing the *Achieving Styles Inventory* (ASI), using a self-descriptive sentence format to measure our nine styles of achieving. The iterative design and redesign involved three steps, based on Brown's (1970) suggested process of combining logical and homogeneous keying and then testing the instrument empirically with defined criterion groups.

First, we developed a pool of items that looked like reasonable indicators of each achieving style and assigned items to their theoretical scale on a rational basis. To encourage respondents to focus on overt behavior, most items were posed in a behavioral mode (I do, I seek, I find) rather than an affective mode (I think, I feel, I believe). By this means, we hoped to minimize social-desirability biases and normative responses.

Second, we administered the instrument to several populations and analyzed the items' intercorrelational matrices. We eliminated items that did not correlate highly with other items on the same scale. Items from one scale that correlated highly with items from another scale were excluded. In the most recent revisions of the test, factor analyses were applied to assure that items belonging to a particular scale loaded highly only on the style being measured and not on other factors. Factor analyses also confirmed that each of the nine subcategories was, in fact, a separate scale.

Third, we empirically examined forms of the instrument on moderately well-defined criterion groups. For example, we hypothesized that persons in service organizations (e.g., nurses and librarians) would score high on vicarious and contributory relational scales, that senior executives would exhibit power and competitive direct styles, and that the engineering students would be high on intrinsic direct and low on vicarious relational scales. When the ASI was administered to such groups, many of our predictions were verified. Using discriminant function analyses, the ASI significantly differentiated among subgroups (e.g., sex and occupation) of the samples. While this procedure was not used to select items for the scales, it was used in the model construction.

The ASI is a 54-item instrument, scored by an unweighted sum-score method. Each scale score is computed by averaging the values of the six items keyed to each scale. The instrument shows adequate reliability, using both Cronbach alphas as estimates of scale reliability and four week test-retest coefficients as estimates of stability. In order to examine the contiguity of adjacent scales in the model, correlational matrices of scale scores were analyzed for several samples of subjects. Results across samples rather consistently demonstrate that the data conform to the conceptualization of a continuum model. That is, correlation coefficients of adjacent scales in the model are moderate, and correlations among scales farther apart on the model diminish. This lends support to our contention that individuals generally have access to contiguous styles.

Over its eight revisions to date, the instrument has been administered to more than 1200 respondents from ten populations: freshmen in psychology and biology, undergraduate engineering students, M.B.A. students, high school students, mid-level executives enrolled in an educational program, those executives' spouses, senior executives in an executive training program and their spouses, supervising nurses, librarians, mid-level technical managers in an R&D organization, and couples of the Mormon faith.[8]

Across this set of populations and across the set of modified instruments, several rather consistent findings have emerged. In older and more traditional populations, for example, sharp sex differences occur, with *men scoring higher on the direct scales* and *women higher on the relational ones. Among engineering and M.B.A. students, however, these differences not only disappear, but indeed seem to be reversed*, with women students appearing more direct and somewhat less relational than men. This latter result seems consonant with our expectations for younger cohorts of women entering

---

[8]A full description of the data described here will be presented in a forthcoming paper.

professional and graduate schools in the wake of the feminist movement.

To summarize, we began the development of the ASI by constructing a strictly logical pool of items. Later, we applied the technique of homogeneous keying of scales and then tested the instrument in samples of empirically-defined criteria groups. At this point, we have a completed nine-style instrument that meets conventional conceptual and methodological criteria for scale development. Investigation of the generalizability of the instrument is the current focus of our research.

Concluding Remarks

We have tried in this research process, to attend to "styles" or "orientations" of achievement rather than "motives" or "needs." This concern with the ways people go about getting what they want provides a process perspective on achievement that hopefully escapes from some of the conceptual and methodological traps surrounding the motivation concept. It also brings us to a step closer (but not all the way) to specific, directly observable behaviors. In addition, the methodology we have used brings into focus the concept of relational achieving, previously masked by the concentration on isolated individuals that marks most achievement motivation research.

While we have worked on *means* rather than ends, we have treated means as rather broad behavioral bands (i.e. "styles"). These behavioral bands are wide enough to encompass (1) a variety of specific *behaviors* and (2) also some attitude-like *propensities* which lie somewhere in the no-person's land between means and ends, between behavior and motives.

The notion of styles also provides a bridge, we feel, between the body of research on achievement motivation and several other research currents not ordinarily seen as closely related. Small group membership and leadership styles, for example, seem to us to be concerned mostly with how people set about getting done what they want (are motivated) to get done. And the growing literature on implementation (e.g., Leavitt and Webb, 1978; Keen, 1977) addresses the issue of how people in larger organizations achieve their ends. An achieving styles perspective focusing on the individual, whether alone or with others, offers a link between traditional achievement motivation and these other bodies of research.

We expect, given a finished instrument, to work empirically in those areas, relating achieving styles of individuals to their gender, occupational choice, and stage in the life cycle. We also plan to study how achieving styles relate to performance as leaders and implementers, as well as to the styles that organizations reward and punish.

REFERENCES

Abramson, L.Y., Seligman, M.E.P., & Teasdale, J.D. Learned helplessness in humans: Critique and reformulation. *Journal of Abnormal Psychology*, 1978, *87*, (1), 49-74.

Adorno, T.W., et al. *The authoritarian personality*. New York: Harper 1950.

Allport, G.W. *Personality: A psychological interpretation*. New York: H. Holt & Co., 1937.

Allport, G.W. Traits revisited. *American Psychologist*, 1965, *21*, 1-10.

Alper, T.G. The relationship between role-orientation and achievement motivation in college women. *Journal of Personality*, 1973, *11*, 9-31.

Alper, T.G. Achievement motivation in college women: A now-you-see-it-now-you-don't phenomenon. *American Psychologist*, 1974, *29*, 194-203.

Angelini, A.L. A new method for evaluating human motivation. *Boletin Faculdade de Filosefia Ciences*, Sao Paolo, 1955, 207.

Atkinson, J.W. Achievement motivation. *International Encyclopedia of the Social Sciences*, 1 (ed.: Sills, D.) San Francisco: The Free Press, 1968.

Atkinson, J.W., & Raynor, J.O. *Motivation and achievement*. Washington: Winston, 1974.

Bales, R.F., & Slater, P. Role differentiation in small decision making groups. in T. Parsons & R. F. Bales, *Family, socialization and interaction process*. Glencoe, Ill.: Free Press, 1955.

Bandura, A. *Social learning theory*. Englewood Cliffs, N. J.: Prentice-Hall, 1977.

Bass, B.M., & Dunteman, G. Behavior in groups as a function of self, interaction, and task orientation. *Journal of Abnormal and Social Psychology*, 1963, *66*, 419-428.

Bem, S.L. The measurement of psychological androgyny. *Journal of Consulting and Clinical Psychology*, 1974, *42*, 155-162.

Bernard, J. *The sex game: Communication between the sexes*. Englewood Cliffs, N. J.: Prentice-Hall, 1968.

Blake, R.P., & Mouton, J.S. *Building a dynamic corporation through grid organization development*. Reading, Mass.: Addison-Wesley, 1969.

Blaxall, M., & Reagan, B. (eds.) *Women and the workplace*. Chicago:

University of Chicago Press, 1976.

Blumen, J. (Lipman) *Selected dimensions of self-concept and educational aspirations of married women college graduates.* Unpublished doctoral dissertation, Harvard University, 1970.

Brown, F.G. *Principles of educational testing.* Hinsdale, Ill.: The Dryden Press, 1970.

Christie, R., & Geis, F.R. *Studies in Machiavellianism.* New York: Academic Press, 1970.

Coser, R.L., & Rokoff, G. Women in the occupational world: Social disruption and conflict. *Social Problems,* Spring 1971, 535-554.

de Charms, R. *Personal causation: The internal affective determinant of behavior.* New York: Academic Press, 1968.

Deci, E.L. *Intrinsic motivation.* New York: Plenum, 1975.

Dweck, C.S. The role of expectations and attributions in the alleviation of learned helplessness. *Journal of Personality and Social Psychology,* 1975, *31,* 674-685.

Dweck, C.S. Children's interpretations of evaluative feedback: The efffect of social cues on learned helplessness. in C. S. Dweck, K. T. Hill, W. H. Reed, W. M. Steihman, & R. D. Parke, "The impact of social cues on children's behavior." *Merrill-Palmer Quarterly,* 1976, *22,* 83-123.

Dweck, C.S., & Reppucci, N.D. Learned helplessness and reinforcement responsibility in children. *Journal of Personality and Social Psychology,* 1973, *25,* 109-116.

Epstein, C.F. *Woman's place.* Berkeley, Calif.: University of California Press, 1970.

Fiedler, F.E. The contingency model and the dynamics of the leadership process. In L. Berkowitz (ed.), *Advances in experimental psychology* (Vol. 11). New York: Academic Press, 1978.

Fleishman, E.A. *Twenty years of consideration and structure.* Paper presented at Southern Illinois University Leadership Conference, April, 1971.

Fleming, J. Predicitive validity of the motive to avoid success in black women. *Humanitas,* May 1977, *13*(2), 225-244.

Ford, Gerald (with John Underwood) In defense of the competitive urge. *Sports Illustrated,* July 8, 1974.

Franklin, B. *Poor Richard's almanac.* Reading, Pennsylvania: Spencer Press, 1936.

French, E., & Lesser, G.S. Some characteristics of the achievement motive in women. *Journal of Abnormal and Social Psychology,* 1958, *68,* 45-48.

Freud, S. *The basic writings of Sigmund Freud.* (A. A. Brill, ed. & trans.) New York: Random House, Modern Library, 1938.

Hochschild, A.R. A review of sex role research. In Joan Huber (ed.), *Changing women in a changing society.* Chicago: University of Chicago Press, 1973.

Hoffman, L.W. Fear of success in males and females: 1965 and 1972.

Journal of Consulting and Clinical Psychology, 1974, 42, (3), 353-358.

Horner, M. *Sex differences in achievement motivation and performance in competitive and non-competitive situations.* Unpublished dissertation, University of Michigan, 1968.

Hull, C.L. *Principles of behavior: An introduction to behavior theory.* New York: Appleton-Century-Crofts, 1943.

Inkeles, A., & Levinson, D. J. National character: The study of modal personality in sociocultural systems. In G. Lindzey & E. Aronson (eds.), *Handbook of Social Psychology* (Vol. 4). 2nd ed. Reading, Mass.: Addison-Wesley, 1969.

Keen, P. Implementation research in OR/MS and MIS: Description versus prescription. Research Paper 390, Graduate School of business, Stanford University, July 1977.

Leavitt, H.J., Lipman-Blumen, J., Schaefer, S., Harris, R. *Vicarious achievement orientation.* Paper presented at annual meeting American Psychological Association. San Francisco, California, August 1977.

Leavitt, H.J., & Webb, E.J. Implementing: Two approaches. Research Paper 440, Graduate School of Business, Stanford University, May 1978.

Lesser, G.S., Kravitz, R.M., & Packard, R. Experimental arousal of achievement motive in adolescent girls. *Journal of Abnormal and Social Psychology*, 1963, 66, 59-66.

Lewin, K., Lippit, R., & White, R.K. Patterns of aggressive behavior in experimentally created social climates. *Journal of Social Psychology*, 1939, 10, 271-299.

Levine, A., & Crumrine, J. Women and the fear of success: A problem in replication. *American Journal of Sociology*, 1975, 80, 964-974.

Likert, R. *New patterns of management.* New York: McGraw-Hill, 1961.

Lipman-Blumen, J. How ideology shapes women's lives. *Scientific American*, 1972, 266 (1), 34-42.

Lipman-Blumen, J. *The vicarious achievement ethic and non-traditional roles for women.* Paper presented at annual meeting, Eastern Sociological Society, April 1973.

Lipman-Blumen, J., & Leavitt, H.J. Vicarious and direct achievement patterns in adulthood. *Counseling Psychologist, 1976, 6(1),* 26-32.

Lipman-Blumen, J., & Leavitt, H.J. *Socialization and achievement patterns in cross-cultural perspective: Japanese and American family and work roles.* Paper presented at International Sociological Association, 9th World Congress, August 1978.

Lipman-Blumen, J., & Leavitt, H.J. Sexual behavior as an expression of achievement orientation. In H. A. Katchadourian (ed.), *Human sexuality.* Berkeley, Calif.: University of California Press, 1979.

Lipman-Blumen, J., & Tickamyer, A. Sex roles in transition: A ten-year review. *Annual Review of Sociology.* Palo Alto, Calif.: The Annual Review, 1975, 297-337.

Lockheed, M. Female motive to avoid success? A psychological barrier or a response to deviancy? *Sex Roles,* March 1975, 41-50.

Maccoby, E.M., & Jacklin, C.N. *The psychology of sex differences.* Stanford, Calif.: Stanford University Press, 1974.

Maslow, A. *Motivation and personality.* New York: Harper, 1954.

McClelland, D.C., Atkinson, J.W., Clark, R.A., & Lowell, E.L. *The achievement motive.* New York: Appleton-Century-Crofts, 1953.

McClelland, D.C. *The achieving society.* New York: Van Nostrand, 1961.

McClelland, D.C., & Burnham, D.H. Power is the great motivator. *Harvard Business Review,* (March/April 1976), *54*(2), 100-110.

Mednick, M.T.S., & Puryear, G.R. Motivation and personality factors related to career goals of black college women. *Journal of Social and Behavioral Sciences,* 1975, *21,* 1-30.

Mednick, M.T.S., & Puryear, G.R. Race and fear of success in college women: 1968 and 1971. *Journal of Consulting and Clinical Psychology,* 1976, *44,* 787-789.

Mednick, M.T.S., & Weissner, H.J. Psychology of women—selected topics. *Annual Review of Psychology,* 1975, *26,* 1-18.

Miller, A.G. Integration and acculturation of cooperative behavior among Blackfoot Indian and non-Indian Canadian children. *Journal of Cross-Cultural Psychology,* 1973, *4,* 374-380.

Miller, J., & Dollard, N. E. *Social learning and imitation.* New Haven: Yale University Press, 1941.

Mischel, W. *Personality and assessment.* New York: Wiley, 1968.

Murray, R.A. *Explorations in personality.* New York & London: Oxford University Press, 1938.

Papanek, H. Men, women and work: Reflections on the two-person career. *American Journal of Sociology,* 1973, *78,* 852-870.

Parsons, T., & Bales, R.F. *Family, socialization and interaction process.* Glencoe, Ill: The Free Press, 1954.

Puryear, G.R., & Mednick, M.T.S. Black militancy, affective attachment, and the fear of success in black college women. *Journal of Consulting and Clinical Psychology,* 1974, *42,* 263-266.

Salancik, G.R., & Pfeffer, J. An examination of need-satisfaction models of job attitudes. *Administrative Science Quarterly,* 1977, *22*(3), 427-456.

Salili, F., Maehr, M.L., & Gilmore, G. Achievement and morality: A cross-cultural analysis of causal attribution and evaluation. *Journal of Personality and Social Psychology,* 1976, *33*(3), 327-337.

Sears, R.R., Whiting, J.W.M., Nowlis, V., & Sears, P.S. Some child-rearing antecedents of aggression and dependency in young children. *Genetic Psychological Monographs,* 1953, *47,* 135-234.

Seligman, M.E.P. *Helplessness: On depression, development, and death.* San Francisco: W. H. Freeman, 1975.

Schechner, R. *Essays on performance theory, 1970-1976.* New York: Drama Book Specialists, 1977.

Simon, H.A. *Administrative behavior.* New York: Macmillan Co., 1957.

Spence, J.T., & Helmreich, R.L. *Masculinity and femininity: Their psychological dimension, correlates, and antecedents.* Austin: University of

Texas Press, 1978.
Stogdill, R.M., & Coons, A.E. (eds.) *Leader behavior: Its description and measurement.* Columbus: Bureau of Business Research, College of Commerce and Administration, Ohio State University, 1957.
Tresemer, D. (ed.) *Sex Roles,* 1976, 2, No. 3. Special Issue on Fear of Success.
Vroom, V.H. *Work and motivation.* New York: John Wiley, 1964.
Vroom, V.H., & Yetton, P.W. *Leadership and decision-making.* Pittsburgh: University of Pittsburgh Press, 1973.
Weiner, B. *Achievement motivation and attribution theory.* Morristown, N. J.: General Learning Press, 1974.
White, R.W. Motivation reconsidered: The concept of competence. *Psychological Review,* 1959, 66, 297-333.

III
Achievement Motivation and Life Span Human Development

Chapter 8

The Graying of America: Implications for Achievement Motivation Theory and Research*

MARTIN L. MAEHR
The University of Illinois, Urbana-Champaign

DOUGLAS A. KLEIBER
The University of Illinois, Urbana-Champaign

Introduction

Anyone involved with education or other social services in the United States is unavoidably, and perhaps very personally, aware of shifting proportions of children, youth, adults and the aged in the general population. Abrupt changes in the size of the youth population are an interesting and unique feature of the social landscape of the U.S. in the twentieth century. The rapid rise in birth and fertility rates, referred to as the "baby boom," produced an extremely rapid increase in the number of children and youth during the years 1956-1976. Just as surely, the recent and very rapid decline in birth and fertility rates will produce an equally steep decline in the relative size of the youth population in the period between 1976-1990 (cf., Figures 1 and 2). Complementary to this, the average life span has continued to increase, eventuating in what has sometimes been called the "Graying of America."

Such drastic changes in age proportions can hardly be ignored. And they have not been! School board presidents and politicians are very much aware of them. It is unclear how aware achievement theorists are of such trends, however; and

---

*Portions of this chapter were presented as a paper in conjunction with an AERA symposium entitled "Recent Trends in Achievement Motivation Theory and Research. The authors are especially indebted to Robert L. Egbert, John Gienapp, John Kelly, Mary Kluender, John Nicholls, and Connie Walker.

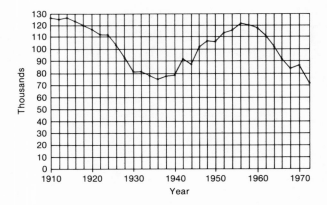

Figure 1. U.S. fertility rate in two-year intervals—1910-1972 (Vital statistics of U.S., 1973. After Egbert, Maehr, and Gienapp, 1968).

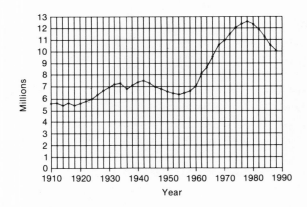

Figure 2. Number of eighteen to twenty-year olds in U.S. population — 1910-1988 (Estimated from 1970 census of population. After Egbert, Maehr, and Gienapp, 1968).

yet, they should be, for such trends suggest new questions and present new opportunities. Although there are a number of possible effects of the "Graying of America," the purpose of this chapter is to examine two implications of these trends so far as achievement motivation theory and research are concerned. First, we will propose that a focus on motivation and achievement in aging is warranted, perhaps even inevitable. Yet, to this point this focus is not evident in psychological research. While increasing attention has been directed in recent years to matters of aging and dying, the characteristics of motivation in later life have been largely neglected. The relative shift in population should have an effect in changing this, but even more significant is the possibility that aging may be associated with qualitatively different patterns of motivation, patterns which, with the prevailing emphasis on achievement in childhood and young adulthood, have gone heretofore unrecognized. Such a focus will likely change our conception of achievement motivation and expand the ways in which we study it.

Second, we will consider broader and more speculative issues about the nature of motivation and achievement where the old predominate. What are the psychosocial implications of these age proportion shifts? To over-dramatize the issue a bit: do such shifts signal the end of the "achieving society" as we now know it? While general attention has been rapidly directed to the drastic shift in age proportions of the population, little attention has been given to the psychosocial impact of these shifts. Yet, it should be obvious that such shifts in age proportions are likely to have an impact on social interaction patterns and therewith on the nature of persons and society.

These shifts relate to changes in family size and one can easily imagine that the size of the family might well have some effect on parents and parenting. The shifts will also affect the capacity, effectiveness and, in some cases, the very existence of institutions which complement the family in the socialization task. Teachers lose valuable contact in classrooms which are overcrowded, but dramatic decreases in enrollment, such as are in evidence in elementary schools around the country at this point in time, threaten to close the schools altogether.

Additionally, the shift in population structure may result in the redistribution of social power. The bulge in the youth population from 1966-1976 conceivably encouraged the development of a "youth culture" which in turn wielded considerable social and political power (cf. for example, Coleman, 1974). It is reasonable to assume, for example, that the large youth cohort of the late sixties was at least in part responsible for lowered voting and drinking ages. The economic implications of a large youth cohort was also fairly obvious in the focus and thrust of advertisements presented in the media. Now, the shift in power is beginning to be evidenced in the budgetary cuts which affect institutions which serve children and youth, especially the public schools, and colleges and universities are recognizing the writing on the wall. Such informal observations can, of course, be multiplied (cf., for example, Egbert, Maehr, & Gienapp, 1978) but this may be sufficient to suggest that population age shifts can have important effects on persons and society as a whole so far as achievement motivation is concerned.

In summary, the "Graying of America" occasions the asking of a number of provocative questions. Somewhat arbitrarily we intend to consider only two issues in this regard. The first issue begins with the fact of demographic shifts and asks: whither achievement motivation theory and research? The second issue asks: whither the achieving society? While these two issues are somewhat unrelated in a theoretical sense, they are certainly both of parallel importance as far as social policy issues are concerned. They both come to mind, almost simultaneously, as one thinks of the implications of recent demographic shifts. Additionally, it should be clear that the second issue can only be dealt with as the first issue is fully understood.

Aging and Achievement Motivation

With the Graying of America it is not altogether unpredictable that someone would finally come along and suggest that the effects of aging on achievement motivation ought to be studied. Given the shifts in population and power already alluded to, this is probably inevitable. Yet, such study is likely to have interesting effects both on how we define achievement motivation and how we study it—and it is that fact which should be of general interest.

While the study of achievement motivation has been extended to varying times and climes (cf., McClelland, 1961; Maehr, 1974; Maehr & Nicholls, in press) it has been largely limited to the study of the young. Indeed, the term achievement motivation itself has a futuristic connotation that seems most applicable to those who have their life and career before them. But possibly this only reflects a kind of ethnocentrism, the kind of ethnocentrism that has been decried in the case of crosscultural studies of achievement behavior (Maehr, 1974; Maehr & Nicholls, in press). In any case, as shifting age proportions in the population make the study of aging relevant, achievement theorists will do well to shift their attention slightly. In so doing, they will need to ask a number of different questions, and the way they answer these questions will likely determine the conceptualization and study of achievement generally.

First, it seems logical to ask—how does achievement motivation change, if at all, in the later years of life? However, in asking such a question, we may find ourselves concerned with rethinking the basic nature of achievement motivation. The predominant theoretical orientation (e.g., Atkinson & Birch, 1970; Atkinson & Feather, 1966; Atkinson & Raynor, 1974, 1977; McClelland, 1961, 1971; McClelland & Winter, 1969) has stressed the importance of characteristic individual differences which remain stable throughout the life span. Thus, it is characteristically assumed that the child who demonstrates high need-achievement in the academic setting of junior high school will be similarly motivated in the context of work thirty years later. Those who stress the importance of role (e.g., Klinger & McNelly, 1969) and social context (e.g., Maehr, 1974, 1977, 1978) may well see the later years as the proving ground for their argument. While there most certainly is evidence that personality and social motivation do not change dramatically during the later years (e.g., Schaie & Parham, 1976) and even that achievement motivation has a remarkable stability over time (Kagan & Moss, 1962,) Veroff (1978) is perhaps begging the question when he contends that:

> Achievement motivation has the same symbolic significance from childhood thorough the adult life span. Particular contexts of achievement motivation naturally vary at different ages, but the quality of motivation will be the same.

Leaving aside the meaning of "symbolic significance" and

"quality of motivation" the viewpoint is one of stability of pattern; but there are, after all, other indications that achievement motivation may undergo substantial changes with age. A willingness to take moderate risks, to compete and make social comparisons, and a future orientation are regularly cited features of achievement-oriented individuals. Yet, there is evidence that older persons are characterized by cautiousness (cf., Botwinick, 1966) and a tendency toward affiliation rather than competition and social comparison (cf., Edward & Wine, 1963; Klein, 1972) and it seems self-evident that a concentration on the future may be qualitatively different for the aged. Certainly, delay of gratification takes on a different meaning for older persons. Is it that they are no longer motivated to achieve or that their motivation to achieve is directed in new and different ways, elicited by differing contexts?

In focusing on achievement behavior of later life adults there will be greater opportunity to consider how role-related, contextual changes affect achievement motivation. The "launching" of children, for example, has been shown to have significant impact on the leisure motivations of adults (Kelly, 1978) and other role transitions, especially those related to retirement, may be expected to affect motivational patterns in general. All in all, one might oversimplify a bit and suggest that considering whether or not achieving orientations and achievement behavior change for older individuals provides an opportunity for achievement research. For one thing, it provides an opportunity to reconfront perplexing trait vs. state issues (cf., for example, Endler & Magnusson, 1976) in a new context and with renewed vigor. Perhaps it might also cause us to reconsider the varied meanings of achievement behavior.

In a recent review of culture and achievement motivation, Maehr and Nicholls (in press) have argued that the cross-cultural evidence forces us to consider achievement in a broader context than it is usually considered. Simultaneously, they have suggested that the different cultural groups do not differ so much in regard to exhibiting achievement behavior. The difference between cultural groups lies rather in how such achievement-related behavior as challenge-seeking, persistent effort in pursuit of excellence, delay of gratification, etc.

is directed to different ends. Generally, individuals pursue ends because attaining them will reflect favorably on themselves. Obviously, many such ends can be and are pursued, but Maehr and Nicholls sugest that three general types of goal striving are properly comprised within the area of achievement motivation. Following Kukla (1978), they suggest that a primary goal is to *demonstrate competence*. Clearly related thereto is the more *intrinsic goal of mastery*. A third goal, perhaps only marginally classified as an achievement goal, is to *gain social approval* for one's striving or one's goals. Maehr and Nicholls' contribution is to incorporate these three goals in one theoretical framework as three possible achievement orientations, each of which leads to a certain style of achievement behavior. Note also that the domain in which such achievement striving is to occur is left open. Presumably, that varies across individuals as well as across cultures. One can strive to be competent in laying bricks, basketweaving, or playing basketball. One can also hold the primary goal of appearing well-intended or conscientious in these tasks in order to gain social approval. Maehr and Nicholls emphasize that such a shift in goals has important effects on the behavior which the person demonstrates. Moreover, a full understanding of achievement behavior requires sensitivity to the fact that apparently similar behaviors may serve different purposes.

Cross-cultural studies prompted Maehr and Nicholls to broaden and redefine achievement motivation in this way. Quite possibly cross-age studies will provide an additional proving ground for this conceptualization. It is intriguing to think about the possibility that striving to attain such achievement goals as outlined by Maehr and Nicholls may be evidence at all age levels, albeit in regard to different task areas. Moreover, it is even more intriguing to consider whether there are certain shifts in preference for the types of achievement goals outlined by Maehr and Nicholls. The goal to demonstrate competence is a goal readily in evidence in academic settings. If one examines students who are on the threshold of proving themselves in the world, who can also readily make social comparisons, and who are surrounded by an ethic which stresses the value of being more competent than someone else, it is natural to stress the goal of competence in achievment striving (cf., Kukla, 1978). But older persons live in a slightly different world, and quite possibly, it is a world where they can go beyond the stress on competence in reference to others to a more intrinsic concern with mastery. Such a shift in focus would be consistent with a gradual disengagement from the world of work where one's competence is defined in terms of salary, status, occupational prestige and social and political power and a movement toward more leisure-based activities where standards of excellence are intrinsically defined and emerge to a great extent out of the activities themselves (cf., Csikszentmihalyi, 1978). Whether or not such a shift characterizes older persons in general, the study of older persons should lead researchers to ask just such questions and this should expand our conceptualization of what achievement motivation is and can be.

The Development of New Approaches

It appears that switching to the study of older persons could provide a profitable proving ground for new hypotheses about achievement behavior. Indeed, there is every reason to believe that the study of achievement patterns in the case of older persons may occasion some surprises and some rethinking in regard to the nature of achievement motivation. It will, most assuredly, force the adaptation of old methodologies and the adoption of new ones. Perusing the numerous studies that have been conducted on achievement motivation, one is impressed with their reliance on highly motivated, approximately-socialized students who have little else to do but play the researchers' silly game. If one is to study older persons, particularly the aged, there is little doubt that this strategy will prove a disaster. Certainly, industrial psychologists have established important precedents for motivational research that can appropriately be followed. But even these may not be adaptable to the study of persons who are no longer working for a company which can command their allegiance, support, and subservience to the whims of a researcher. It is hard to predict the methodology that will evolve in response to such problems. In our own initial, struggling efforts (Maehr, Kelly & Kleiber, 1979) we have resorted to interviews, case histories and ethnography. That seemed to be a logical first step, but it can be only a transitional phase on the way to solving a wide range of procedural problems. In any case, the call for an experimental anthropology of motivation expressed earlier in regard to cross-cultural research (Maehr, 1974) might be every bit as relevant in cross-age studies of the type we are proposing. It is, of course, important that those who would move to study motivational processes in the aging recognize the inherent methodological problems and take advantage of established techniques that have been discussed by those who have been working in aging and life span development for some time (e.g., Schaie & Baltes, 1978). Inevitably, as older persons are studied, techniques will be shunned and the armamentarium of research techniques will expand. That, too, is no mean outcome stemming from the "Graying of America."

The Graying of America and The Achievement Ethic

A second implication of the Graying of America reaches beyond effects on achievement motivation theory and research in any strict sense. Given the fact that certain demographic shifts are occuring, what effects might these have on society as a whole? How might this affect societal values associated with work, play, and achievement? With more people retiring and retiring earlier and with increasing attention to retirement, leisure lifestyles and a leisure ethic must be considered as viable reorientations. Might the Graying of America eventuate then, in the demise of the achivement ethic? These are questions which inevitably push us beyond the typical domain of achievement motivation research. Answering these will likely involve extrapolation, questionable generalization, and... sheer speculation. Yet, they are questions which are properly asked, even if they can only be improperly answered. Are there answers in current work which speak to these issues? Quite possibly, at least, it may prove to be an interesting enterprise to view current work in such futuristic terms.

In projecting what societal effects might ensue with the

Graying of America, there are a number of lines of thought that can be pursued. A first and reasonably obvious one is to focus on the fact that a certain age group, which may have its own values, goals, styles, and behavioral preferences, is predominant. As the age proportions shift, will power and influence shift? If it does how will overall societal values, goals, and behavioral patterns change? The study of older persons, already proposed, may suggest answers here.

One possibility which might be considered is that the expansion in the numbers of older individuals will lead more quickly to an art, science and technology of aging and that a greater "age consciousness" will pervade all levels of society. And one effect of this may be a divergence of societal models of competence, varying from the traditional patterns of those who gain success in occupational contexts to those who find fulfillment in personally satisfying life styles. The whole concept of achievement may revolve more around personal definitions of mastery and excellence. Intrinsically satisfying, expressive pursuits may increase in importance. In embracing these latter behavioral styles, a young individual's achievement orientation and pattern of achievement behavior are less likely to be defined by goal-setting, a future orientation, and delay of gratification as by the pursuit of intrinsically-satisfying experiences, present centeredness, and emergent motivation based on task-specific challenges (see especially, Csikszentmihalyi, 1975, 1978; Maehr & Nicholls, in press).

At a more metaphysical level Sarason (1977) takes to task the prevailing ideology that one can and should find significance in life in a single career and vocation. He bases his analysis to some extent on the writings of Ernest Becker (1973), who discusses the tendency to deny an awareness of death by maintaining an illusion of immortality in some "character lie." Sarason suggests that at least in Western culture, such self-deceit is frequently embodied in the social significance of an achieved occupational position. But he contends that such illusions are less and less tenable as our awareness of aging expands and that a more reasonable alternative is to see aging as growing and a variety of work situations (we might say achievement contexts) as growth opportunities. However conscious individuals are of those problems of existence, or to whatever extent such consciousness is raised in the future, it is safe to say that achieving orientations are ultimately based on a sense of time and mortality, concepts with changing meaning in an aging population.

In addition to these impressions of age and aging on the young, demographic changes may affect them even more directly. The Graying of America is not only a function of lengthening life span, of course; an equally, if not more important, contribution to the overall picture is that fewer children are being born. A drastically decreased birth rate following a period of expanding population creates a peculiar situation as far as age proportions in the population are concerned. Essentially, the children born today are born into an adult world—more so than ever before. Moreover, whereas previous generations could always hope that if they just worked hard enough there would be "room at the top" for them, this generation must recognize that their elders—an unusually large group—have a stronghold on the top and may not wish to vacate it. What are the possible implications of such a state of affairs? In order to illustrate that there are important implications for an achievement ethic here, two complementary possibilities will be considered.

Early Socialization Experiences in a Gray Society

A preliminary basis for believing that population shifts may have an overall effect on achieving patterns in a society can be found in the work of Zajonc (1976; cf. Markus & Zajonc, 1977; Zajonc & Markus, 1975). Since this work is rather generally known (though not always fully accepted[1]), a brief summary should suffice. Essentially, Zajonc has argued (and demonstrated to a substantial degree) that aptitude test scores decline as a function of family size. The larger the size of the family, the lower aptitude scores are likely to be. Thus, in periods when

---

[1] It may be noted, for example, that it apparently did not figure strongly in the final conclusions of the *Advisory Panel on the Scholastic Aptitude Test Score Decline* (1977).

larger families are in vogue, a general lowering of achieving patterns can be expected, culminating in generally lower college aptitude scores. In explaining this correlation, Zajonc has stressed likely variations in the cognitive enrichment of the socializing environment, particularly in the varying intellectual experience and teaching processes that probably characterize smaller and larger family units. The fewer the competing siblings and the older (and wiser) the existing siblings are, the lower the demand on and depletion of existing intellectual resources.

Perhaps motivational factors are involved as well. Population shifts create socialization patterns that could well affect achievement motivation as well as cognitive growth. Children born in this period of reduced birth rate will likely be raised in socialization units where the adult/child ratio will be greater. If one reasons with Egbert and his colleagues (Egbert, 1978; Egbert & Kluender, 1977; Egbert, Maehr, & Gienapp, 1978) that this will also eventuate in greater (adult) attention paid to the individual child, this might have motivational as well as cognitive and general achievement effects. On a priori grounds one might surmise that the increased adult/child ratio could eventuate in either of two possibilities. One the one hand, greater attention paid to the individual child might effect a greater sense of personal worth and perhaps also a greater sense of *personal causation* (deCharms, 1968, 1976), leading to an increased orientation toward achievement. Alternatively, there is the possibility that the greater adult/child ratio could overwhelm the child, creating a kind of dependence and eventuating in a decreased sense of personal causation and/or achievement motivation.

Not surprisingly, the achievement motivation literature to this point has not dealt excessively with the effects of size-determined socialization patterns on achievement motivation. There are, however, scattered pieces here and there in the literature which may move this discussion one small step beyond speculation toward a credible hypothesis, one worth putting to the test. For example, Rosen (1961) presented a line of reasoning not unlike that presented here in conducting one of the early (and one of the few!) studies on the effects of family size on achievement motivation development. Briefly, Rosen argued that the large family is more likely to value responsibility, conformity, and obedience over individual achievement. The demands of handling a larger group tend to require this. While children raised in a larger family are perhaps expected to be more self-reliant, this is likely to be in areas of basic caretaking and not in areas where clear standards of excellence are readily and repeatedly brought to bear. Thus, one might surmise that the child brought up in a larger family is not so likely to be presented with a sequence of appropriate challenges, at least in those areas valued by adult caretakers. Rosen's results indicated an overall negative relationship between size of family and need for achievement, even when socioeconomic status was controlled; but a closer look at the data revealed that this relationship might depend somewhat on social class. It was in the case of middle-class subjects that the relationship was most clearly evident, with some significant reversals in the case of upper- and lower-class subjects. Thus, the results were somewhat more complex than one might wish in providing clear evidence for the hypothesis that the social dynamics that are likely to be associated with smaller families will likely lead to higher achievement motivation. Clearly the issue must be pursued further, and attention should also be given to how varying adult/child ratios in other socializing institutions may affect the development of achieving orientations.

When the "Children of the Sixties and Seventies" Hit the Job Market

Assuming that small families raise children to have a greater sense of their own importance, a greater sense of their own self-worth, and possibly, a greater orientation toward achievement—or even assuming that such is not the case—it is interesting and important to consider what might ensue.

Here attention might be called to one critical situation which the "children of the sixties and seventies" will probably face. These children who were and are born into and raised by small families, will enter a job market where entry-level jobs are likely to be readily available. (Their cohort group will be small and we assume that societal needs will remain relatively constant). One might expect that their initial experience with the job market should be a successful one, therewith also raising their level of career aspiration. However, one may note that, considering population proportions alone, upward mobility may prove difficult (the cohort groups produced by the "baby boom" would presumably dominate upper level positions). The question becomes—what is likely to happen under such conditions?

First, what will happen will depend significantly on the kinds of persons we will have. Thus, the study of the (personality) effects of being reared in small families is a critical step in making reasonable projections.[2] Of equal importance is how various personality types are likely to relate to lack of upward mobility or career progression. In piecing together answers to these two questions one can imagine at least two scenarios:

(1) Assuming that socialization in small family units increases achievement motivation, one might expect the "children of the sixties and seventies" to be a frustrated generation. Given an age cohort composed of individuals who tend to have a strong sense of their own importance, of their own self-worth and possible high achievement motivation, they will probably expect to accomplish great things. Additionally, this expectation would presumably be reinforced by their initial job/career experiences. But an analysis of population trends indicates that *movement to upper level positions may be severely retarded* (recall that the immediately elder age group is large and presumably it will dominate upper level positions perhaps for even a longer period than current age-associated promotion and retirement practices allow). What might happen under such conditions is difficult to surmise. Some might argue that a rebellion against the establishment would occur. However, our line of reasoning suggests a slightly different possibility. Following

---

[2]Associated with such study, of course, must be parallel work on the complementary or counter effects of other socialization units. In particular, attention must be given to the possibility that increased use of "day care facilities" may modify the effects of being born into small family settings.

the analysis presented by Egbert (1978; Egbert & Kluender, 1977; Egbert, Maehr, & Gienapp, 1978), this cohort is not likely to band together *as a group* and lash out at the external factors that are the source of their distress. As *individuals*, they could acquire and may retain the belief that they personally are not the cause of such failure to achieve and instead blame the situation. As a result, it would be reasonable for them to direct their efforts away from achievement as "employees" and move toward achievement outside the market place, finding satisfaction in avocational rather than vocational contexts, for example. For the individuals involved, that may be a hopeful possibility. However, it does not necessarily bode well for a society which looks to the young as a source for continued economic, industrial, and technological growth and revitalization. It has also been suggested that this scenario could lead to "active economic warfare between established companies staffed (more and more feebly) by aging employees and active competitive young companies run by the young and staffed by the young" (Evans, 1979).

(2) The second scenario hardly has an optimistic side. Assuming that socialization in small units has the opposite effect; that is to say, that it tends to overwhelm the individual and therewith reduce his/her sense of ability and competence, giving him/her a sense of powerlessness. This scenario—far more than the first—promises a peaceful coexistence between generations. But peace may well have been won at a high price. Motivation to excel, exhibit excellence, and do well not only at a job or in a career, but at anything, might be severely reduced for a generation or so.

One may choose and defend either scenario at this point. They are, in vague outline, equally plausible. However, in either case, the point is that motivation and achievement patterns, as we know them, may be drastically changed by population trends. Indeed, given changing opportunities, the goals of achievement may have to be adjusted, if not redefined. That, of course, is the point we wish to stress.

And Where Do We Go From Here?

Let us first reconsider the crucial points of this exercise in crystal ball gazing. Some drastic and perhaps dire consequences of current demographic shifts on achieving behavior have been suggested. While these specific suggestions are highly speculative, the important point is that it may not only be theoretically interesting but also of practical significance to observe population shifts when and wherever they occur. Moreover, learning something about the effects of population on social psychological behavior may be increasingly useful in the world in which we live. One should note in this connection that drastic birth rate changes such as have been experienced in the U.S. (as well as in the USSR, Germany, Japan, and France) are likely to be experienced by other developing countries in the near future as well. We are not merely dealing with the prediction of a one-time phenomenon. Societal experience with age proportion shifts and distinct age groups may occur repeatedly. And even if they do not, the virtues of scholarship dictate that there are benefits in simply establishing relationships between demographic trends and social psychological variables and organizing the diverse data for meaningful interpretation.

Finally, one may note that by attending to such over-arching global issues, priority is given to areas of research on achievement that have been less than thoroughly studied. The effects of varying size of socializing unit deserve our interest. What effects does increasing the adult/child ratio have—on behavior in general as well as on achievement motivation in particular? Are any effects of family adult/child ratio trends neutralized by possible counter trends, such as placing children in day care settings? How are parental motivation patterns and family stability affected by these changes? These are examples of the kinds of questions that this inquiry should prompt. At the very least, looking into the future as we have in this analysis points up how little we know about the socialization of achieving orientations, about the role of socializing institutions and socializing situations, about the endurance of achieving orientations through the life-span and about the ultimate implications for society. Perhaps also this exercise has suggested that those of us interested in achievement motivation could, if we put our minds to it, say something significant to those who cannot avoid predicting, planning for, and deciding about the future, i.e., government planners and policy makers.

Conclusion

In this chapter we have moved rapidly, and some would say precipitously, from a micro- to a macro-analysis of achievement. More dangerously still, we have jumped from a limited amount of evidence to some rather bold interpretations. Yet, such boldness may be justified. We cannot reliably know the future, of course, but there are certain signs of the future in the present and these deserve attention. As a matter of course, policy makers, planners, and many social scientists see it as their special business to consider options of the future. It is our impression that psychologists have been especially remiss in this regard: possibly because of the constancy that they see in the person, they tend to ignore the whirl of changing social events which surround persons.

In any case, several things ought to be clear from any cursory reading of the current social scene. We are fast becoming a nation of older persons. That has direct implications for our society. Whether or not we are concerned with relevance it will affect the kind of research we are encouraged to do. But such demographic shifts will also, in all probability, have numerous indirect effects as well. We have speculated about several of these, choosing to focus particularly on how the Graying of America will affect the next generation or so. While it is obviously imprudent to suggest that everything be put aside in order to chase after futuristic questions, projecting to the future does suggest issues in the present that might be given greater attention, and that, we believe, is the major value of such a focus.

But above all, the point of this chapter is to suggest that a new future lies ahead for achievement motivation researchers. The Graying of America presents new possibilities for those who are interested in "hard data" and may entice a few to become intrigued with the "softer" issues of social history, policy, and planning.

REFERENCES

Advisory Panel on the Scholastic Aptitude Test Score Design. *On further examination.* N.Y.: College Entrance Board, 1977.
Atkinson, J.W., & Birch, D. *The dynamics of action.* N.Y.: Wiley, 1970.
Atkinson, J.W., & Feather, N. (Eds.). *A theory of achievement motivation.* N.Y.: Wiley, 1966.
Atkinson, J.W., & Raynor, J.O. (Eds.). *Motivation and achievement.* N.Y.: V.H. Winston and Sons (Wiley), 1974.
Atkinson, J.W., & Raynor, J.O. *Personality, motivation and achievement.* N.Y.: Hemisphere, 1977.
Becker, E. *The denial of death.* N.Y.: Free Press, 1973.
Botwinick, J. Cautiousness with advanced age. *Journal of Gerontology,* 1966, *21,* 146-158.
Coleman, J.S. (Chairman). *Youth: Transition to adulthood—report of the panel on youth of the President's Advisory Committee.* Chicago: University of Chicago Press, 1974.
Csikszentmihalyi, M. Beyond boredom and anxiety. San Francisco: Jossey-Bass, 1975.
Csikszentmihalyi, M. Intrinsic rewards and emergent motivation. In M. Lepper & D. Greene (Eds.), *The hidden costs of reward.* Hillsdale, N.J.: Lawrence Erlbaum Associates, 1978.
deCharms, R. *Personal causation.* N.Y.: Academic Press, 1968.
deCharms, R. *Enhancing motivation.* N.Y.: Irvington Publishers (Wiley), 1976.
Edward, A., & Wine, D. Personality changes with age: Their dependency on concomitant intellectual decline. *Journal of Gerontology,* 1963, *18,* 182-184.
Egbert, R.L. Social impact of the changing sizes of youth populations. George A. Miller Lecture, University of Illinois, January, 1978.
Egbert, R.L., & Kluender, M. The potential effect of quantitative discontinuity on post-secondary attendance in the 1980's: A feasibility study. (Report prepared for the U.S. Office of Education, Department of Health, Education, and Welfare). Mimeo, University of Nebraska, 1977.
Egbert, R.L., Maehr, M.L., & Gienapp, J.C. Youth in transition: A study of the decreasing youth population and its impact on society. A proposal for research. Mimeo, University of Nebraska, 1978.
Endler, N.S., & Magnusson, D. Toward an interactional psychology of personality. *Psychological Bulletin,* 1976, *83,* 956-974.
Evans, R. Personal Communication. April 12, 1979.
Kagan, J., & Moss, A. *Birth to maturity.* N.Y.: Wiley, 1962.
Kelly, J.R. Family leisure in three communities. *Journal of Leisure Re-*

search. 1978, *10*, 47-60.
Klein, R. Age, sex and task difficulty as predictors of social conformity. *Journal of Gerontology*, 1972, *27*, 229-236.
Klinger, E., & McNelley, F.W. Fantasy need achievement and performance: A role analysis. *Psychological Review*, 1969, *76*, 574-591.
Kukla, A. An attributional theory of choice. In L. Berkowitz (Ed.), *Advances in experimental social psychology* (Vol 11). N.Y.: Academic Press, 1978.
Maehr, M.L. Culture and achievement motivation. *American Psychologist*, 1974, *29*, 887-896.
Maehr, M.L. Sociocultural origins of achievement motivation. *International Journal of Intercultural Relations*, 1977, *1*, 81-104.
Maehr, M.L. Sociocultural origins of achievement motivation. In D. Bar-Tal & L. Saxe (Eds.), *Social psychology of education: Theory and research*. N.Y.: Wiley, 1978.
Maehr, M.L., Kelly, J., & Kleiber, D. Elements of motivation in later-life adults. Mimeo, University of Illinois, Urbana-Champaign, 1979.
Maehr, M.L., & Nicholls, J.G. Culture and achievement motivation: A second look. In N. Warren (Ed.), *Studies in cross-cultural psychology* (Vol. 3). N.Y.: Academic Press, in press.
Markus, G., & Zajonc, R.B. Family configuration and intellectual development: A simulation. *Behavioral Science*, 1977, *22*, 137-142.
McClelland, D.C. *The achieving society.* N.Y.: Free Press, 1961.
McClelland, D.C. *Motivational trends in society.* N.Y.: General Learning Press, 1971.
McClelland, D.C., & Winter, D.C. *Motivating economic achievement.* N.Y.: Free Press, 1969.
Rosen, B. Family structure and achievement motivation. *American Sociological Review*, 1961, *26*, 574-585.
Sarason, S.B. *Work, aging and social change.* N.Y.: Free Press, 1977.
Schaie, K., & Baltes, P. *Life span developmental psychology.* N.Y.: Academic Press, 1978.
Schaie, K., & Parham, I. Stability of adult personality tracts: Fact or fable. *Journal of Personality and Social Psychology*, 1976, *34*, 146-158.
Veroff, J. Social motivation. *American Behavioral Scientist*, 1978, *31*, 709-730.
Zajonc, R.B. Family configuration and intelligence. *Science*, 1976, *192*, 227-236.
Zajonc, R.B., & Markus, G.B. Birth order and intellectual development. *Psychological Review*, 1975, *82*, 74-88.

Chapter 9

Motivational Determinants of Adult Personality Functioning and Aging*

JOEL O. RAYNOR
State University of New York at Buffalo

Theoretical and Empirical Background

The present theoretical effort is an outgrowth of earlier conceptual analyses of the role of future orientation as a determinant of achievement-related motivation (Raynor, 1969, 1974a) and of the determinants of motivation for career striving (Raynor, 1974b). The earlier effort—to understand the determinants of immediate achievement-oriented activity—was primarily concerned with situations where immediate success/failure is believed by the individual to determine the opportunity for subsequent striving to attain some future goal. Thus, immediate success/failure was conceived as "important" by the individual because immediate success earned the opportunity to continue and hence might lead on to future success, but immediate failure meant future failure through loss of the opportunity to continue. The conceptual analysis of such contingent path situations led to elaboration of the basic equations of theory of achievement motivation and provided a more general theory which now derives the earlier model as a special case of the more general one. The later effort—to

---

*This paper represents an initial effort to extend and then to apply to the problem of aging the expectancy x value theory of motivation that has evolved in the study of achievement-oriented activity (cf. Atkinson, 1957, 1964; Atkinson and Feather, 1966; Raynor, 1969, 1974a). A more complete treatment of the topic is presented elsewhere (Raynor, forthcoming) based upon this initial effort, with additional theoretical assumptions and discussion, and with additional emphasis on substantive areas other than achievement-oriented activity.

understand the determinants of motivation for career striving—began to focus on "important future goals" per se, as well as contingent paths that lead on to them. We introduced the consideration of the future sense of self or self-image which might serve as a goal of immediate activity, whose attainment would indicate to the individual that "I have become who I always wanted to be."

The major impetus for the present analysis came initially from application of the equations of the more general theory of achievement motivation to contingent path situations where a succession of immediate successes allowed continued striving for the future goal, or when continued success led the individual to an increased positive evaluation of competences believed to be prerequisites for eventual career success. In this way we were able to derive the effects of successive successes on motivation for (each new) subsequent activity along the career path. In fact, by using the elaborated theory of achievement motivation in conjunction with the concepts of open and closed contingent career paths, we were able to derive some rather (at the time) startling conclusions concerning the effects of success on subsequent career striving. We discovered that motivation to achieve some (important) future goal could be expected to eventually decrease as a function of continued success if that future goal had initially been seen as the final or ultimate goal of the career path, and if no subsequent achievement goals had been added on to the path as a function of earlier successes in getting closer to it. Such a closed contingent path with a final or ultimate goal was then seen as representing a commonly perceived situation for individuals who either have concrete future goals which they are successfully approaching, or a situation where the individual now believes he is in a "dead-end career" with no prospects for subsequent advancement beyond a certain level. It became obvious that aging would always be positively correlated with this progression from earlier to later steps of a career path. Thus aging could be seen as a process through which the psychological effects of a series of successes (or failures) have their impact on the determinants of subsequent motivation of immediate activity, both directly as they affect subjective probability of success for later steps of the path, and indirectly as they affect the individual's perceptions of his own competence (e.g., the extent to which the individual believes he possesses the prerequisite competences for eventual success along that career path). Thus the analysis of motivation and career striving laid the conceptual groundwork for extension of theory of motivation to a theory of adult personality functioning and change—in striving for future career success. And the consideration of a "future self-image" or "future sense of self" whose attainment is contingent upon achievement of the final or ultimate goal of the career path provided the conceptual impetus for linking future goal striving to "feeling good about oneself." Thus the earlier analysis anticipated later developments to be described below.

The discovery of a positive relationship between the rated necessity of doing well now to earn the opportunity to strive for future success (future-importance) and to feel good about oneself (self-importance) (Raynor, Atkinson, and Brown, 1974) suggested the assumption that earning the opportunity to continue was important for positive self-evaluation—because attainment of the future goal that was contingent upon immediate success is anticipated to provide positive feelings of self-worth. Findings that indicated that "important" as opposed to "unimportant" future goals were seen as of much greater motivational significance (Raynor, 1976) further strengthened the view that attaining some important future goal involved more than just an additional concrete achievement that one could feel good about because it was a difficult accomplishment per se, or because it meant greater financial rewards, power, or security, per se, although such interpretations were also possible. Rather, it also suggested and reinforced the view that "becoming a future sense of self" was also implicated in future goal striving when "important" future goals were involved. "Feeling good about oneself" would be contingent upon immediate success in earning the opportunity to continue to strive for the future goal because attainment of the "important" future goal itself provided a substantial source of "feeling good about oneself."

The finding of a positive interrelationship between

future-importance, self-importance, and the extent of perceived possession of a competence (self-possession) (Raynor and English, forthcoming) focused attention upon the role of possession of a prerequisite competence, and changes in an individual's competence judgments, as a function of continued success along the contingent career path. It now became necessary to consider the individual both *before* the results of an evaluation of prerequisite ability were known, *at the time* the individual was about to undertake test-taking behavior (which had been done by the theory of achievement motivation) and *after* the results of such an evaluation, when the individual might make inferences about how much competence was possessed in this career area by virtue of having successfully moved through a contingent gate toward the final goal of the path. It now became apparent that consideration of the resultant effects of successful striving on beliefs concerning the perceived possession of valued competences, whose possession had been validated by the process of successful movement toward the future goal, should also implicate the motivational effects of past success in determining motivation of immediate activity. Thus it appeared reasonable to consider, in the absence of any data bearing on the issue, that an individual might also be motivated to maintain past successes as a means of feeling good about himself, or be motivated to maintain his belief in a high level of possession of prerequisite competences (which result from past successes along the path). In fact, in an earlier study (cf. Atkinson, 1969), where retrospective reports were obtained concerning important vs. unimportant past events, results suggested that past successes and failures could be conceptualized as having motivational impetus in a manner analogous to anticipated future success/failure. While the evidence was certainly not conclusive, or even strongly supportive of such a possibility, the sugestion of discovering that maintenance of a "past sense of self" contingent upon maintenance of past successes/high evaluation of prerequisite competences lead directly to consideration of the past time-linked sense of self which might provide esteem-income to the individual and hence serve as motivating influences on immediate activity. In particular, it became apparent that as an individual moved successfully through a series of contingent steps in a career path toward the final or ultimate career goal, and as the predicted decrement in (positive) motivation to attain that final goal became substantial, the individual might lose interest in attaining the future goal (future sense of self in career striving) and become increasingly more motivated to maintain past successes (past sense of self) and in that way still provide esteem-income in approaching the final stages of that particular career path. This change in locus of motivation from the future to the past not only seemed plausible as a means of accounting for a wide variety of previously unrelated movitational phenomena in a number of different contexts, but it explicitly focused attention on the need for a variable of personality organization that would go beyond the categorization of behavioral potentials (e.g. motive) or the categorization of the effects of success/failure on judgments about one's own abilities (e.g. competence) to one that could represent the motivational impetus to be derived from "feeling good about oneself" that seemed to be contingent upon the attainment of important future goals and might very well be contingent upon the maintenance of important past goals. Thus we extended the tentative conceptual analysis of the self-image or an individual's self-concept as a *future sense of self* to consideration of the self-concept as a *past sense of self* and a *present sense of self,* and assumed that taken together they provide for a self-system whose collective motivational significance lies in their function as a source of esteem-income to the individual. Our conceptual effort then focused directly on the individual's self-image, its future, past, and present components, as a means of insuring feeling good about oneself. The concept of individual cultural value was then added as a means of differentiating different future goals so as to be able to determine why some future goals might provide a possible sense of self to a greater extent than others.

We now assume that striving to feel good about oneself, or to insure positive esteem-income, is a primary goal of adult life, and that use of this assumption as the underlying motivational impetus provides a theory of adult personality functioning and change which can integrate the study of personality and the study of motivation using explicit theoretical assumptions to derive testable and disprovable hypotheses while focusing on "important life behaviors" as well as the more general problems of self-identity, psychological morale, psychological integration, psychological health, and identity-crisis. The resulting conceptual analysis can be considered both as a further extension of theory of achievement motivation, and as a general theory of adult personality functioning and change. It is applied not only to occupational-educational career striving, where the concept of "career" represents a combination of a perceived self-image and a sequence of contingent steps prescribed by society (e.g., role) but to any substantive area. Thus the theory applies for any significant source of esteem-income where immediate activity can bear upon success/failure in attaining a criterion of performance which is contingently related to the opportunity to continue to see oneself in that particular way, and hence which is contingently related to the level of esteem-income that can be anticipated and/or retrospected by an individual pursuing such activity. We have pursued our conceptual analysis focusing only on outcomes contingent upon an individual's skill and/or effort in immediate activity over time, reserving for another paper the application of the present conceptual scheme to that part of the self-system which is impervious to change as a function of immediate success/failure, but rather which is assumed by the individual to represent a pattern of enduring dispositions, traits, and/or abilities which are believed by him to be fixed and immutable and hence immune to feedback from immediate activity.

In subsequent sections we present the initial results of this conceptual effort.[1] We do so by first presenting the assumptions of the theory, which are then applied to closed and open careers. We then derive some of the implications of the theory for these kinds of careers, and finally we apply the implications to relevant substantive issues.

---

[1] See Raynor (forthcoming) for a more complete statement of the theoretical assumptions and implications of this position.

Overview of the Theory

In this theory it is first assumed that a primary goal of adult life is to feel good about oneself, or put another way, to *maximize positive esteem-income*.[2] Second, it is assumed that a primary source of positive esteem-income comes from *positive self-evaluation*. Thus the individual's self-image or sense of self is assumed to be an important motivational variable. There are usually three senses of self that are pursued simultaneously. In answer to the question, "Who am I?", individuals often indicate an occupational, sexual, and family identity, but not necessarily in the order just given, and many others are also possible. A self-image functions as an outcome or consequence of immediate activity when attainment or maintenance is contingently related to the outcome of immediate activity. Third, it is assumed that there are three time-linked senses of self—the *future sense of self*, or "who I am becoming," the *past sense of self*, or "who I have been," and the *present sense of self*, or "who I am now." An individual can have one, two, or three time-linked senses of self to serve as sources of esteem-income, for any self-image. Fourth, it is assumed that all compatible sources of esteem-income combine additively to produce the total esteem-income from a particular substantive sense of self and for all self-images. Senses of self can differ in substantive focus, and an individual can see himself as having several different substantive senses of self at once, or none at all. Thus in the usual case there can be 3×3 or 9 substantive time-link self-images which can provide additive amounts of esteem-income to produce a total amount of "feeling good about oneself."

Esteem-income represents the anticipated and/or retrospected amount of positive thinking about himself in context of his circumstances in life. However, esteem-income does not directly influence the direction, vigor, or persistence of immediate action. Esteem-income reflects the emotional tone or

---

[2]In the more complete treatment (Raynor, forthcoming) individuals are assumed to be motivated to maximize positive value and/or to minimize negative value, with two kinds of value—information value and affective value—distinguished. The present analysis refers only to affective value, and does not consider information value.

"morale" of the individual at a given point in time. The concept of "career" is used to relate the time-linked senses of self to action. Thus the fifth assumption of the theory concerns the "career," conceived as a joint function of the individual's sense of self or self-image and a particular opportunity for action sometimes referred to as "role" but here identified as the series of steps in a path to a concrete attainment goal. However, "career" is a general term that is not limited to occupational pursuits, but rather reflects any substantive self-image that is related to action opportunities that are seen as contingently related to attainment and/or maintenance of that self-image. Thus "career" is the psychological variable that links the internal view of self and the external view of role (as seen by others) so that "careers" are *behavioral opportunities for self-identity*. When attempting to predict action, we refer to "careers," whereas when referring to value in the self-system we refer to esteem-income. When analyzing the cultural opportunities that are usually perceived by individuals we could refer to "roles," but rather we refer to sequences of steps which may be seen as a contingent or noncontingent path.

Most individuals in our culture can pursue at least three simultaneous senses of self—careers in the "occupational" area related to earning a living, careers in the sexual area related to masculinity/feminity and sexual functioning, and careers related to their family identity—as husband/wife and/or father/mother. We will restrict attention for the moment to what we have termed the educational/occupational careers, since these are most clearly related to the conceptual analysis of motivation and career striving for which previous theoretical analysis and empirical research are most relevant (cf. Raynor, 1974b). Later we will consider other "careers" that traditionally have not been conceived as "achievement-related" but will here be conceptualized in similar terms as involving immediate activity that contingently bears on attainment or maintenance of the self-image and therefore involves esteem-income and goal-striving.

We have already noted that esteem-income can be anticipated as a function of becoming a future sense of self, that it can be retrospected as a function of becoming a future sense of self, that it can be retrospected as a function of maintaining a past sense of self, and/or it can be obtained as a function of seeing oneself as one presently is. Thus "becoming," "having been," and "being" are used here as terms to refer to these three time-linked sources of esteem income.

Value and Esteem-Income

We now turn to consideration of the particular sources of esteem-income of each time-linked sense of self. Here we build upon and extend expectancy-value theory of motivation. We assume that *value to the individual* (sometimes referred to as valence or attractiveness) of a goal or threat also applies to value of the self-image, and that esteem-income is directly proportionate to value in the self-system. Thus the attractiveness of an outcome is used as a basis for determining esteem-income of the self-image. There are assumed to be four[3] sources of value to the individual for each outcome/consequence, where a goal refers to a positive outcome and a threat refers to a negative outcome.

The *difficulty value* of an outcome refers to the attractiveness/repulsiveness derived from the perceived chances of attainment of the outcome. This is similar to the concept of incentive value in theory of achievement motivation (cf. Atkinson, 1957, 1964, 1974; Atkinson & Feather, 1966), when applied to "being" or "becoming," and is assumed to be inversely related to the subjective probability or expectancy of success or goal attainment; the easier the task outcome, the less its positive difficulty value, the harder the task, the greater its positive difficulty value. Thus the assumption implies that for the self-system, an individual anticipates greater esteem-income contingent upon success on a difficult task that defines "who I am becoming" than an easy task—that is, he expects to feel better about himself if he succeeds on a task that he believes offers a low subjective probability of success, than if he succeeds on a task that he believes offers a high subjective probability of success. Or, an individual retrospects about some past success that defines "who I have been" and feels better about himself if he

---
[3]Intrinsic value is also taken into account in the more complete treatment (Raynor, forthcoming) so that five rather than four sources of value are considered.

recalls the task as one that was difficult to achieve success in, that if he recalls the task as one for which success was easy. For both anticipated and retrospected difficulty value it is the individual's present belief about the difficulty of some future goal or some past goal, rather than either what others think the difficulty should be for that individual, or what that individual thought at the time of past action, that is relevant. For example, as individual might have faced a task with the belief that it was very difficult, and therefore anticipated feeling very good about himself contingent upon success. However, he then finds out that the taks is in fact very easy for him. When retrospecting he might recall the task as initially very difficult, and then feel very good about himself, or he might recall the task as in fact having been very easy, and then feel very little positive esteem-income contingent upon that success. The individual's recall of past level of difficulty is assumed to be the critical variable for determining amount of esteem-income derived from the difficulty value of the past sense of self.

Another source of value is the *instrumental value* of an outcome. Instrumental value refers to the number of opportunities for action that the attainment of an outcome is believed to guarantee (in the future) or to have guaranteed (in the past). Future instrumental value applies to behavior in contingent paths, where immediate success is believed to earn the opportunity to try for some number of future successes and immediate failure is believed to guarantee future failure through loss of the opportunity to try for future successes (cf. Raynor, 1974a). Thus the instrumental value of immediate success is greater the number of steps in a path to the final goal of a contingent path, or put another way, the instrumental value of an immediate outcome varies with the length of the contingent path of which it is part of the first step. It follows that single outcomes not related to other activities-outcomes have no future instrumental value. Similarly, the final or ultimate goal of a contingent path has no instrumental value for activity in that path. It also follows that as the number of steps in a contingent path decrease as an individual moves successfully through the path, the future instrumental value of the immediate step also decreases. However, the instrumental value of a past success correspondingly increases as a function of success along the path, for the individual can perceive that some past success has guaranteed an increasingly greater number of later opportunities for success.

The third source of esteem-income comes from what is termed *extrinsic value*, or the "extrinsic incentives" in previous theory of achievement motivation. These are sources of feeling good (about oneself) that derive from rewards that are contingent upon goal attainment, but do not relate to the difficulty value of the task per se. Money, approval, power, security, etc. are common sources of extrinsic value, and are assumed to function to provide positive value (esteem-incòme in the self-system) contingent upon immediate success when they are perceived to be appropriate and/or usual outcomes of that activity (cf. Kruglanski et al., 1976).

Finally the *individual culture value* of an outcome refers to the extent to which an individual has acquired the belief that attainment of that outcome is good/bad in his cultural context. In the self-system it refers to how good or bad a person the attainment of that skill, outcome, or self-image will make or has made the individual. It implicates a moral-evaluative source of esteem-income. Thus, the *consensual cultural value* associated with success is most often positive in our culture, and for many individuals in our culture their *individual culture valuation* of success is also positive. A "successful person" is believed to be a ' good" person; so is an "intelligent" person. Most professional/occupational outcomes are highly positively valued along this good/bad dimension in our culture (consensual cultural value), and many individuals share that evaluation so that their individual cultural value for "doctor," "lawyer," etc. is also positive. Individual cultural value may also be negative, as might be the case for "failure," and "alcoholic," etc. Individual cultural value is not previously represented in expectancy-value theory of achievement-motivation, and is here introduced particularly to refer to a source of esteem-income that, while not uniquely tied to the self-system, is most often potently aroused when an individual is involved in pursuit of the "self-image—contingent path" that we refer to as a "career." *We assume that senses of self that can be used as primary means of self-identity are most often ones that provide substantial cultural value to the individual.* The variable of cultural value provides an additional source of "feeling good about oneself" that was not considered in earlier conceptual analyses, and provides a link between the task-oriented outcomes of difficulty value and instrumental value, on one hand, related as they are to striving per se, and to evaluation of final or ultimate goals, on the other hand, related as they are to "important" life outcomes involving big extrinsic rewards and *large cultural valuation.*

Note that these four sources of value can provide esteem-income for each of the three time-linked senses of self, and for each substantively different "career." Thus there are in this theory at least 4 × 3 × 3 = 36 potential, additive sources of positive esteem-income for the individual, and an equal number of sources of negative esteem-income. These can *simultaneously* provide esteem-income for individuals having clear-cut careers in the occupational, sexual, and family areas that have future, past, and present senses of self. Of course, the individual may have fewer, or more, simultaneous careers, or the individual may have only one or two time-linked senses of self for any of these, and any of the four sources of positive (or negative) value may provide a large or small quantity of that particular value. Or the individual may have no clear-cut sense of self, so that no esteem income is derived from the self-esteem. Thus the theory can represent extremely varied sources and amounts of esteem-income that a particular individual might experience at any particular time.

One of the critically important aspects of the present theory concerns the implications of the above assumptions for changes in esteem-income as different careers become available or old ones no longer are psychologically available, as different time-linked senses of self become part of the self-system or become unavailable as sources of esteem-income, and as the different sources of value change in amount of esteem-income they provide.

Psychological Careers

Thus far we have assumed that a person's time-linked senses of self can provide (positive or negative) esteem-income. We now need to relate these assumptions to behavior in the real world. We assume that a future sense of self can be obtained by successful striving along the steps of a contingent path, so that the self-image can be represented as the final or ultimate goal of a closed contingent path. Recall that a contingent path is defined as a series of steps (activity-outcome pairings) where immediate "success" earns the opportunity to continue along the path (i.e., to strive for some number of future successes along the path) but immediate "failure" guarantees future failure through loss of the opportunity to continue along that path. The combined linkage of the self-image and the steps along a path has been termed the "career," which is the psychological variable that relates the self-system to behavior in the real world. As noted earlier, we prefer "contingent path" to the concept of "role" because the former spells out those particular activities and outcomes that are expected by the individual to occur, and in what order of occurrence, so that it is both a more specific concept than the more general notion of "role," and represents the individual's perception (internal viewpoint) rather than some judge's view of normative social behavior (external viewpoint).

It is worthwhile to give the reader some feel for the kinds of issues and problems that the theory has been applied to. We do so in the following examples with the precautionary note that the implications of the theory will follow from manipulation of algebraic statements of the assumptions made above, which will then be given psychological meaning. However, in this paper we have dispensed with formal derivations in order to move the reader on to some substantive issues that the theory addresses and to avoid premature specifications. At a later time we plan a formal presentation and derivation of these and other implications of the theory. At the present time we need research studies to guide choice of specific mathematic representations, particularly about what will be termed "have been," or the effects of the retrospected past.

Becoming, Being, and Having Been
Along a Closed Contingent Career Path

We are now ready to apply the theory in a general way to a

typical sequence of career striving. We use this as one possible application of the theory to illustrate how its assumptions lead to implications about the time-orientation of the individual, his level of esteem, possible "identity crises" that might be faced, and whether striving to attain goals through instrumental action in the real world or cognitive reorganization to defend and maintain past accomplishments might be the primary strategy for feeling good about oneself.

We first introduce a means of representing the assumptions of the theory in graphic form, which eventually will be used as the basis for stating the mathematical model. Figure 1 shows a "time line" which represents the person in the immediate present at point "0," his anticipated future steps as positive numbers (+1, +2, +3, etc.) with the furthest anticipated future step as the final or ultimate step in the anticipated future (+N), and his retrospected (recalled) past steps as negative numbers (-1, -2, -3, etc.) with the furthest recalled past step as the final or ultimate step in the retrospected past (-N). Note that so far this time line does not refer to real events as we follow an individual from one step to another step, but rather represents the individual's thoughts about (e.g., his cognitive representation of) possible future steps, and recalled past steps, as the individual anticipates and retrospects at point "0" in the here-and-now.

Let us assume that we are depicting a college student who as a freshman decides (for whatever reason) that he "wants to be a doctor," who knows little about his own capabilities concerning the prerequisites needed to become a doctor, but who has checked the undergraduate catalog of his University and some graduate school brochures so that he knows the required sequence of events that must transpire to eventually earn his

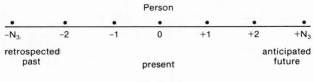

Figure 1.

medical degree. Figure 2 depicts this student at the start of his career. Note that the anticipated steps along the path are represented as part of a closed contingent path with +N as "becoming a doctor" by earning the M.D. and the steps to that final goal are the courses to be taken, coupled with the good grades that must be earned, to insure continued progression through the undergraduate program to earning a spot in a medical school, to successfully progressing through the graduate years with its contingent gates until finally at last the M.D. is awarded. Because the student took no "pre-med" courses in high school, and engaged in no activity contingently related to "becoming a doctor" before enrolling in introductory biology (the first step of the path), Figure 2 does not show any retrospected steps back into the past as the student takes the introductory course. We would describe this career as "future-oriented" and the student as "becoming." Since the student believes it is difficult to make it through all of the steps to the M.D., the difficulty value of the future goal is relatively high. Since the student knows he must do well in introductory biology in order to continue, many opportunities for future success are at stake, so the instrumental value of the first step is relatively high. Since the student believes that "it's good to be a doctor" (doctors are intelligent, educated, devoted helpers of their fellow man, etc.) the individual cultural value of the future goal is relatively high. And since the student believes that he can earn a good living, find security, and influence others by being a doctor, the extrinsic value(s) of the future goal is relatively high. Thus the individual anticipates a relatively large amount of "feeling good about himself" (positive self-esteem) as he looks forward to "becoming a

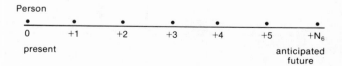

Figure 2. The start of career striving in a closed contingent path: becoming.

doctor" by successfully moving through the contingent steps of this career. His sense of self is defined primarily in terms of "who he is becoming" and right now, as he anticipates his future, this makes him feel very good about himself. In other words, his present self-esteem is high because the value of his career as "becoming" is high.

Let us restrict ourselves for now to the student who successfully moves through the first several steps of this contingent career path by attaining those necessary prerequisites that allow himself to psychologically accept the fact that he has met the stated requirements while at the same time allowing the external gatekeepers of the University to also certify that he has met the requirements to become a "biology major." Figure 3 depicts the student at this middle stage of career striving. There are several important changes that have taken place. First, the individual can now see himself *both* in terms of "becoming" and in terms of "having been." He is still striving to earn the M.D. by trying to do well in upper class courses to earn the college degree to get into medical school, to do well in medical school, etc. At the same time, he can now look back to his past successes in negotiating the requirements to get this far along the path and "feel good about himself" as a function of the difficult value of his past successes ("that was a tough introductory biology course—only 15 A's out of 100 and I got one of them!"), the instrumental value of his past successes (e.g., doing well in introductory biology earned the chance to move on, and so did doing well in intermediate biology, and "without doing well in them, I wouldn't be where I am today"), the extrinsic value of his past successes ("being an upper-classman lets me impress the under-classmen in the

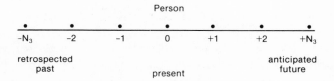

Figure 3. The middle of career striving in a closed contingent path: becoming, being, having been.

Undergraduate Biology Association), and the cultural value of his past successes ("it makes me a good person to have been a successful biology/pre-med student thus far"). His esteem-income is no longer restricted to the anticipated future. He has a history of career success which when retrospected provides a substantial amount of esteem-income.

In addition, by successfully negotiating the various gates and by attributing his success to internal factors such as *high ability* to do biology well, and *hard work* in this area, the student has developed a much clearer present sense of self in this career due to an increment in perceived degree of possession of those prerequisites which he believes have made success possible. The amount of esteem-income that present sense of self can provide because of the increase in value produced by upward assessment of ability is increased. We refer to an individual's assessments of his own ability as competence, adopting the term proposed by Moulton (1967), and can now refer to the various sources of value derived from "being"— that is, from seeing oneself as "intelligent" and "hardworking" in this area. Difficulty value increases because the individual sees himself as being competent—one of the increasingly few who have progressed to this point since only 50 out of the original 100 in introductory biology are now biology majors; the instrumental value of "intelligence" in the area of psychology also increases because he believes that possessing this competence will allow him to successfully negotiate the later steps of the path; the extrinsic value of being "intelligent" is increased because possessing more of it allows for getting summer jobs as a research assistant with one of his instructors who was impressed with his intelligence in the course, and asked him if he would work on his summer project at $5000/month for two months. Cultural value of "being" has increased because it's "good to be intelligent, and the more intelligent, the better it is."

We therefore see that in this arbitrary example, there are now three sources of esteem-income (past, present, and future) rather than only one (future) or possibly two (present and future) that existed at the beginning of this particular career. Other things equal, we expect an increase in total esteem-income because of this increase. Note, however, that the increase is due primarily to an increase in esteem-income due to the "present sense of self" or "being." This is because as the past sense of self became a source of esteem-income and it increased in value with each successful step along the path, the future sense of self correspondingly provided less esteem-income, *assuming the path has remained a closed contingent one*, with no new goals added on to the anticipated future. The reasons for this are important, and basic to the implications of the theory. First, since the number of steps has remained anchored by the goal of "becoming a doctor" through earning the M.D., and the individual has moved successfully through several steps, the instrumental value of the (next) first step along the path has decreased, because there are fewer future opportunities that the (next) immediate success can earn. That is, there are fewer steps between the individual and the final goal of the closed contingent path. In addition, the student has increased his confidence in eventual success along the path due to both an increase in perceived competence in this area and the fact that there are fewer contingent steps remaining, so that the difficulty value of the final goal has decreased. This means that if the cultural and extrinsic values of the final goal have not changed (let us assume that they have not, although they in fact may have increased or decreased), the total value of the path to the final goal, the anticipated future sense of self, has decreased, with a concomitant decrease in anticipated esteem-income generated by "becoming" along this career path. In our example, the increase in esteem-income produced by an increase in "having been" along this path is counterbalanced by the decrease in esteem-income produced by the decrease in "becoming" along this path, although in fact it may be less than, equal to, or exceed that lost from "becoming."

Let us now follow our successful student to the end stage of career striving along this particular career path. This student has just successfully passed his last examination and thus has "become" a doctor by earning his M.D. But because this a closed path the individual still has not at all considered what the next step, if there is to be one, along this career path might be. Figure 4 shows that there is no anticipated future, so that

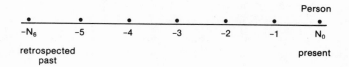

Figure 4. The end of career striving in a closed contingent path: having been, being.

there is no esteem-income to be derived from "becoming." On the other hand, there is a substantial value due to the past sense of self, based on the entire series of successes along the path, and an increase in perceived competence concerning "having had what it takes" to successfully negotiate the contingent path. Both the past and present senses of self can provide substantial esteem-income. But note while the student at the beginning was "becoming," at the end he is now a "has been." That is, the relative impact on esteem-income on the individual has shifted from future-oriented to past-oriented, with a continued increase in the role of the present self-image as "competent in this area" to increase (perhaps) or at least not decrease total esteem-income. Note, however, that along the way the individual might have perceived a limit to his continued upward evaluation of prerequisite competences, some upper point beyond which he has not continued to upwardly evaluate his competences, either because real-world feedback has suggested such a limitation or because he has stopped attributing his successes to his own ability and/or effort. We might expect such a limitation to occur at some *middle point* of career striving, although it might occur earlier or not at all. If this were the case, we would then say that the major shift that has taken place from the beginning to the end of this closed career path has been one from "becoming" to "having been."

The example presented here is an arbitrary one but it includes what we think may represent a frequently occurring sequence, one that precedes a major career "crisis" and one that can be reasonably used to characterize some previously observed differences as a function of aging.

The "typical" kinds of changes that this particular example illustrates may be summarized as follows: (1) a shift from "becoming" to "having been" as a function of success in a closed career path, produced by a decrease in the difficulty value of the future goal and the instrumental value of the immediate next step leading to that future goal, on the one hand, and an emerging past sense of self which forms the basis for sources of value to provide esteem-income whereas at the start no such past sense of self existed for this new career path; (2) an increase in "being" based upon upward evaluation of competences which are seen as being validated by the very successes that move the individual along the career path, up to some point, where the individual reaches some asymptotic assessment of competence.

Aging and Career Striving

Of particular significance for an understanding of adult personality functioning and change is the correlation which might be assumed between aging and the various stages of striving outlined above. While by no means perfect, we expect some correlation between "youth" and the beginning stage of career striving, middle age with the "middle stage" of career striving, and old age with the "final stage" of career striving. We do not mean to rule out a 60-year-old facing the first step of a new career path, nor a 30-year-old faced with the last step of a closed career path. But generally, given certain facts of life to be discussed in greater detail in later sections, we expect the "usual" pattern of career striving to result in such a correlation, although the variations of these patterns in fact allow us to conceptualize such phenomena as "second careers," "new careers," and "retirement careers." This means that we can derive that young people will be characterized primarily as "becoming," whereas older people will be characterized primarily as "having been," on the average, for a given group of individuals, other things equal. This allows us to understand phenomena concerning what has been termed the "generation gap" from a new perspective, particularly why older individuals sometimes appear to younger ones to be "trying to maintain the status quo" of past successes, while the younger ones are trying to "make it" in the real world through future successes. In fact, when no further action is possible after attaining the final goal of a closed career path, we predict that the individual is left with the task of insuring continued esteem-income primarily by maintaining by cognitive means the various sources of perceived value of past successes. Recall that no further action is possible in the final stage of a closed contingent career because there are no additional steps to engage in. This means that "becoming" is no longer possible.

Maintaining the Values of Past Success

The above analysis allows us to predict that so long as the individual is "becoming" to some extent, he will be motivated to attain those goals remaining in the contingent path. Whether that enhances positive or negative motivation will depend upon individual differences in strength of positive and negative motivation for the kinds of outcomes in question. However, once the individual can no longer strive because there are no future goals, then he is forced to maintain esteem-income through other, cognitive means—that is, by trying to maintain the various sources of value of past successes at their highest magnitudes. Cognitive rearrangement and cognitive defensiveness will be the only means of maintaining esteem-income. So long as the individual can act, he can receive the value of the present sense of self by utilizing his skills, or by trying to validate at a higher level the various abilities that he believes are the necessary prerequisites for immediate success. But again, once striving is no longer possible maintenance of the present (as well as the past) sense of self is dependent upon cognitive means rather than action. We then see why individuals, when faced with the "end" of their careers, are expected not only to be past-oriented to a large extent, but actively motivated as a means of maintaining self-esteem to distort, exaggerate, and otherwise engage in cognitive activity whose aim is the maintenance of the value(s) of past success—

whether it be to maintain the difficulty value by insisting on how "tough" it was to succeed in the past, or to recall the large number of opportunities that past success earned so that instrumental value is maintained, etc. When "becoming" is no longer possible as a means of feeling good about oneself, and present skills cannot be utilized, "having been" can still make the individual feel good so long as he can continue to see the value(s) of his past successes, other things equal. Note, however, that the availability of alternative careers will be an important factor in predicting whether or not these defensive maintenance reactions occur, as well as whether a closed career can turned into an open one (see below).

One factor that we have thus far neglected in our identification of "becoming" with the young or younger, and "having been" with the old or older, as a rough correlation, is the fact that *as a function of age* individuals sometimes come to perceive a leveling off or even a decrement in the degree of possession of previously valued competences which, although they are still seen as prerequisites, are now also seen as skills left behind in one's youth. This perceived decrement in prerequisite competence has the consequence of diminishing esteem-income that can come from the present sense of self, so that it often happens that as one approaches the final goal of a closed career both the future and present senses of self provide diminishing esteem-income, with only the past sense of self left to provide it.

The critical assumptions of the present theory concerning age and aging can be seen from the above to be that success/failure experiences along a contingent career produce changes in various motivational variables which influence esteem-income and level of motivation for activity; (1) success/ failure influences perceived competence, which influences present sense of self and esteem-income derived from it; (2) success/failure influences confidence, which changes the difficulty value of the future sense of self; (3) success reduces the number of steps remaining along a closed contingent path, which influences the instrumental value of the immediate next step along the path and hence esteem-income from "becoming" through immediate activity; (4) success increases "has been" based on previous success, so that esteem-income from the past sense of self plays an increasingly greater role.

Career Striving in an Open Contingent Path

An important distinction is made between a closed and an open career path. Recall that the closed path is one where the individual has a final or ultimate goal (+N) whose attainment will mark the end of striving along the path because the last goal of the career is fixed at the outset and remains unchanged as a function of success in moving toward it. On the other hand, an open path is one where the individual may initially have a final or ultimate goal, but an immediate success suggests one or more new goals that add on to the end of the path whose length has now remained the same or even increased. The implications of this open path for esteem-income are seen in Figure 5, which first shows the closed career path in the early stage (Figure 5a), and then after several successes have moved the individual to the middle stage of the closed career (Figure 5b) and the open career (Figure 5c). Note that as in the closed path, both the retrospected past and the anticipated future can contribute to esteem-income in the open path. But there has been no decrease in "becoming" for the open path as "having been" increases as is the case with the closed path. In the open path the individual can build up a source of positive esteem-income as a function of past successes along the path while still retaining the initial impetus for "becoming" as a consequence of new additional goals which continually become apparent as a function of successful immediate striving. In fact, so long as the path remains open—that is, so long as new goals become apparent as a function of continued immediate success—there will be no "late" or "final" stage of the career because the individual does not approach an ultimate or final goal. In an open path the "final" goal does not exist as a fixed target, whose attainment would signal the end of "becoming" in this career. Thus the open path has extremely important implications for life striving because it provides a means of understanding the difference between individuals who remain psychologically young through

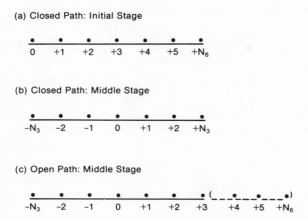

Figure 5. Early (a) and middle (b) stages of striving in a closed path, compared to the middle stage of striving in an open path (c).

continued "becoming" and those who become psychologically old through exclusive dependence upon "having been" to feel good about themselves.

The distinction between open and closed careers provides a powerful tool for the analysis of the apparently paradoxical situation where relatively successful individuals, who would be expected to be satisfied and "fulfilled" as a function of their success, are in fact restless, bored, uninterested, and "lost" at the pinnacle of their careers—a phenomenon that Sheehy's (1974) *Passages* identifies and which we believe to be the inevitable consequence of career striving along a closed as opposed to an open career. First, esteem-income from "becoming" is lost as a function of success in the closed career. In fact, the future sense of self is lost upon attainment of the final goal of the closed career path. The individual no longer knows "who I am becoming." This precipitates an "identity-crisis" (Erickson, 1968) since the individual has previously (in this career path) seen himself primarily as "becoming." We expect that the individual who has been striving successfully along a closed contingent path begins to notice the loss of attractiveness of the final goal only gradually, as each success slightly diminishes its anticipated esteem-income. But the impact of this loss becomes strikingly apparent after successful attainment of the final goal itself, for now there is literally "nothing to look forward to" in the career, and while the loss of value is a gradual process, no longer having any future goal at all is seen and experienced as qualitatively different. The individual wakes up one morning to find he has no reason to continue to pursue the career. In fact, he might realize that he can just as well stay in bed and retrospect about the past in the career, and assess his present competences as a function of his past success, and feel just as good about himself as if he "went through the motions" of immediate activity which has lost its instrumental and difficulty values for attainment of a future sense of self.

On the other hand, the individual faced with the open path never loses the anticipation of "becoming" because there is always a new sense of self to be attained as a function of the ever-changing (new) final goal of the career. At any given time the individual is pursuing a final goal of the career path, but this final goal continually changes. Renewed esteem-income can come to be anticipated and loss of the future sense of self avoided, while at the same time the backlog of past successes builds a source of positive "having been" and a positive sense of "being" as is the case for the closed career. If the "cake" is "becoming" and the attainment of those goals the "eating," then the person in an open career path "can have his cake and eat it, too," whereas the person in a closed career path can have the cake, or eat it, but not both. For in closed careers, becoming turns to having been, whereas in the open path becoming produces having been while still remaining becoming.

Note also that success in a closed path increasingly ties the individual to his past as a primary means of self-identity and feeling good about himself, whereas success in an open path provides a continued link between the individual's past, present, and future senses of self as the means for self-identity and esteem-income.

Interaction of Open-Closed Careers

In the real world it is rare to find an individual whose entire life is characterized as career striving along an open path in any of the substantive (occupational, sexual, familial, etc.) identities people usually have. More often we expect to find a path remaining open for some period of time, after which either a change in the impact of some dominant situational (path) feature of their career, such as a change in the nature of the occupation or family situation, or the impact of some internal change, usually the failure to use success as a signal for continued upward evaluation of competences, produces a change in the effect of success on the perceived new opportunities for continued striving along that career path. That is, sooner or later, in the usual case, we expect open paths to become closed paths, either because the world imposes some final or ultimate goal that was not initially perceived—because the individual was too far away from that goal in the earlier stages of career striving for it to have any psychological impact, or because the individual realizes that a limit to his own competences in this area sets some limit to his realistic prospects for continued advancement along the path—at which time the new final goal of the previously open path becomes the actual final goal of the now closed contingent path. When immediate success fails to provide for the usual new possibilities, the open path becomes closed and a reduction in "becoming" sets in. We have been mathematically precise about the predicted interrelationships between competence, length of path, and the transition from open to closed path (cf. Raynor, 1974a).

It is also possible that a closed path can become open. Thus while early successes might not suggest new possibilities to add on to the path, later successes might begin to, so that the closed path becomes open. Again, the apparent length of path will interact with the perception of a final or ultimate goal of the path to determine whether the path will function as open or closed at a given point in time.

The implications of the above are that the distinction between open and closed paths, while critical for an understanding of adult personality functioning, should not be taken as fixed as a given individual since an open path can become closed and a closed path can become an open one. However, if and when these changes occur we can predict their effects. Such changes are believed to be critically important in adult personality functioning.

Maintaining vs. Attaining
as Time-Linked Orientations to Esteem-Income

We began the present analysis with the assumption that feeling good about oneself is an important goal of adult life. That is, we make this assumption because we believe it will help us understand adult functioning in a better way, and will lead to predictions concerning adult life that would not be systematically derivable without such an assumption. We also assumed that an important source of esteem-income is the individual's image or sense of self, which can provide more or less (positive or negative) esteem-income as a function of the four different sources of value defined earlier. Implicit in the combination of these two assumptions is the notion that the individual consciously perceives a past, present, or future sense of self, or, at the very least, becomes aware or is made aware of time-linked sense of self—or has to become aware of it—in order to feel good about himself as a function of its perception. We now want to make this explicit so as to dintinguish between the effects of the past attributable to the automatic carry-over of unsatisfied motivation from past situations where motivation has been aroused—the concept of inertial motivation (see Atkinson & Birch, 1970, 1974)—and the esteem-income and motivation of action attributable to the *retrospected* past, or the consciously recalled/remembered past as it relates to the self-system and past success/failure. We assume that inertial effects can influence immediate action without the mediation of consciousness or awareness, while we assume that such conscious attending to the past is the necessary ingredient for the self-system to function so that the past can influence the individual. (We assume a similar need for conscious attending for both becoming and being to influence the individual.) Thus inertial effects and retrospected effects are seen as operating independently of each other.

It follows that "becoming" means a focused attention on the anticipated future, and in fact large total value associated with future goals is expected to shift the focus of the individual to the anticipated future so that the individual will be increasingly preoccupied with attaining future success and therefore with his self-identity as "becoming." That is, *an increase in anticipated value focuses attention to its source*, making available a valued self-image to be used as a means of self-identity. Thus in the final stage of a closed path contingent career the individual is expected to be primarily concerned with his identity as "having been," and to an increasingly great extent, as its value increases. We expect that this shift will produce an important shift in the conscious factors experienced by the individual as sustaining the career, and how he relates to that career, and in fact on the determinants of immediate action in that career.

When an individual's esteem-income is primarily from past success, as when at the final stage of closed career striving, we expect the individual's thoughts to be primarily past oriented. We expect him to report that "the best part of my life (career)" is in the past, if such a statement could be made without negative evaluative connotation (but see below), when faced with the late stages of closed career striving, whereas when faced with early stages of career striving, whether open or closed, the individual should report that the "best part of my life (career) is in the future."

The individual who is "becoming" can continue to receive esteem-income by seeing himself as successfully moving through steps to future success, and in order to do so he must be primarily concerned with the task requirements for success along the path. His goals in life within this career are those of the anticipated future and he is dominated by thoughts of how his present actions will translate the anticipated future into present reality.

The individual who "has been" successful can continue to receive esteem-income by seeing himself as having been successful in that career, and particularly, by maintaining the values of past success at levels which provide sufficient esteem-income. Thus the individual is expected to be primarily concerned with the maintenance of the values of past success, such as recalling how difficult past success was, how a given success is important in earning the chance the person is now confronted with, how much money (etc.) the past success produced, and how good that success makes him as a person. Note that this maintenance does not involve any action in the real world, but consists of recall and reminiscing about past success and cognitive work, thinking about the circumstances of past success and its value as now seen in the present. Failure to do this might result in the erosion of esteem-income from the past. On the other hand, the cognitive work involved in "becoming" consists of daydreams of future success, and cognitive work to anticipate how valuable that series of future successes along the anticipated career path will be and the planning/doing required for immediate success. In other words, we expect the thoughts and daydreams of individuals who face early vs. late steps in a closed contingent career path to be quite different in their time orientation—future-oriented vs. past-oriented—and that their emotional outlook (esteem-income) should be closely tied to this cognitive work. When the "psychologically young as becoming" thinks about the future he feels good about himself, but when he thinks about the past, that feeling disappears. When the "psychologically old as has been" thinks about the past he feels good about himself, but when he thinks about the future that feeling disappears. The "psychological morale" of such individuals will depend upon whether they are engaged in their usual cognitive elaboration of their stage of career striving—and will change when they focus on a time orientation that provides a different level of esteem income.

The Link Between Esteem-Income and Action

To the extent that attaining vs. maintaining involves thought exclusively, we can predict that early vs. late career striving in a closed path will be dominated by thoughts of anticipated future success and what it takes to attain it, on the one hand, or by thoughts of retrospected past success and what it was like. However, in a contingent path, immediate activity is always a requirement so long as the final goal of the past has not as yet been attained. We believe that dominant thoughts, linked as they are to the stage of striving, can provide the primary determinants of immediate activity—or in the language of the dynamics of action (Atkinson & Birch, 1970), can provide additional instigating force. That is, the stage of striving will define the purpose for which the individual engages in the immediate (next) step of the contingent career. Earlier conceptual analyses have developed how this works for "becoming" (cf. Raynor, 1974a). We will want to conceptualize the effects of "having been," which has until now received no attention within this conceptual scheme, and eventually provide a mathematical model to account for its effects on immediate behavior.

Maintaining vs. Attaining as Determinants of Immediate Action

One of the important new orientations to grow out of the extension of theory presented here is the re-emphasis on the interrelationship between *action in the real world* to attain future success, and reorganization of the cognitive field, the *maintenance of the values of past success* (cf. Lewin, 1938), as alternative means of insuring continued esteem-income. These provide for different purposes in adult life. At the beginning of career striving attainment of goals through action perceived to be instrumental to their attainment will be seen by the individual as the primary means of obtaining esteem-income. Expectancy-value theory of action as previously defined will predict best. At the end of career striving, with becoming no longer a possibility, and being limited to evaluation of abilities which cannot be further utilized because there are no future accomplishments to which they can be applied, the individual is left with retrospection about past successes and maintaining their value as the primary means of obtaining esteem-income. Theories addressed to such cognitive work will predict best, and Expectancy-value theory needs to be extended to take such factors into account. In middle career striving, anticipation of future goal attainment, retrospection about past goal attainment, and evaluation of present levels of prerequisite abilities (respectively, becoming, having been, and being) will all be seen as contributing to feeling good about oneself in this career. Theory concerned with both *attaining* and *maintaining*, as well as *evaluating* competence through seeking information and making attributions to various causes of success, will all be relevant, but each if used alone will be limited in its success because all three kinds of theory are necessary since all three time-linked processes are expected to be occurring at once. In fact, we need theory that considers the interrelationships between orientations toward the future (attaining), the past (maintaining), and the present (evaluating), something that until now has been missing because of the failure to recognize that the dominant theories of motivation/action in psychology have dealt with different aspects of the time-linked behavioral situation. Expectancy-value theory has been primarily concerned primarily with "becoming." Cognitive consistency theory and theories of psychological defenses have been concerned primarily with "having been." Trait theories of personality, self-attribution theory, and information-seeking theory are concerned primarily with "being."

We expect that time-linked sources of esteem-income will produce quite different reactions to a situation where the individual is challenged to justify feeling good about himself as a function of a particular career, either by himself or by others. We expect that in the beginning of striving the individual will take the attitude that, "if I work hard to fulfill those prerequisites that will allow me to move on toward the next steps of the career path, I can justifiably feel good about myself because I will be becoming who I want to become." There is little psychological defensiveness that might involve cognitive distortion in order to maintain esteem-income from past successes because there are none to "defend." The individual is open to information concerning "How I am doing" to meet the stated prerequisites, and is open to both positive and negative feedback that might by useful in attainment of immediate success and therefore earning the opportunity to continue to "become." However, the individual at the end of striving in a closed contingent path is predicted, in comparison to the individual at the beginning of striving in a closed contingent path, to be more motivated to use psychological defenses and cognitive work to insure continued positive valuation of his past successes in order to feel good about himself. This individual will be more likely to "define his record of past accomplishments" so as to insure that their values which provide esteem-income are maintained at present value or enhanced. At this time this is the only way the individual can continue to feel good about himself. There is no possibility for further accomplishment, so the individual believes, because all goals have been attained. Now the task is maintenance of the "historical record," the set of accomplishments, particularly if others were to attack that record. Such an "attack" might try to belittle the record by indicating that past successes were trivially easy (reducing difficulty value if believed by the individual under attack); that past successes had little to do with later opportunities—the person would have gotten the chance even if they had failed (thereby reducing instrumental value if believed); that times have changed and the current generation does not value such kinds of accomplishment (such as technological absolescence reducing cultural value), etc. Such defensive reactions that are expected will be to protect past success from losing their ability to provide esteem-income. *The primary goal is to feel good about oneself*, however that can be brought about. As Lewin (1938) implied with regard to the issue of locomotion vs. cognitive reorganization, both attaining goals that have value, or maintaining the values of previously attained goals, can function for the individual to provide esteem-income. *We explicity predict that action will be preferred to cognitive work so long as further action is possible*, so that the probability of a defensive reaction will be smallest in early careers, moderate in middle careers, and greatest in late careers, when these careers are perceived as closed. It follows that if a closed career becomes an open one, defensiveness should decrease, and the decrease should be proportionate to the extent of increase in "becoming" (i.e., length of future path).

For individuals in the middle of careers of a closed contingent path, we expect some defensiveness, but since the past record can be justified by immediate success and the utilization of already-acquired skills, we expect this to be limited. The individual can act to attain future success. *Since in our society the cultural value of "becoming" is expected to be greater for the majority of individuals than that for "having been,"* striving to attain valued future goals is expected to provide greater esteem-income that maintains the values of past goals, holding the particular magnitudes of difficulty, instrumental and extrinsic values constant. Put another way, in Western culture most individuals learn that others value more the forward-looking, open stance with its prospects for continued advancement, improvement, and success, than the backward-looking, closed stance with its necessity for argumentation to defend the status quo. Various measures of openness vs. closedness to new information, negative information, and use of psychological strategies of defense to distort reality should yield a consistent picture indicating greater openness for new information, negative information, and lack of defensiveness, for the individual who is primarily "becoming" as described here, in comparison to the individual who is primarily "having been" as described here. Change from the end of a closed contingent path to a new contingent path, or change of a closed to an open contingent path, should produce greater openness for new information, for negative information, and less defensive reaction to maintain values of past successes.

Concluding Remarks

The preceding gives a general indication of the application of an expectancy-value theory of adult personality functioning and change to the problem of motivation and aging. However, not considered here are the relation of expected value of the past to immediate action, the effects of failure (rather than success), the role of negative value and negative esteem-income, and most importantly, the role of individual differences in degree of positive and negative motive dispositions of the individual for a particular substantive career area. The present discussion has assumed that individuals are more or less positively motivated within a given substantive career, which need not be the case. In the domain of achievement-oriented activity, this corresponds to the assumption that individuals are relatively stronger in the motive to achieve success that the motive to avoid failure (i.e., that they are success-oriented). Additional treatment of individual differences in degree of positive and negative motive dispositions (Raynor, forthcoming) suggests that different effects are to be expected for individuals who either are negatively motivated in a particular substantive career and who therefore would be conceptualized as equivalent to an individual in the achievement-oriented domain of activity in whom the motive to avoid failure is stronger than the motive to achieve success (i.e., a failure-threatened individual), or who have equal strengths of relevant positive and negative motives.

REFERENCES

Atkinson, J. W. Motivational determinants of risk-taking behavior. *Psychological Review*, 1957, *64*, 359-372.

Atkinson, J. W. *An introduction to motivation*. Princeton: Van Nostrand, 1964.

Atkinson, J. W. Measuring achievement-related motives. Unpublished final report, NSF Project GS-1399, University of Michigan, 1969.

Atkinson, J. W. The mainsprings of achievement-oriented activity. In J. W. Atkinson & J. O. Raynor (Eds.), *Motivation and achievement*. Washington, D.C.: Hemisphere Publishing Corp., 1974.

Atkinson, J. W., and Birch, D. *The Dynamics of Action*. New York: John Wiley & Sons, 1970.

Atkinson, J. W., and Feather, N. T. (Eds.), *A theory of achievement motivation*. New York: Wiley, 1966.

Erikson, E. H. *Identity, youth and crisis*. New York: W. W. Norton, 1968.

Kruglanski, A. W., Riter, A., Amitai, A., Margolin, B. Shabtai, L., & Zaksh, D. Can money enhance intrinsic motivation?: A test of the content-consequence hypothesis. *Journal of Personality and Social Psychology*. 1975, 31, 744-750.

Lewin, K. *Conceptual representation and measurement of psychological forces*. Durhan, N. C.: Duke University Press, 1938.

Moulton, R. W. Motivational implications of individual differences in competence. Paper presented at the meetings of the American Psychological Association, Washington, D. C., September, 1967. Also in J. W. Atkinson & J. O. Raynor (Eds.), *Motivation and achievement*. Washington, D. C.: Hemisphere Publishing Corp., 1974.

Raynor, J. O. Future orientation and motivation of immediate activity: An elaboration of the theory of achievement motivation. *Psychological Review*, 1969, 76, 606-610.

Raynor, J. O. Future orientation in the study of achievement motivation. In J. W. Atkinson & J. O. Raynor (Eds.), *Motivation and achievement*. Washington, D. C.: Hemisphere Publishing Corp., 1974. (a) Also in J. W. Atkinson and J. O. Raynor, *Personality, motivation, and achievement*. Washington, D. C.: Hemisphere Publishing Corp., 1978.

Raynor, J. O. Motivation and career striving. In J. W. Atkinson & J. O. Raynor (Eds.), *Motivation and achievement*. Washington, D. C.: Hemisphere Publishing Corp., 1974. (b) Also in J. W. Atkinson & J. O. Raynor (Eds.), *Personality, motivation, and achievement*. Washington, D. C.: Hemisphere Publishing Corp., 1978.

Raynor, J. O. Future orientation, self-evaluation, and motivation for achievement. Unpublished research proposal, State University of New York at Buffalo, 1976.

Raynor, J. O. Motivation and aging: A theory of adult personality functioning and change. In J. O. Raynor & E. E. Entin (Eds.), *Motivation, career striving, and aging*. Washington, D. C.: Hemisphere Publishing Corp., forthcoming.

Raynor, J. O., Atkinson, J. W., & Brown, M. Subjective aspects of achievement motivation immediately before an examination. In J. W. Atkinson & J. O. Raynor (Eds.), *Motivation and achievement*. Washington, D. C.: Hemisphere Publishing Corp., 1974.

Raynor, J. O., & English, L. D. Relationships between self-importance, future-importance, and self-possession of personality attributes. In J. O. Raynor & E. E. Entin (Eds.), *Motivation, career striving, and aging*. Washington, D. C.: Hemisphere Publishing Corp., forthcoming.

Sheehy, G. *Passages: Predictable crises of adult life*. New York: E. P. Dutton, 1974.

Chapter 10

A Developmental Perspective on Theories of Achievement Motivation*

DIANE N. RUBLE
Princeton University

Research in the area of achievement motivation is concerned ultimately with predicting the strength of desire to engage and persist in achievement-related activities. There have been numerous theoretical orientations guiding this research and several important components of achievement behavior have been identified. However, notably absent from theoretical formulations of achievement (with only a few exceptions—see Smith, 1969; Weiner, Kun, & Benesh-Weiner, in press) has been a *developmental* perspective—a model of the emergence of the central components of motivation and an analysis of the antecedents and consequences of such developmental changes.

A developmental perspective is important, in part, because theories of achievement motivation have provided the justification for many educational enhancement programs and for some educational policies, such as decisions regarding homogeneous vs. heterogeneous classroom groupings; and yet the age level of children involved has typically not been a major factor in these policy decisions. However, as discussed below, recent research concerning social and cognitive development suggests that the nature and determinants of achievement striving may change considerably during the course of childhood; and that because of such changes, educational decisions based on current theories may not only be ineffective at some age levels but may, in some cases, have effects opposite to those intended (e.g., Boggiano & Ruble, in press).

---

*Portions of this chapter also appear in D. N. Ruble, & A. K. Boggiano, Optimizing motivation in an achievement context, in B. K. Keogh (Ed.), *Advances in Special Education*, (Vol. 1), Greenwich, Conn.: JAI Press, in press.

A developmental perspective may also make substantial contributions to a more basic theoretical understanding of achievement processes. Obviously age-related differences are important for concerns with the generalizability of any theory. However, in addition, the study of achievement processes in young children may provide insight into which variables are central in the sense of early emergence. It should also lead to a more in-depth and complete conceptualization of central constructs, such as effort and ability, a need identified in recent writings by achievement theorists (e.g., Weiner, 1979). Furthermore, a developmental analysis may lead to a more dynamic, change-oriented conception of achievement behavior more generally. That is, this perspective may provide insight into the process of entering new areas of achievement (e.g., different athletic endeavors) at any point in the lifespan.

In this chapter, the possibilities for relating developmental processes to theories of achievement motivation will be explored. A specific goal is to provide a framework within which to study the types of social and cognitive developmental changes most relevant to variables central to theories of achievement motivation. The chapter consists of four sections. The first represents a brief description of a few of the central variables and principles that have emerged from theories of achievement motivation, and which, thus, seem particularly important to consider developmentally. Second, previous approaches relevant to a developmental analysis are discussed; and the alternative conceptualization of a developmental perspective, as represented in this paper, is presented. In the third and fourth sections, the proposed framework for a developmental analysis of theories of motivation is presented; Section 3 describes cognitive-developmental changes, and Section 4 describes social-environmental changes.

Central Constructs in Theories of Achievement Motivation

Although basic models of achievement motivation have undergone numerous revisions, extensions, and shifts in emphasis since the seminal work of McClelland and his colleagues in the late 1940s and 1950s (McClelland, Atkinson, Clark, & Lowell, 1953), there does appear to be, at least on a surface level, considerable agreement across theories about what constitutes the major determinants of achievement orientation. (See Ruble and Boggiano, in press, for a review of these theories). Although the specific relationships among variables vary across models, each contains, either implicitly or explicitly, three types of factors: (1) individual differences, (2) expectations or probability of success, and (3) incentive value of outcomes.

The *individual difference variable* in the early theories (e.g., Atkinson, 1957) consisted of a relatively stable, unconscious motive to achieve, which was assumed to be determined by early socialization experiences. Recently, however, a number of theorists have questioned the usefulness of the global motivation concept (e.g., Heckhausen, 1977) and have proposed instead that research should focus on more cognitively-oriented constructs, such as self-perceptions of competence and the perception of personal control over outcomes. (See Weiner, 1972, and Heckhausen and Weiner, 1974, for conceptual and empirical justifications for this shift toward cognitive theories of motivation.) A focus on self-perceptions is crucial to the main concern of this paper with developmental changes because it emphasizes the need to examine cognitive-developmental variables as well as socialization processes in understanding the origins of achievement motivation.

The specific definition of *expectancy or probability of success* varies considerably. For example, it may refer either to the individual's subjective estimates of success at a given task or to a more objectively-based probability of success based on social norm information. An additional problem with the construct is that predictions based on expectations for success may be dependent on other interactive factors included in the models, such as: (1) a likely relationship between perceptions of *competence* (the individual difference factor) and expectations (Kukla, 1972; Moulton, 1974; Weiner, 1974), such that, for example, individuals believing they are incompetent at a task may have a relatively low subjective expectancy of success even if normative data indicates the task is quite easy; and (2) a likely correlation between the level of *effort* expected or intended at a task and expectations for success (Kukla, 1972; Revelle & Michaels, 1976). Because of these theoretical problems, it seems important to begin conceptually to separate the motivational effects of real task difficulty (e.g., enhancement of effort following failure) from subjective expectations of success or level of confidence. A developmental analysis may provide some insights into these relationships (e.g., children may be able to evaluate the difficulty of a task before they have a stable sense of their own level of competence). Most importantly, however, it is clear that expectations of success are related to achievement striving (e.g., Crandall, 1969; Feather, 1966; Weiner, 1974); and thus the emergence of this concept and the factors that relate to it (e.g., perceptions of effort; future goal orientation) are important foci for a developmental analysis of achievement.

The *incentive or reinforcement value* of an outcome refers, in a general sense, to whatever contributes to positive affect (e.g., pride) or value of achieving a goal. Early definitions of this construct referred primarily to the challenge value of a difficult task. More recent cognitively oriented theorists have used the term more broadly to include other intuitively important aspects of incentive, such as the significance of the particular area of achievement for the individual (Crandall & Battle, 1970) or its informational value for self-evaluation (Weiner, 1974). A developmental analysis of this construct would include examining conceptual changes in what constitutes a valued or challenging activity as well as corresponding socialization variables involved in defining valued activities.

In summary, an analysis of points of integration across theories does suggest some consensus regarding three major determinants of achievement motivation. However, definitional ambiguities and likely confoundings across variables suggest that we cannot consider these as three totally separate constructs that make independent contributions to predicting achievement behavior. These basic theoretical problems complicate somewhat our task of applying a developmental analysis to achievement motivation. Nevertheless, it seems reasonably safe to conclude that a better understanding of motivational processes in children is dependent upon examinations of: (1) the emergence of concepts related to the central constructs described above, such as perceptions of ability and challenge; (2) developmental changes in processing or integrating information relevant to these concepts; and (3) socialization factors that affect such processes.

Alternative Approaches to the Development of Achievement Motivation

It should not be concluded from the preceding discussion that a concern with developmental issues has been absent from the achievement motivation literature. Indeed, a major goal underlying the initial ideas in the area (McClelland et al., 1953) and some of the early research was concerned with the socialization of achievement motivation (e.g., Rosen & D'Andrade, 1959; Winterbottom, 1958). With the hope of identifying the optimal conditions for individual achievement, numerous studies subsequently pursued such questions as what kinds of parental characteristics or child-rearing practices relate to achievement striving and behavior (cf. Crandall, V. J., 1963; Crandall, V. C., 1967). In general, this early research could be described in terms of two major characteristics: (1) a focus on individual differences — e.g., what factors seemed to produce high vs. low achievers; and (2) a passive view of children in that they were assumed, in a sense, to be molded by socializing agents (usually the parents) to be motivated to achieve or not.

The socialization of individual differences in achievement continues to be an issue of major interest and importance. However, the position taken in this paper is based on a more cognitive-developmental approach (e.g., Kohlberg, 1969) that an equally important but neglected aspect of the socialization of achievement concerns the more universal processes of development and a view of the child as taking a more active role in such processes. That is, the rules, standards, and values that underlie achievement motivation may be viewed as cultural conventions that are common to the learning experiences of children within a culture. For example, children must learn that standards of success are often based on relative outcomes in much the same way as they learn cultural definitions of sex-appropriate behavior. Children are active participants in this process in the sense that they are presumed to be motivated toward predictability within their environment and thus to seek information relevant to norms or self-definition. In addition, the interpretation of such information is mediated by the cognitive-developmental level of the child.

One reason to turn to this alternative developmental approach is that, although a few generalizations may be derived from the numerous studies on socializing individual differences, the results on the whole have been disappointing; the correlations are typically low and frequently inconsistent across studies (Crandall, 1963; Smith, 1969; Zigler & Child, 1969). One possible reason for such inconsistencies is a frequent failure to systematically consider developmental level in examining the child-rearing antecedents of achievement motivation. A developmental perspective on these relationships would suggest that children's motivational orientations may be highly susceptible to some kinds of social information (such as peer comparison) and socialization influences (such as rewards for meeting standards of competence) at particular ages. Indeed, data from longitudinal research does suggest that the developmental timing of parental demands is a critical factor in making predictions of later achievement (Crandall, 1972; Kagan & Moss, 1962).

A second reason to incorporate a cognitive-developmental approach is that theories of achievement motivation have increasingly relied on constructs involving cognitive mediation. Thus, age-related cognitive structural changes or information processing shifts may dramatically alter the nature of achievement motivation in children at different ages. There are, in fact, several indications of important developmental changes in variables related to achievement motivation. For example, several studies have shown that children younger than 6-7 years prefer easy tasks, while older children tend to choose moderately difficult tasks (e.g., Halperin, 1977; Veroff, 1969). In addition, longitudinal research has shown that although there are some consistencies in achievement striving over time, these are very dependent on the developmental levels being compared and the specific type of achievement behavior examined (Crandall, 1972; Kagan & Moss, 1962).

To date, there has been little research on cognitive and social developmental changes related to achievement motivation. Thus it is not yet possible to provide a set of conclusions about why such shifts occur. Instead, the purpose of this chapter is to present a possible framework for describing developmental variables most likely to affect children's achievement-related perceptions and behaviors, which is summarized in Figure 1. This framework will include both cognitive-developmental factors (next section) and developmental factors that are more social-experiential in nature (final section), such as age-related shifts in the messages received from socializing agents.

Cognitive Components of a Developmental Framework

Although previous investigators have recognized the fairly obvious point that the developmental level of a child may signal important shifts in the nature of achievement motivation (e.g., Veroff, 1969; Weiner, 1974), there has been little attempt to begin a systematic description of the nature of the different types of shifts involved. In order to illustrate the enormity of the task of relating developmental processes to motivational theories in a comprehensive way, a simplified model of achievement is presented in Figure 2, modified from Weiner's (1974) current attributional theory of achievement motivation. To exemplify the model, consider an individual with the following *information available:* She has generally

Cognitive Components		Social-Environmental Components
Concepts	Processing and Integrating	
Ability	Forming Evaluative Judgments	The Composition of the Environment
Success and Failure	Inferences about Outcomes Causes	Messages Received from Socializing Agents
Challenge	Integrating Sequential Information	Opportunities Available to Exhibit Behavioral Consequences

Figure 1. A framework for describing developmental changes relevant to theories of achievement motivation.

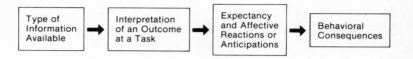

Figure 2. An illustrative model of achievement processes (based on Weiner's, 1974, attributional theory of achievement motivation).

been successful at mathematical games and is about to attempt a game known to be quite difficult. She succeeds, and *interprets* her success as due to her high ability (Frieze & Weiner, 1971). This interpretation results in a high *affect* and an *expectation* for continued success, which in turn leads to a *behavioral response* to persist if she fails on a subsequent trial (Weiner, 1974).

Developmental processes can and probably do enter into and affect predictions at every step on the model. In order to simplify the presentation of a descriptive framework for cognitive-developmental influences on achievement, the analysis will be divided into two types of categories: (1) changes in children's *concepts* of the central constructs of achievement, and (2) changes in the way the constructs are processed or integrated.

Changes in Children's Concepts of Achievement

The first category is represented by the boxes in the model and would consist of developmental changes in the meaning of concepts such as ability, success, or challenge, which would affect the achievement process. For example, if young school-aged children lack the concept of relatively stable abilities, then they may tend not to interpret their outcomes in terms of the presence or absence of competence; or, if they do, an attribution to ability may not have the predicted relationship to subsequent expectations because it is perceived as unstable. In this section three concepts will be discussed as illustrations of the types of developmental changes that may be crucial to understanding the nature of achievement striving in children of different ages.

The concept of ability. Inferences concerning ability have emerged as a major determinant of achievement orientation, both in influencing responses to previous outcomes and in goal orientation and expectations for the future. Yet, the concept of ability is somewhat complex, involving inferences concerning abstract internal states and perceptions of some degree of stability over time and situation. To what extent do young school-age children think of themselves and others as possessing relatively stable abilities?

A number of studies examining children's perceptions of other people suggest that there are fairly dramatic changes in this process occurring during the early years of school. In general, the literature suggests that with increasing age, descriptions of other people become more abstract and complex, involve more inferential concepts, focus less on overt characteristics (e.g., appearance), and tend to indicate perceptions of greater consistency or stability of other people's qualities (Livesley & Bromley, 1973; Peevers & Secord, 1973; Shantz, 1975; Rholes & Ruble, 1978). Similar trends have also been reported regarding conceptions of the self (Montemayor & Eisen, 1977).

As Shantz (1975) notes, one of the more interesting aspects of these findings is that the developmental changes do not appear to be gradual, but rather show quite radical shifts at a particular age level—sometime between 7 and 8 years of age. For example, Livesley and Bromley (1973) found that the differences between 7 and 8 year olds, were often greater than differences between 8 and 15 year olds. These findings suggest that the first two or three years of school may be critical times in the development of children's conceptions of themselves and others, and that research focusing on processes underlying the development of dispositional concepts may find this age group particularly productive.

Some recent research has focused specifically on developmental changes in children's concepts of abilities. For example, Nicholls (1978) has suggested that 5-13 year olds' concepts of ability and effort can be described in terms of four developmental levels. At *level 1* (5-6 year olds), effort, ability, and outcome are not distinguished from each other, and only by *level 4* (after age 10-11) is ability used systematically as a capacity which may be clearly differentiated from effort in predicting outcomes. These levels, however, are based on children's conceptions of ability in only a limited sense—i.e., perceptions of the abilities of a pair of children working on math problems in which levels of effort and outcome were varied. Additional analyses in this study revealed that a different aspect of the concept of ability (i.e., recognizing that more difficult tasks require more ability) showed the major developmental shift between 6 and 8 years of age. Clearly, additional work is needed on developmental changes in this very basic aspect of achievement orientation.

The concept of success and failure. Children's standards of success or concepts of attainment appear to shift developmentally in two different, though interrelated ways. The first concerns the perception of control or personal responsibility for outcomes. Several investigators have suggested that an important change occurs in young children's achievement orientation once they become aware of their role in producing goal-oriented outcomes (Bialer, 1961; Crandall, Katkovsky, & Crandall, 1965; Heckhausen, 1967; Veroff, 1969). For example, according to Bialer (1961), very young children respond to all goal attainments with pleasure and to nonattainments with displeasure, but there is no real evaluation of success and failure because the child does not associate the outcome with his/her own behavior. The exact nature of this shift, the approximate age levels at which it occurs, and its causes have not been made explicit in these articles, partly because the concept must be inferred from diverse behaviors. However, these questions deserve systematic empirical attention since the ability to perceive personal causation is central to cognitive theories of achievement motivation; and, in fact, "true" achievement motivation may not develop until this point (Heckhausen, 1967).

Second, standards of success appear to shift developmentally in the extent to which they are based on normative (social comparison) versus absolute standards. According to one theory of developmental changes in achievement (Veroff,

1969), there are two kinds of achievement motivation: (1) autonomous, which is based on *internalized* standards or comparisons of the self with the self, and (2) social, which concerns standards based on *social comparison*. According to the theory, the second type does not develop until the early school years, after "considerable reinforcement, usually from siblings or parent" (Veroff, 1969, p. 50). Recent empirical research generally supports these ideas: (1) there seems to be little use made of social comparison information for self-evaluation until at least second grade (e.g., Boggiano & Ruble, in press; Ruble et al., 1976b; Ruble et al., 1979); (2) children's appraisals of their class standing, as compared to their teachers' ratings, were found to be considerably more realistic after age 8 (Nicholls, 1978); and (3) there appears to be an increase in interest in comparing one's performance with other during the early school years (Dinner, 1975; Ruble, Feldman & Boggiano, 1976).

The Concepts of Challenge

Central to the idea of achievement striving is the concept of challenge—selecting activities that involve some uncertainty about success in order to maximize the affective value of succeeding or to increase self-knowledge concerning competence (Atkinson, 1957; Harter, 1974; Weiner, 1974). Numerous findings have suggested, however, that an orientation toward challenge does not seem to characterize the achievement behavior of preschool or kindergarten-aged children. For example, several studies have indicated that younger children tend to prefer to repeat tasks at which they have experienced previous success, while older children prefer to repeat tasks at which they have failed (cf. Crandall, 1963; VanBergen, 1968). Similarly, young children tend to prefer to work on easy tasks, while older children prefer tasks of moderate or high levels of difficulty (Halperin, 1977; Veroff, 1969). Finally, older children have been found to spend a longer time at insoluble than soluble tasks, while younger children do not seem to differentiate among the tasks (Harter, 1975).

Thus, there appears to be an important change during the

early school years in children's orientation toward or awareness of the challenge value of tasks. Although the specific mental age at which such shifts occur are difficult to designate, given the numerous differences across studies in ages of subjects and tasks used, the major changes seem to appear between the ages of 6 to 9 years (cf. Veroff, 1969). There is thus a correspondence in developmental changes among concepts of challenge and the concepts of ability and attainment. This correspondence may indicate that the changes in behavioral response to challenge reflects an increasing awareness of the incentive value of difficult tasks as the concept of ability and its relationship to outcomes on difficult tasks emerges (Nicholls, 1978).

A few limitations of the above conclusions should be noted. First, the developmental differences in children's task preferences may reflect something other than an increased desire for mastery or challenge. For example, in one recent study, no developmental trend was found in the number of children choosing to repeat an interrupted task when the difficulty level of the task was matched to the child's grade level (Young & Egeland, 1976). Thus, previous developmental differences found using the same task may have been partly due to differential perceptions of the task by children at varying age levels. Second, findings that young children prefer easy tasks or continue to work on a task already solved does not necessarily imply an absence of mastery motivation. For example, Harter (1975, 1978) suggests that the young children's continued interest in an already solved task may reflect a different kind of mastery motivation. That is, the task itself may represent an interesting sensory event that they can control on their own; and, as such, correctness as a goal may be less important. A similar analysis may apply to the finding by Weiner et al. (in press) of preschoolers' greater interest over time than 7-8 year olds in a task involving a contingent relationship between effort and outcome.

Changes in Processing and Integrating

The second category of developmental influences on

achievement is represented by the arrows in Figure 2 and refers to changes in the way information is sought, attended to, and processed, and to changes in how the constructs are related. Such information processing changes would affect the predictions derived from the achievement model even if the constructs were conceptualized as intended by the adult model. Thus, for example, young children may have a stable notion of ability as a causal category but may fail to use it according to logical predictions because of a tendency not to integrate over time previous outcomes or because of different methods of causal analyses or causal schemata. A few examples of apparent changes in achievement-related information integration and their possible implications will serve to illustrate the importance of this developmental factor in the study of achievement motivation.

Forming evaluative judgments of others. One line of research involving information integration has examined children's evaluations of others as a function of information about effort, ability and outcome. Consistent with most findings in the moral judgment literature, the youngest age groups (4-6 years) primarily distribute rewards received on the basis of outcome, while the impact of the explicitly volitional variable (in this case, effort) increases with age, at least until about 10-12 years of age (Salili, Maehr, & Gillmore, 1976; Weiner & Peter, 1973). As in moral judgment research the magnitude and timing of the shift may be partly dependent on artifactual aspects of the task, such as a recency effect of the outcome information (e.g., Feldman et al., 1976). Nevertheless, recognizing the importance of effort and integrating it with outcome information to form evaluative judgments are developmental processes of major interest to motivation theorists, and further study of the emergence of these processes is clearly needed.

Inferences concerning outcomes or their causes. A second line of research has been concerned with developmental changes in the perception of the interdependent relationship among effort, ability, the difficulty of the task, and outcome. It is reasonably well established that even first-grade children can integrate different elements of achievement information in making judgments and that they have some idea of the compensatory relationships between possible causes of outcomes—e.g., that a child with low ability will have to try harder than a child with high ability to achieve the same outcome (Karabenick & Heller, 1976; Kun, Parsons, & Ruble, 1974; Shaklee, 1976; Weiner & Kun, in press). However, it is also the case that the ability to make finer discriminations and more complex judgments increases with age (Karabenick & Heller, 1976; Nicholls, 1978; Shaklee, 1976; Weiner et al., in press). More interesting perhaps is that there also seems to be shifts in the perception of the logic underlying relationships among elements. For example Kun (1977) presents findings from five year olds showing an asymmetry in the judgments of inferring effort from ability information. She describes the findings in terms of a "halo schema" for ability inferences and suggests that young children may perceive themselves as competent only if they try very hard, not if they succeed with little effort. Similar findings are reported by Nicholls (1978).

Integrating sequential information. A third example of the developmental changes in integrating information about achievement concerns inferences made on the basis of events separated in time. Children's abilities to accumulate information about a series of success and failure outcomes increase with age both in terms of inferences about the difficulty of the task and the ability required (Shaklee, 1976), and in terms of changing expectations about their own successes (Parsons & Ruble, 1977). Furthermore, for both types of inferences, consistent significant effects of success/failure outcome information were shown only after the preschool age level.

Social-Environmental Components of a Developmental Framework

Another possible source of developmental differences in achievement motivation is social experimental in nature. Children probably become more familiar with making evaluative judgments, with success and failure outcomes, and with comparing their performance with others, once they enter school, in addition to changes in basic cognitive structures of information processing capacities. Furthermore, it seems likely that socializing agents expect and provide reinforcement differentially across age levels for achievement activities. These types of differences may be divided into three categories: (1) Changes in the environment, (2) Changes in messages received from socializing agents, and (3) Changes in opportunities available to exhibit behavioral consequences. The discussion of these developmental differences and their possible effects will be brief, since there is little direct evidence relevant to these points.

Changes in the Environment

The major universal change relevant to achievement processes is the *entrance into school*. The most obvious impact of this change, in terms of the achievement model in Figure 2, is a shift in the type of information available relevant to interpreting an outcome. For example, the school environment provides children with a large number and range of comparison others performing similar tasks, thereby allowing a basis for assessments of competence that did not previously exist. Certainly, the increase in use of peer comparison information for self-evaluation after the second grade level (e.g., Ruble et al., 1979) is suggestive of the importance of this type of environmental change. Since young children have been shown to utilize social comparison information in other contexts (cf. Masters, 1971), it may be that achievement-related group experiences are crucial to applying comparison information to self-evaluations of competence. Perhaps an earlier use of social comparison would be observed in children with earlier group experiences. Other aspects of the entrance into school that merit study regarding their impact on the achievement processes include: (1) the introduction of a teacher in a child's life— i.e., a significant and powerful adult, other than parents; and (2) the dramatic increase in the amount of time devoted to structured achievement-related activities.

Changes in Messages Received from Socializing Agents

One fascinating aspect of socialization, about which even

the most basic descriptive information is unavailable, is an age-related change in the ways socializing agents attempt to directly influence children. That is, not only are achievement processes affected by changes in the composition of children's environment and their perceptions of it, but also by changes in the nature of the expectations, reinforcement, and punishment by significant others. Indeed, it would be strange if, for example, parents did not modify achievement-related demands and feedback as children's capabilities mature. The usefulness of recognizing that socializers may provide differential information and reinforcement on the basis of categorical distinctions has been seen in the study of sex differences in achievement (cf. Dweck & Goetz, 1977; Parsons et al., 1976). However, changes in socialization behavior as a function of age-level categories have been neglected.

One implication of this type of developmental process applies to the work on individual differences described earlier; that is, the timing of increasing demands in relation to changing capabilities may be a critical determinant of maintaining optimal striving, as others have suggested (e.g., Veroff, 1969; Weiner, 1974). Additional interesting issues that have, to date, received virtually no attention include: (1) to what extent and when external standards of success change with age—e.g., do parents typically shift from focusing their children's attention on task mastery or completion to how well they did relative to others; and (2) do demands become particularly stringent or punishment for lack of motivation particularly frequent at certain age levels, such as during the acquisition of reading skills.

Changes in Opportunities Available to Exhibit Behavioral Consequences

The self-evaluative processes described in Figure 2 are presumably manifested in achievement behaviors, such as intensity of effort or choice of task (Weiner, 1974). However, even this very simple step in the model may also be influenced by age-related social changes. For example, the capacity to choose freely among multiple categories of activities may be a relatively late developing phenomenon, in that the activities of children are likely to be quite closely monitored by adults at least through the early years of school.

Summary and Conclusions

Theories of achievement motivation have changed in many ways since the initial efforts of McClelland and his colleagues in the early 1950s. Models of achievement have incorporated new variables and assumptions to adapt to a growing body of empirical data. In addition, basic philosophies about the nature of achievement have shifted from an emphasis on unconscious motives toward an emphasis on active information-seeking, with cognitions about achievement being viewed as basic determinants of achievement striving. Given this apparent breadth and flexibility of theoretical formulations in the area of achievement, it is surprising that the relationship of developmental processes to these models has generally been neglected. Not only is this perspective essential to understanding the determinants of achievement striving at different age levels, but it is also likely to provide new insights about the changing nature of achievement-related behavior as new activities are being mastered. For example, shifts over time in the self-evaluation processes of adults acquiring new skills may show close parallels to changes in the bases of self-evaluation over age.

The goal of this chapter was to suggest one way of conceptualizing the link between models of achievement and development. Three categories of developmental processes—changes in achievement-related concepts, information processing, and social experience—seem most central to this link; and a review of the available literature indicates that dramatic changes in each of these processes occurs during the early years of school. Additional research is needed not only to describe and explain more fully the nature of developmental changes in achievement, but also to examine the specific implications of such developmental trends for making accurate predictions from existing achievement models.

REFERENCES

Atkinson, J. W. Motivational determinants of risk-taking behavior. *Psychological Review*, 1957, *64*, 359-372.

Bialer, I. Conceptualization of success and failure in mentally retarded and normal children. *Journal of Personality*, 1961, *29*, 303-320.

Boggiano, A. K., & Ruble, D. N. Competence and the overjustification effect: A developmental study. *Journal of Personality and Social Psychology*, in press.

Crandall, V. C. Achievement behavior in the young child. In Hartup, W. W. (Ed.), *The young child: Reviews of research*, Washington, D.C.: National Association for the Education of Young Children, 1967.

Crandall, V. C. Sex differences in expectancy of intellectual and academic reinforcement. In C. P. Smith (Ed.), *Achievement-related motivation in children*. New York: Russell Sage Foundation, 1969.

Crandall, V. C. The Fels Study: Some contributions to personality development and achievement in childhood and adulthood. *Seminars in Psychiatry*, 1972, *4*(4), 383-397.

Crandall, V. C., Katkovsky, W., & Crandall, V. J. Children's beliefs in their own control of reinforcements in intellectual-academic situations. *Child Development*, 1965, *36*, 91-109.

Crandall, V. C., & Battle, E. S. The antecedants and adult correlates of academic and intellectual achievement effort. In J. P. Hill (Ed.), *Minnesota Symposia on Child Psychology* (Vol. IX). Minneapolis: University of Minnestoa Press, 1970.

Crandall, V. J. Achievement. In H. Stevenson (Ed.), *Child psychology: Sixty-second yearbook of the National Society for the Study of Education*. Chicago: University of Chicago Press, 1963.

Dinner, S. H. *Social comparison and self-evaluation in children*. Unpublished doctoral dissertation, Princeton University, 1976.

Dweck, C. S., & Goetz, T. E. Attributions and learned helplessness. In J. H. Harvey, W. Ickes, & R. F. Kidd (Eds.), *New directions in attribution research* (Vol. 2), Hillsdale, N.J.: Lawrence Erlbaum Associates, 1977.

Feather, N. T. Effects of prior success and failure on expectations of success and subsequent performance. *Journal of Personality and Social Psychology*, 1966, *3*, 287-298.

Feldman, N. S., Klosson, E., Parsons, J. E., Rholes, W. S., & Ruble, D. N. Order of information presentation and children's moral judgments. *Child Development*, 1976, *47*, 556-559.

Frieze, I., & Weiner, B. Cue utilization and attributional judgments for success and failure. *Journal of Personality*, 1971, *39*, 591-606.

Halperin, M. S. Sex differences in children's response to adult pressure for achievement. *Journal of Educational Psychology*, 1977, *69*, 96-100.

Harter, S. Pleasure derived by children from cognitive challenge and mastery. *Child Development*, 1974, *45*, 661-669.

Harter, S. Developmental differences in the manifestation of mastery moti-

vation on problem-solving tasks. *Child Development,* 1975, *46,* 370-378.
Harter, S. Effectance motivation reconsidered. Toward a developmental model. *Human Development,* 1978, *21,* 34-64.
Heckhausen, H. *The anatomy of achievement motivation.* New York: Academic Press, 1967.
Heckhausen, H. Achievement motivation and its constructs: A cognitive model. *Motivation and Emotion,* 1977, *1,* 283-329.
Heckhausen, H. & Weiner, B. The emergence of a cognitive psychology of motivation. In P. C. Dodwell (Ed.), *New Horizons in Psychology 2.* Baltimore, MD.: Penguin Books Ltd., 1974.
Kagan, J., & Moss, H. *Birth to maturity.* New York: Wiley & Sons, 1962.
Karabenick, J. D., & Heller, K. A. A developmental study of effort and ability attributions. *Developmental Psychology,* 1976, *12,* 559-560.
Kohlberg, L. *Stages in the development of moral thought and action.* New York: Holt, Rinehart & Winston, 1969.
Kukla, A. Foundations for an attributional theory of performance. *Psychological Review,* 1972, *79,* 454-470.
Kun, A. Development of the magnitude-covariation and compensation schemata in ability and effort attributions of performance. *Child Development,* 1977, *48,* 862-873.
Kun, A., Parsons, J. E., Ruble, D. N. Development of integration processes using ability and effort information to predict outcome. *Developmental Psychology,* 1974, *10,* 721-732.
Livesley, W. J., & Bromley, D. B. *Person perception in childhood and adolescence.* New York: Wiley, 1973.
Masters, J. C. Social comparison by young children. *Young Children,* 1971, *27,* 37-60.
McClelland, D. C., Atkinson, J. W., Clark, R. A., & Lowell, E. L. *The achievement motive.* New York: Appleton-Century-Crofts, 1953.
Montemayor, R., & Eisen, M. The development of self-conceptions from childhood to adolescence. *Developmental Psychology,* 1977, *13,* 314-319.
Moulton, R. W. Motivational implications of individual differences in competence. In J. W. Atkinson & J. O. Raynor (Eds.), *Motivation and achievement.* New York: Winston, 1974.
Nicholls, J. G. The development of the concepts of effort and ability, perception of academic attainment and the understanding that difficult tasks require more ability. *Child Development,* 1978, *49,* 800-814.
Parsons, J. E. & Ruble, D. N. The development of achievement-related expectancies. *Child Development,* 1977, *48,* 1975-1079.
Parsons, J. E., Ruble, D. N., Hodges, K. E., & Small, A. Cognitive-developmental factors in emerging sex differences in achievement related expectancies. *Journal of Social Issues,* 1976, *32,* 47-62.
Peevers, B. H., & Secord, P. F. Developmental changes in attributions to descriptive concepts to persons. *Journal of Personality and Social Psychology,* 1973, *27,* 120-128.
Raynor, J. O. Future orientation in the study of achievement motivation.

In J. W. Atkinson & J. O. Raynor (Eds.), *Motivation and achievement.* Washington, D.C.: Winston, 1974.

Revelle, W., & Michaels, E. J. The theory of achievement motivation revisited: The implications of inertial tendencies. *Psychological Review,* 1976, *83,* 394-404.

Rholes, W. S., & Ruble, D. N. The understanding of dispositional causes of behavior: A developmental study. Unpublished paper, 1978.

Rosen, B. C. & D'Andrade, R. The psychological origins of achievement motivations. *Sociometry,* 1959, *22,* 185-218.

Ruble, D. N. & Boggiano, A. K. Optimizing motivation in an achievement context. In B. K. Keogh (Ed.), *Advances in special education Vol. 1,* Greenwich, Conn.: JAI Press, in press.

Ruble, D. N., Boggiano, A. K., Feldman, N.S., & Loebl, J. A. A developmental analysis of the role of social comparison in self-evaluation. Manuscript submitted for publication, 1979.

Ruble, D. N., Feldman, N. S., & Boggiano, A. K. Social comparison between young children in achievement situations. *Developmental Psychology,* 1976, *12*(3), 192-197.

Ruble, D. N., Parsons, J. E., & Ross, J. Self-evaluative responses of children in an achievement setting. *Child Development,* 1976, *47,* 990-997.

Salili, F., Maehr, M. L., & Gillmore, G. Achievement and morality: A cross-cultural analysis of causal attribution and evaluation. *Journal of Personality and Social Psychology,* 1976, *33,* 327-337.

Shaklee, H. Development in inferences of ability and task difficulty. *Child Development,* 1976, *47,* 1051-1057.

Shantz, C. N. The development of social cognition. In E. M. Hetherington (Ed.), *Review of Child Development Research* (Vol. 5.) Chicago: University of Chicago Press, 1975.

Smith, C. P. *Achievement-related motivation in children.* New York: Russell Sage Foundation, 1969.

Van Bergen, A. *Task interruption.* Amsterdam: North-Holland Publishing Co., 1968.

Veroff, J. Social comparison and the development of achievement motivation. In C. P. Smith (Ed.), *Achievement-related motives in children.* New York: Russell Sage Foundation, 1969.

Weiner, B. *Theories of motivation: From mechanism to cognition.* Chicago: Rand-McNally, 1972.

Weiner, B. (Ed.), *Cognitive views of human motivation.* New York: Academic Press, 1974.

Weiner, B. A theory of motivation for some classroom experiences. *Journal of Educational Psychology,* 1979, *71,* 3-25.

Weiner, B., & Kun, A. The development of causal attributions and the growth of achievement and social motivation. In S. Feldman & D. Bush (Eds.), *Cognitive development and social development.* New York: Erlbaum, in press.

Weiner, B., Kun, A., & Benesh-Weiner, M. The development of mastery, emotions, and morality from an attributional perspective. *Minnesota Symposium on Child Development* (Vol. 13). New York: Erlbaum, in press.

Weiner, B. & Peter, N. V. A cognitive-developmental analysis of achievement and moral judgments. *Developmental Psychology,* 1973, *9,* 290-309.

Winterbottom, M. The relation of need for achievement to learning experiences in independence and mastery. In J. Atkinson (Ed.), *Motives in fantasy, action, and society.* Princeton: Van Nostrand, 1958.

Young, E., & Egeland, B. Repetition choice behavior as a function of chronological age, task difficulty and expectancy of success. *Child Development,* 1976, *47,* 682-689.

Zigler, E. & Child, I. L. Socialization. In G. Lindzey & E. Aronson (Eds.), *The handbook of social psychology* (Vol. 3). Reading, Mass.: Addison-Wesley, 1969.

IV
Attributions and Achievement Motivation

Chapter 11

Attributional Style, Task Selection and Achievement*

LESLIE J. FYANS, JR.
The University of Illinois and the Illinois State Board of Education

MARTIN L. MAEHR
The University of Illinois, Urbana-Champaign

The role of causal attributions in determining motivation to achieve has been the object of intensive study with generally interesting and valuable results (Dweck & Goetz, 1978; Weiner, in press). Thus, it seems quite clear that causal attributions play a critical role in determining the perception of success and failure as such (cf. Maehr & Nicholls, in press) and also mediate responses to these perceived events. Regarding the latter, a primary focus has been on certain behavioral patterns like persistence (see, for example, Andrews & Debus, 1978) at a task in the face of success, failure, or a moderate degree of challenge. Of course, it is of major importance to learn that certain causal attributions affect persistence at a learning task. Such attention to learning tasks not only determines immediate performance levels but ultimately should determine the course of intellectual growth (cf. Rosenshine & Berliner, 1978).

But there is another facet to this matter that has been given considerably less attention, even though it is likewise of significance as far as achievement patterns are concerned. That facet is concerned with individual differences in attributional biases as these determine behavioral preferences related to achievement. Previous research certainly indicates that

---

*Reprinted by permission of authors and publisher of *Journal of Educational Psychology*, 1979, Vol. 71, 499-507. Copyright (1979) by the American Psychological Association.

individuals hold different beliefs about personal causation in achieving situations, including the school in particular (Crandall, Katkovsky, & Crandall, 1965). It seems quite clear that some are likely to believe that success in school is generally attributable to their ability, while others are likely to believe that it is simply a matter of luck. That in itself is a matter of some interest, particularly since social class and ethnicity appear to be related to such patterns (cf. Coleman, Campbell, Hobson, McPartland, Mood, Weinfeld, & York, 1966; Katz, 1967; Friend & Neal, 1972). But this study asks a further question: How do such biases affect task preferences? Kukla's (1978) recent reconsideration of the Atkinson (1957; Atkinson & Feather, 1966; Atkinson & Raynor, 1974) risk-choice model in terms of attribution theory suggests interesting answers in this regard. Briefly, Kukla's analysis implies that individuals who believe that they achieve because they are able, will seek out situations which put this ability to the test. Conversely, individuals who hold a low regard for their ability will tend to avoid such tasks (cf. also Nicholls, 1975, 1976; Sohn, 1977).

A first purpose of this study is to test these derivations from Kukla's attributional analysis of achievement choices. But this study also goes a step beyond the Kukla formulation. Whereas, this formulation focuses on approaching or avoiding tasks where ability is to be demonstrated, we consider simultaneously whether individuals, who attribute achievement to effort, likewise choose tasks where doing well is dependent on effort. Also considered, is whether those who assume that success is outside their control, choose tasks where success is externally controlled, in this case a matter of luck. A theoretical rationale for choices which complement such attribution biases is not as readily available as in the case of ability biases. But, it seems reasonable enough to suggest that beliefs in the efficacy of effort or luck will be associated with the perceived attractiveness of tasks where outcomes are primarily determined by either of these two factors. In the case of effort, it has also been pointed out (Maehr & Nicholls, in press) that some individuals will approach achievement tasks with the purpose of demonstrating good intentions. That is, rather than demonstrating that they are competent, their goal is to demonstrate virtue in trying hard. This might suggest that those who essentially believe in the efficacy of effort should choose opportunities to demonstrate effort both because it is believed to be a way to success and because it is a valued course of action.

In any case, the study asks a very basic quesiton: Do individual differences in achievement attributions relate to task choices that individuals make? But while that question may seem simple, the answers could well have profound significance as far as understanding the nature of intellectual growth is concerned. If beliefs about personal causation are indeed associated with task preferences, it should be of special interest to education. Indeed it would mean that achievement attributions are not only important in the limited setting of the classroom but are part and parcel of a more general achievement syndrome. When beliefs about one's ability are complemented by preferences and choices which serve to enhance such ability, a kind of reciprocal determinism (cf. Bandura, 1978) may evolve which could have important long term effects on achievement.

Aside from these major goals, the study also investigated the influence of grade level, sex and sociocultural identification upon attribution and task selection. Clearly each of these factors is likely to affect attributional processes (cf. Salili, Maehr, & Gillmore, 1976; Salili, Maehr, & Fyans, 1978). The question is, do such factors also modify any task selection biases? Additionally, the stability of individual differences in attribution patterns was considered across situations. The specific concern was to determine whether immediate success or failure on an achievement task might abrogate any relationship between attributional bias and task selection.

In outline, three major research questions guided the investigation.

(1) The first research question was a *congruence* question: How well can task selection be predicted from an individual's attributions? One would expect good prediction if the attributions and selections were congruent.

(2) The second question was the *consistency* question. The major interest here was—will any match-up (or congruence) between attributions and task selections occur repeatedly? Also

of importance here is the influence of outcome (success or failure) in the performance on the task upon subsequent levels of congruence.
(3) The final question was the *generalizability* question. This question was concerned with whether or not the level of congruence would be stable across educational levels (5th through 12th grades), sexes, and socially diverse (rural and urban) groups.

Method

Subjects

A total of 743 students in grade levels ranging from 5th through 12th grades participated in this study. The students were from 7 different schools: 3 elementary, 3 junior high, and 1 senior high. It should be specifically noted that students were not self-selected into the experiment. Virtually all students in the selected classrooms participated and the four who did not apparently did so for reasons irrelevant to the study. To further facilitate the investigation of the generalizability of this study's results across differing social groups, the students were drawn from both a predominantly rural school system and from an urban school district. Table 1 presents the samples of subjects taken from each grade level, their sex, and urban/rural background.[1]

Materials

Causal Attribution Questionnaire. A specially composed questionnaire, based on such measures as the IAR (Crandall, Katkovsky, & Crandall, 1965) was employed to assess the student's success attributions. Each item had a stem describing

---

[1] The inequalities in the N size of Table 1 reflect the population available. However, the variance component analysis procedures used to complete the generalizability coefficient accounted for the factor in producing the results.

Table 1
Breakdown of Subject Sample Relative to
Grade, Sex, and School System

| | School System | | | |
| | Urban | | Rural | |
Grade Level	Male	Female	Male	Female
Grade Five	38	49	54	53
Grade Six	48	41	56	54
Grade Seven	31	30	30	38
Grade Eight	30	30	30	32
Grade Nine	19	5	6	22
Grade Ten	12	12	--	--
Grade Eleven	8	6	--	--
Grade Twelve	9	9	--	--

a performance outcome for which the student would attribute the cause. Following the stem were three phrases each containing an attribution (to ability, to effort, or to task-difficulty-luck). The task-difficulty-luck attributions were combined into a single external attribution item, since pretesting indicated no ascriptions to luck alone. One of the items is presented below as an example:

When you do well on a test at school, is it more likely to be:
a. Because you study for it,
b. Because you are smart,
c. Because you are lucky and got an easy test.

Five items describing successful outcomes[2] were presented and students were asked to complete the sentences by relating the attributional phrase "they liked best." They were told that there were not right or wrong answers for the questions and that their teachers would not see their papers.

Procedure

Assessment of Attributional Style. Subjects were adminis-

---

[2] It was originally intended that failure attributions would also be considered. However, preliminary work indicated that failure attribution responses in this questionnaire did not allow for the assumption of a normal distribution or scaling and, more generally, presented a curious and uninterpretable picture.

tered the forced-choice attribution questionnaire to assess their attributional style in classroom settings. All assessment measures were administered by a twenty-two year old male graduate student.

The following explanations were given to all groups:

> "My name is _____ and I'm here today because I'm interested in what students your age think about some things. I wrote some questions here (showing questionnaire) and your answers will give me the information I need. Here are some sentences to finish and some questions to answer. Decide which is *most* like you and draw a circle around it. This is not a test; your teacher will not see your papers so there are no wrong answers. You can choose the answer that is most like what you think and draw a circle around it. (Pass out the questionnaires). You will have about 15 minutes. *Now* you can begin."

Later, the students' responses to the five items were averaged to determine the student's attributional style.

Task Selection. Two weeks after the assessment of their attributions, an experimenter took each subject individually to a room, introduced himself (or herself), and presented the subject with three boxes. Each box contained identical angle matching tasks which required the subject to match an angle printed on a card to one of the five remaining angles. In actuality none of the angles matched exactly.

The instructions given to each child were as follows:

> "I have brought you to this room so you could play some games. Here are three boxes (point to boxes); each box has a game in it. No one game is harder than the others. (Experimenter points to each game as he explains starting at left.) In this game you do best by *trying* hard. In this game you do best by *knowing* what to do. In this game you do best because you are *lucky*. In the game you choose, I will put five cards on the table in front of you. Then I will give you one more card. You should find one of the cards in front of you that matches the last card, but you cannot put your card down beside the others or move them in any way. After you choose a game, I'll open the box and you can play. Now you can choose a game."

The student was then asked to select which of the remaining games he liked second best. The experimenter recorded the student's choices, opened the box chosen first, and removed the other two boxes.

The students were randomly assigned success or failure independent of their performance upon the angle matching task. If they were to receive success they were told "That's good; you have done very well." If they were assigned failure they were told "That's bad; you have not done very well."

The experimenter then replaced the cards in the box, removed it, and put three new boxes on the table. The instructions were the same as with the first presentation. Again the subjects gave their first and second choices. The experimenter recorded the student's two choices; he then opened the box chosen first while removing the other two boxes. Following this the students were all assigned success relative to the angle matching performance. The experimenter took the student back to his (or her) classroom.

Eight experimenters were used to collect the task preference data. To insure against variability in procedure, each experimenter participated in five concentrated training sessions. As evidence of the effective training sessions, no significant differences between the data obtained among any of the eight experimenters were found. Of course, experimenters were not given any prior information on the purpose of the study nor were they aware of scores on the attribution measures.

Analyses

Congruency. Initially, product moment correlations (ϕ) were run between a student's attributions and task selections and averaged across all selections.[3] However, these correlations, as with all correlations, give only the *general* level of association between any two variables x and y. Furthermore,

---

[3]While each student selected two games, there were three possible groupings of games which could be matched with success-attributions. Thus, there were three prediction situations in which to predict selections from attributions. One prediction situation was the first game selected by each student. However, before the second game selection, each student was assigned success or failure. Thus, there were two possible prediction situations of second selections: those selections of student assigned success outcomes and those selections of students assigned failure.

they are symmetric (i.e., rxy = ryx). What was clearly needed was a measure, *not* of general association but of *predictive* associations, much like a regression weight in a regression equation (i.e., By.x). This index of predictability would be asymmetric (like a regression weight By.x) and would indicate how much error in predicting y is reduced when given information from x. Thus, this predictive index, for the present data would indicate how well one can predict task selections from attributions. This level of predictive associations was ascertained through the use of the statistic lambda (λ) as described by Hays (1973). The interpretation of lambda indicates the percentage of error in predicting task selections when given knowledge of a student's attributions. A mean lambda value corresponding to each λ for each grade level was thus calculated.

Consistency. An ANOVA was calculated employing as the independent variable the outcome (success or failure) which was randomly assigned after subjects performed upon their first selection. The dependent variables were the subjects' second selection and difference scores reflecting any discrepancy between a subject's first and second selection. These analyses were conducted to determine whether the immediate experience of failure or success would modify the relationship between the presumably more enduring attributional orientations and task preferences.

Generalizability. To answer the issues addressed by this question, several factorial ANOVAs each employing grade level, sex, and culture as independent variables were calculated. The dependent variables were scores developed to express the level of congruency between each student's attribution and subsequent task selections. Thus, for example, if the game the individual ranked first exactly matched that individual's attributions, that individual was assigned a score of 3. If an individual's game selection which matched his (or her) attribution was ranked second, that individual was given a congruency score of 2. However, if the game selection ranked first or second by the subject did not match his/her attributional style, that subject was given a congruency score of 1. Using these congruency scores as dependent variables, the variance components for the independent variables were then transformed into generalizability coefficients following Cronbach, Gleser, Nanda, and Rajartanam (1972) and Golding (1975) in order to show directly the generalizability of attributional self-selection across grade levels, sex, and culture.

RESULTS

Congruency

The product moment correlations calculated between a student's attribution and task selections are given in Table 2 for each grade level. As can be seen from Table 2, the correlations between attributions and task selections are relatively strong across grade levels, having an average ϕ of .43. However, as mentioned earlier, there are indicators of the general, symmetric relationship between attributions and task selections. The actual predictive associations when predicting ability tasks from ability attributions, effort tasks from effort attributions, and luck tasks from luck attributions are given by the lambda coefficients. The lambda coefficients expressing this predictive association at each grade level are given in Table 3.

Table 2
Average Correlations Between Attributions and Task Selections For Each Grade Level

Grade Level	
Grade Five	.32
Grade Six	.19
Grade Seven	.40
Grade Eight	.46
Grade Nine	.42
Grade Ten	.45
Grade Eleven	.64
Grade Twelve	.59

Table 3
Mean Lambda Values (λ) For Each Grade Level

Grade Level	
Grade Five	.16
Grade Six	.38
Grade Seven	.53
Grade Eight	.30
Grade Nine	.44
Grade Ten	.16
Grade Eleven	.56
Grade Twelve	.38

It should be remembered that the coefficients shown in Table 3 are conceptually similar to that of a regression weight for predicting specific tasks from specific attributions. Numerically, they reflect the percentage decrease in error in prediction when one has knowledge of a student's attributions. The lambdas shown in Table 3 appear relatively strong, having an average of .36 across grade levels. Thus, across grade levels, knowledge of a student's specific attributions reduces the error in predicting that student's specific task selection by an average of 36%. To attain such a reduction in error with a correlational statistic one would have needed an initial correlation of rxy = .60. Obviously, this represents a sizeable reduction in prediction error. This finding may be clarified even further, with an example of a particular contingency table between attributions and task selections.[4] An illustrative case is given for 12th grade students for the first selection situation and is displayed in Table 4. This table demonstrates further the interpretation of the lambda and correlation coefficients of Tables 2 and 3 and indicates that one can readily predict student game selections given knowledge of their attributions.

Consistency

The extent to which the above congruency attributions and game selections will occur consistently is presented in Table 5.

The ANOVA results presented in Table 5 show that the level of congruency between attributions and game selections was

Table 4
An Example of the Contingency Between Student
Attributions and Task Selections in the
First Selection Situation for 12th
Grade Male Students

Attributions	Task Selections		
	Ability	Effort	Luck
Ability	3	1	0
Effort	1	4	0
Luck	0	0	0

---

[4]Many contingency tables were computed for this study and are summarized by the correlation and lambda coefficients in Tables 2 and 3. The original contingency coefficients are available on request from the first author.

Table 5
Effects of Randomly Assigned Outcome (Success or Failure) Upon Level of Second Task Selected and Difference Scores

Dependent Variable	SS	df[a]	Mean Square	F Ratio	P
Second Task Selected:					
Between Groups	.01	1	.01	.01	.90
Within Groups	341.69	554	.62		
Difference Score:					
Between Groups	.11	1	.11	.09	.75
Within Groups	298.18	554	1.17		

[a] The smaller df relative to the initial sample size reflects students with missing data.

not significantly different between the first and second selections. Likewise, this congruency was consistent regardless of whether success or failure outcomes were experienced by the students after completing the first angle matching task. None of the F-ratios for any of the dependent variables (as shown in Table 5) were statistically significant, even at p = .05 level.

Generalizability

The generalizability coefficients for the independent variables of grade level, sex, and culture for each congruency dependent variable are presented in Tables 6 and 7.

Before describing the results as shown in Tables 6 and 7, it may be helpful to review the meaning of generalizability coefficients. A large generalizability coefficient for a particular independent variable indicates high *within-level* correlations among the congruency scores for that independent variable. Thus, a high generalizability coefficient for the independent variable of "grade level" would mean that *within each grade level* (5th, 6th, etc.) there was a high degree of similarity of congruency scores, but that there were large differences in congruency scores *between grade levels*. All discussions or statements concerning attribution-game selection congruency, then, must be specific to each grade level. A high generalizability coefficient would indicate that there was a high degree of

Table 6
Generalizability Coefficients for Congruency Scores
At First Selection Situation

Grade Levels	Sex	Rural/Urban	Attributional Style
.007	.003	.002	.93

Table 7
Generalizability Coefficients for Congruency Scores
At Second Selection Situation

Grade Levels	Sex	Rural/Urban	Attributional Style
.033	.004	.002	.96

generalizability of congruency across all other non-grade level factors included in the study (e.g., sex). A low generalizability coefficient for a particular independent variable (e.g., grade level) would indicate a *lack* of congruency score variance accounted for by the particular independent variable.

The results presented in Tables 6 and 7 indicate that there is no large amount of congruency score variance accounted for by grade level, sex, or urban/rural differences. In fact only a very minimal amount of dependent variable variance is accounted for by these variables. Thus, any statements concerning the congruency between attributions and game selections can be generalized across the various grade levels, sexes, and differences represented in this study.

A question could be raised as to the apparent contradiction between these generalizability results and those expressed by the correlations and lambdas of Tables 2 and 3. However, the dependent variable for the generalizability analysis was a congruency score, not just a task selection. Thus, this score reflected both the strength and weaknesses in prediction of tasks from attributions. This was apparently quite similar for different grade levels.

However, the results presented in Tables 6 and 7 clearly indicate that the major portion of dependent variable variance is accounted for by individual differences in attributional styles. Eighty-five percent of the congruency score variance was associated with the student's pre-treatment attributions. Further, it may be noted from the generalizability coefficients that, while there were large differences in task selections between attribution groups, there was a great deal of task preference similarity within such groups. Thus, congruence was specific to each student's attribution. Moreover, this congruency between individual attributions and task preferences was highly generalizable across grade levels, sex, and urban/rural.

Discussion

In accord with Kukla's hypothesis, students who believe that they achieve because they are able, choose tasks where skill or ability could be demonstrated. Conversely, students who attribute success to factors other than ability tend to avoid such tasks. The educational implications of such tendencies deserve to be stressed. When individuals select tasks requiring competence, it is then that they are likely not only to test but also to enhance their ability. When, however, students characteristically select tasks whose outcome is simply a matter of external factors such as luck, it is not likely that their intellectual competence will be enhanced. Games of chance are not noted for improving academic skill.

Interestingly, this general finding accords nicely with the results of a recent study (Shiffler, Lynch-Sauer, & Nadelman, 1978) on the relationship of self-concept and task orientation in informal classrooms. In this case, a generally positive concept of self was associated with on-task, achievement-related behavior. The broader perspective that may be emerging in these studies is that the kind of freedom of choice that is granted in so-called "open" classrooms is warranted for certain kinds of students: those who have a sense of competence. These students can be trusted to do the kinds of things that will facilitate intellectual growth with minimal direction and control on the part of the teacher. Indeed, given such a positive orientation toward achievement it would be dangerous to risk ruining a good thing by exercising anything more than minimal control over their behavior (cf. Maehr, 1976, 1977). Quite clearly, for other children such freedom may have distinctly negative effects. Specifically, those students who lack a sense of competence are not likely to attend to that which facilitates intellectual growth. These students can hardly be left to their own devices.

The findings discussed thus far relate specifically to the Kukla hypothesis. However, the results suggest something more than anticipated by that hypothesis, namely that students do more than simply approach or avoid opportunities to demonstrate ability: *they exhibit preferences in accord with their belief about the reasons for success.* That is, beliefs in the causal role of ability, effort, or luck lead to ability, effort, luck, task choices. While the Kukla hypothesis accounts for ability and luck choices, it does not relate in any direct or simple manner to effort choices. However, following Maehr and Nicholls (in press) it may be reasoned that individuals can have different goals in an achievement situation. Not only do some basically wish to demonstrate competence and others avoid situations where their competence may be put into questions, in line with Kukla's hypothesis, some clearly believe in the efficacy of effort, act accordingly, and perhaps also demonstrate virtue as they do so (cf. Nicholls, 1976). Certainly, there is no denial of personal responsibility for achievement in the case of these students, as might be so in the case of luck-oriented students. These students may, however, be exhibiting a different approach to achievement, one that likewise has important effects on intellectual growth. Whereas the ability-oriented student may need the challenge, the new test of his/her skill and be turned off when it is not there, the effort-oriented student is more likely to accept the burden of "doing his daily stint."

It seems likely that attributions guide task preferences which in turn affect achievement patterns. Students who hold different beliefs about the reasons they succeed in school, will tend to exhibit continuing motivation, interest patterns, and activity preferences which are differentially effective in further encouraging academic achievement and intellectual growth. Moreover, it is of paramount significance that individual orientations in achievement attribution were found to be associated with choices, regardless of age or urban/rural background and irrespective of immediate feedback experience. This reinforces the pervasive importance of individual attributions as determinants of achievement behavior. All of this suggests that further attention should be given to the *sources of these beliefs*, not only in the classroom (see, e.g., Andrews & Debus, 1978; Dweck et al., 1978) but in the home, and extra-school, sociocultural environment. In any event, the relationship between beliefs and task selection revealed in this study suggest the existence of motivational syndromes which operate to enhance or inhibit school achievement in significant ways. It now remains to specify these syndromes further and clarify antecedent-consequent relationships in this regard.

Superficially, the task situation employed in this study is not unlike the kinds of situations characteristically employed in studies of achievement behavior. However, the task was termed "a game" for the specific purpose of deemphasizing the external pressure and with a view to giving "intrinsic interests" full opportunity to be realized. In reviewing these facts, it may first of all be stated that the patterns of these results are most likely to be applicable to what Maehr (1978) has defined as "play" situations: situations where external evaluations and extrinsic rewards are minimal and where also intrinsic factors are dominant. This in no way suggests that the present experimental situation represents an unimportant one as far as human behavior is concerned. Indeed, as far as achievement in school related tasks is concerned, it is perhaps most appropriate to focus on how one chooses and performs when there are few external pressures (cf. Maehr, 1976, 1977). What does a person choose to do on his or her own? What kind of "play" does he or she select? Do such choices serve to further intellectual growth? As most educators recognize, it is when students develop autonomous and independent interests in learning or in tasks which facilitate intellectual growth, that the goals of schooling are best achieved. Thus, it logically becomes important to consider how students choose and perform when evaluation is minimized—as was the case in this particular study. As a matter of fact, the present results are most interesting as they exhibit close relationships between beliefs about school achievement and the choices made when the students are, so to speak, at play.

REFERENCES

Andrews, C.R., & Debus, R.L. Persistence and the causal perception of failure: Modifying cognitive attributions. *Journal of Educational Psychology.* 1978, 70, 154-166.

Atkinson, J.W. Motivational determinants of risk-taking behavior. *Psychological Review*, 1957, 64, 359-372.

Atkinson, J.W., & Feather, N.T. (Eds.). *A theory of achievement motivation.* N.Y.: Wiley, 1966.

Atkinson, J.W., & Raynor, J.O. (Eds.). *Motivation and achievement.* Washington, D.C.: Winston, 1974.

Bandura, A. The self system in reciprocal determinism. *American Psychologist.* 1978, 33, 344-358.

Coleman, J.S., Campbell, E.Q., Hobson, C.J., McPartland, J., Mood, A.M., Weinfeld, F.D., & York, R.L. *Equality of educational opportunity.* Washington, D.C.: U.S. Department of Health, Education and Welfare, 1966.

Crandall, V.C., Katkovsky, W., & Crandall, V.J. Children's beliefs in their own control of reinforcements in intellectual-academic achievement situations. *Child Development*, 1965, 36, 91-109.

Cronbach, L.J., Gleser, G.C., Nanda, H., & Rajartanam, J. *The dependability of behavioral measurements: Theory of generalizability for scores and profiles.* N.Y.: Wiley, 1972.

Dweck, C.S., Davidson, N., Nelson, S., & Enna B. Sex differences in learned helplessness: II—The contingencies of evaluative feedback in the classroom and III—An experimental analysis. *Developmental Psychology*, 1978, 14, 268-276.

Dweck, C.S., & Goetz, T.E. Attributions and learned helplessness. In J. H. Harvey, W. Ickes, & R.F. Kidd (Eds.), *New directions in attribution research* (Vol. 2). Hillsdale, N.J.: Lawrence Erlbaum Associates, 1978.

Friend, R.M., & Neal, J.M. Children's perceptions of success and failure: An attributional analysis of the effects of race and social class. *Developmental Psychology*, 1972, 7, 124-128.

Golding, S.L. Flies in the ointment: Methodological problems in the analysis of the percentage of variance due to persons and situations. *Psychological Bulletin*, 1975, 82, 278-288.

Hays, W.L. *Statistics for psychologists.* N.Y.: Holt, Rinehart, & Winston, 1973.

Katz, I. The socialization of academic achievement in minority group children. In D. Levine (Ed.), *Nebraska Symposium on Motivation* (Vol. 15). Lincoln: University of Nebraska Press, 1967.

Kukla, A. An attributional theory of choice. In L. Berkowitz (Ed.), *Advances in experimental social psychology* (Vol. II). N.Y.: Academic Press, 1978.

Maehr, M.L. Continuing motivation: An analysis of a seldom considered educational outcome. *Review of Educational Research*, 1976, *46*, 443-462.

Maehr, M.L. Turning the fun of school into the drudgery of work: The negative effects of certain grading practices on motivation. *UCLA Educator*, 1977, *19*, 10-14.

Maehr, M.L. Sociocultural origins of achievement motivation. In D. Bar-Tal & L. Saxe (Eds.), *Social psychology of education: Theory and research*. N.Y.: Hemisphere Publishing Corporation, 1978.

Maehr, M.L., & Nicholls, J.G. Culture and achievement motivation: A second look. In N. Warren (Ed.), *Studies in cross-cultural psychology* (Vol. 3). N.Y.: Academic Press, in press.

Nicholls, J.G. Causal attributions and other achievement-related cognitions. *Journal of Personality and Social Psychology*, 1975, *31*, 379-389.

Nicholls, J.G. Effort is virtuous, but it's better to have ability: Evaluative responses to perceptions of effort and ability. *Journal of Research in Personality*, 1976, *10*, 306-315.

Rosenshine, B.V., & Berliner, D.C. Academic engaged time. *British Journal of Teacher Education*, 1978, *4*, 3-16.

Salili, F., Maehr, M.L., & Fyans, L.J., Jr. The development of moral and achievement judgments: A study of the interaction of social, cultural, and cognitive developmental factors. Unpublished research report, Institute for Child Behavior and Development, University of Illinois, Urbana, Illinois, 1978. (In press, *International Journal of Intercultural Relations*.)

Salili, F., Maehr, M.L., & Gilmore, G. Achievement and morality: A cross-cultural analysis of causal attribution and evaluation. *Journal of Personality and Social Psychology*, 1976, *33*(3), 327-337.

Shiffler, N., Lynch-Sauer, J., & Nadelman, L. Relationship between self-concept and classroom behavior in two informal elementary classrooms. *Journal of Educational Psychology*, 1977, *69*, 349-359.

Sohn, D. Affect-generating powers of effort and ability self attributions of academic success and failure. *Journal of Educational Psychology*, 1977, *69*, 500-505.

Weiner, B. A theory of motivation for some classroom experiences. *Journal of Educational Psychology*, in press.

Chapter 12
A Re-examination of Boys' and Girls' Causal Attributions for Success and Failure Based on New Zealand Data

JOHN G. NICHOLLS

The University of Illinois at Urbana-Champaign*

Almost any study one conducts on achievement motivation is likely to turn up sex differences, provided subjects of both sexes are included. At an earlier time, these differences seemed to represent nuisance value for achievement motivation reseachers. The scene has changed dramatically. The cry is now, "vive la difference." We actively seek sex differences, though the ultimate aim is often their elimination.

Research on causal attributions for success and failure is no exception. Many studies show males to be more likely than females to attribute their successes to high ability and less likely to attribute their failures to low ability. This difference is held to mediate a more active achievement orientation in males than females. The studies I have conducted with New Zealand children generally confirm this pattern and developmental changes in the concept of ability appear likely to amplify this sex difference in older children. However, this sex difference

---

*When this chapter was written John Nicholls was Visiting Associate Professor, Department of Psychology and Educational Psychology, University of Illinois. Present address, Department of Educational·Psychology, Purdue University, West Lafayette, IN., 47907.

This chapter is a revision of a paper presented at a conference on cross-cultural sex differences in achievement motivation at the Meeting of the American Educational Research Association, Toronto, March 31, 1978.

The studies including Polynesian children were supported by a grant from the Nuffield Foundation, London, to John Nicholls at Victoria University of Wellington, New Zealand. The contribution of Rosemary Sutton to this work is gratefully acknowledged as is Margaret Collinge's assistance with the second study described. The cooperation of schools and officers of the Wellington Education Board is also acknowledged with thanks.

was, in one case, reversed. This finding raises new questions about the important factors in the socialization of achieving tendencies. Further, several studies reveal a sex difference in causal attributions of a type different to that generally noted in the literature. At certain ages, *girls appear less logical in making causal attributions than do boys.* The search for an explanation for this new finding, made it apparent that *much of the previous research has been conducted in the context of a masculine rather than a feminine conception of success and achievement.* The outline of a more adequate attribution theory approach to the study of achievement motivation is, therefore, presented.

Attribution Theory and Achievement Behavior

The attribution theory approach to achievement motivation rests on the assumptions that people generate explanations for their achievement outcomes and that these explanations mediate their affective reactions to outcomes, their expectations for future outcomes of behavior, and the affective value of expected future outcomes. Within this general framework a variety of different conceptions are developing (e.g., Kukla, 1978; Weiner, 1974; 1979).

The position adopted here is that children who attribute high ability to themselves or who have high self-concepts of ability will be more actively achievement-oriented than those who have low self-concepts of ability. Children with high self-concepts of ability will generally expect to succeed on achievement tasks and will attribute success to ability. Attribution of success to high ability will produce a strong sense of accomplishment (Nicholls, 1976a: Sohn, 1977) and expectation of future success (Weiner, Nierenberg, & Goldstein, 1976; Valle & Frieze, 1976). When they fail, these same children are likely to attribute this to something other than ability, such as luck, lack of effort, or poor teaching. It follows that children with high self-concepts of ability will prefer difficult tasks, persist in the face of difficulty or lack of success, and, more generally, tend to choose achievement situations. That is, they will choose tasks offering the possibility of displaying the high ability they believe they have.

Children with low self-concepts of ability, on the other hand, will expect failure and attribute this to low ability. Success, not being expected, will be attributed to factors like good luck, extra effort, or teacher assistance rather than to high ability. These children, therefore, will find success less rewarding and it will provide less basis for expecting future success. In failure, they will find confirmation of their belief that they have low ability and thus expect further unpleasant failures. Such children are unlikely to seek achievement tasks, especially likely to avoid difficult tasks, and to give up when faced with difficult tasks (Kukla, 1972; 1978; Nicholls, 1979).

The general trend of research findings in North America is that boys display the high self-concept of ability constellation of causal attributions and behaviors more than girls do (Deaux, 1976; Dweck & Goetz, 1978; Frieze et al., 1976). As will now be shown, this pattern is also commonly found among New Zealand children. It is argued that, as a consequence of changes in the nature of achievement tasks and of the concept of ability, this sex difference has a larger impact on the achievement of males and females after than during the elementary school years. Important though the sex difference in perceived ability is, it will be noted that it does not always occur. Finally it will be argued that this description of the sex difference in achievement orientation may be more than a half truth, but it is certainly not the whole truth.

Boys are More "Achievement-Oriented" Than Girls

Almost all the relevant studies I have conducted in New Zealand produced results indicating more active achievement-orientations in boys than girls. For example, among high SES 9 to 13-year-olds, boys were more likely than girls to choose the most difficult of three unseen puzzles: $X^2(2) = 6.81$, $p < .05$ (Nicholls, 1978; see Table 1). This result is consistent with Veroff's (1969) findings and suggests higher self-concepts of ability for boys than girls (Kukla, 1978). Thus, this evidence appears consistent with the evidence, now to be discussed, of sex differences in causal attributions.

Table 1.
Level of Aspiration Choices of Boys and Girls:
Numbers of Boys and Girls Choosing Each Puzzle

Sex	Difficulty of Puzzle Chosen		
	Easy	Moderate	Difficult
Female	3	27	10
Male	3	16	21

Firstly, bright upper SES 13-year-old children were given manipulated success experiences on an angle-matching task (Crandall, V. C., 1963; Nicholls, 1975). Among other things they were asked to attribute success to proportions of ability, effort, luck, or difficulty using an adjustable pie-graph (Nicholls, 1975). There was a non-significant tendency for boys more than girls to attribute success to ability and a significant trend for girls more than boys to attribute success to luck, $F(1,96) = 4.55$, $p<.05$. There were 10 items on the angle-matching task, two failure and the rest success. Facial expressions in response to feedback were rated after each item. Boys' expressions were more positive after success and more negative after failure than girls'. This trend was significant for three success items and one failure item, all of which were early in the sequence. Both the attributional and affective results accord with the picture of boys as more achievement-oriented than girls.

In a different study, 12-year-old lower SES children, 71% Polynesian and 29% Pakeha (white), were taught material on Thailand in their regular classrooms. On completion of the 30-minute lesson, which required students to gather information by studying pictures of Thailand, a short test on the lesson content was presented. After children had scored their test answers, they selected explanations for the number of items they got correct and those they got wrong using a paired-comparison procedure. Though boys and girls gained similar scores on the test, girls were less inclined to attribute success on test items to ability, $F(1,198) = 16.27$, $p<.001$, and more likely to attribute success to luck, $F(1,198) = 4.23$, $p<.05$. Boys were more inclined to attribute failure to lack of effort, $F(1,198) = 4.70$, $p<.05$, and uninteresting work, $F(1,198) = 7.73$, $p<.01$.

Girls attributed failure to bad luck more than boys $F(1,198) = 9.24$, $p<.01$. Ethnicity had only one effect on attributions in spite of clear ethnic differences in standardized achievement test scores. Polynesians attributed success to luck more than Pakehas (whites) did, $F(1,198) = 6.10$, $p<.05$.

Two studies of perception of own academic attainment and causal attributions for academic success and failure also suggest girls are less likely than boys to see themselves as high in ability. Perception of own attainment was measured by presenting children with 30 schematic faces lined up vertically on a page. The top one represented the class member who performed the best in reading (or schoolwork generally) and so on. Children indicated how they perform relative to others by choosing one of the faces. Causal attributions were measured by asking children to indicate the relative importance of possible causes of success and failure in reading or schoolwork generally using a paired comparison procedure. The causal factors presented included effort, ability, luck and difficulty, but others were added in one study.

In the first of these two studies, 12-year-olds of mainly middle-class backgrounds were asked to rate their general academic attainment and explain occasions when they performed well and poorly in schoolwork (Nicholls, 1976b). Boys' and girls' perception of own attainment did not differ, but boys attributed success to effort more than girls, $t(262) = 4.75$, $p<.001$, and girls attributed success to effort more than boys $t(262) = 4.75$, $p<.001$, and boys' and girls' achievement test scores and grades were similar in this sample.

In the second study, 6, 8, 10 and 12-year-olds (mainly middle-class) were presented the same questions, in this instance about reading attainment and success and failure in reading (Nicholls, 1979). Girls in this sample gained higher reading grades than boys, $F(1,532) = 8.51$, $p<.005$. Their perception of own attainment was also higher than boys' but the effect was not as marked as for grades, $F(1,532) = 4.33$, $p<.05$. Causal attributions differed in only one of the eight possible instances. Despite higher perceived and actual attainment, girls more than boys attributed failure to lack of ability, $F(1,532) = 3.95$, $p<.05$. Thus, whether girls perform as well or better than boys and even when they see themselves as performing better than boys, they are less likely than boys to explain their performance in a fashion that makes for continuing active achievement behavior.

The sex difference in ability attributions reported in this section is of special interest because ability attributions are particularly important as determinants of expectancies for future performance and affective responses to task outcomes. When the achievement situation provides consistent and relatively unambiguous outcome cues, the difference in ability attributions may not have especially unfortunate consequences for girls' achievement behavior. This appears to be the case for much schoolwork where history of similar performance would generally lead to similar perceived attainment and similar expectancies for future achievement in boys and girls. On the other hand, when new tasks have to be faced and there are fewer unambiguous cues on which expectancies might be based (Crandall, V. C., 1969; Nicholls, 1975) girls' less positive ability attributions could have much more negative effects: avoidance of difficult options and giving up in the face of difficulty (Dweck & Goetz, 1978). In the move from school to college or work, past performance provides a relatively uncertain basis for estimating future performance. Thus, the transition from school to work or college is especially likely to make motivational differences between the sexes more marked than they are during the school years (Nicholls, 1978). The attributional differences discussed above might, therefore, help explain the fact that girls' achievement in most spheres other than the domestic one falls behind boys' in the years after formal schooling (Maccoby & Jacklin, 1974).

Contrary to the General Trend

Perception of own reading attainment and attributions for success and failure in reading (paired comparison) measures were administered to 8 and 10-year-old lower SES children; 55% Pakeha (Caucasian), 28% Maori (N.Z. Polynesian) and 17% Pacific Island (Polynesian). There were no sex differences on standardized reading tests, but girls perceived their attainment as higher, $F(1,312) = 8.16$, $p<.01$. Girls were more prone to attribute success to good ability $F(1,312) = 5.24$, $p<.05$ and less prone to attribute failure to poor ability, $F(1,312) = 3.76$, $p<.05$. Boys more than girls attributed failure to not usually trying, $F(1,312) = 10.24$, $p<.01$ and girls more than boys attributed failure to bad luck $F(1,312) = 11.58$, $p<.001$. There were few significant ethnicity effects or sex × ethnicity effects. These attributions (though not achievement test scores) suggest stronger motivation for girls than boys. The finding that the girls of this sample were on-task significantly more than boys during reading is consistent with this suggestion (Sutton & Nicholls, 1977).

It is of interest that though there were clear sex differences in attributions and classroom behavior in this study, there were few ethnic differences. This is consistent with other indications that race is often not a powerful predictor of personality characteristics (Edwards, 1974) whereas sex differences are commonly found.

These sex differences are so clearly opposed to the majority of others that they demand attention. One of the studies cited earlier used a similar method for measuring attributions for success and failure in reading with middle SES children, but revealed more negative ability attributions for girls than boys. It seems, therefore, that these results cannot be explained purely in terms of the sex-typing of reading. Sex-typing of reading could, however, be a factor within low SES groups.

Explaining Sex Differences in Causal Attributions

The common trend, in middle and upper-middle SES samples, was for girls to make attributions and behave in a manner suggesting less active intellectual achievement orientations. The "self-socialization" process proposed by Kohlberg (1966) could, as Parsons et al. (1976) note, account for this sex difference. Teacher feedback patterns described by Dweck et al. (1978) could also produce the sex difference in attributional bias. They found that negative feedback boys and girls receive from teachers differed markedly. Boys' negative feedback included a greater proportion of disapproval for aspects of behavior or performance irrelevant to the intellectual quality of their work than did the feedback girls received. More of girls' negative feedback involved criticism of the intellectual quality of their work. Dweck et al. (1978) went on to show that the pattern of feedback girls receive produces more self-derogatory ability attributions than does the feedback boys receive. The reasons for the different teacher treatment of boys and girls are not established. Boys' greater disruptiveness could, for example, elicit more conduct-related negative feedback. In any event, these differences in teacher feedback would increase or maintain the sex difference in attributions. A study by Neale with middle-class New Zealand children replicated the essential findings of the Dweck et al. (1978) study. Thus, the teacher feedback patterns described by Dweck et al. (1978) probably contribute to the sex differences found here.

The reversed sex trends in attributions with the lower SES sample could reflect different teacher treatment of these children. This appears improbable, however, because the observational data show that the boys in this sample were significantly more disruptive and off-task than the girls: behavioral differences likely to elicit the above differences in pattern of teacher feedback to boys and girls. It is also unlikely that teachers of the different samples differed in any relevant way.

My hypothesis is that the teacher-feedback patterns are similar in middle and lower SES classrooms and the reversal of sex differences in attributions *reflects SES differences in the perceived sex-role appropriateness or attainment value of reading*. If this proves to be the case, it will suggest that the effect of differences in feedback received by boys and girls depends on the perceived sex-role appropriateness or attainment value of the activity concerned. It is likely that middle class boys and girls value attainment in academic activities, and reading in particular fairly highly, but the lower SES males do not value these activities so highly. This proposal is based only on informal observations (made during time as a grade-school teacher and six summers in various unequivocally low status manual occupations). Lower SES males appear to value academic excellence less than females or higher

SES males and to value physical strength and skill more. Thus, even if low SES boys receive feedback that is similar to that received by middle-class boys, they may be less likely to "take advantage" of this to infer good ability at reading if they see this ability as of relatively little value or marginally sex-role appropriate.

The finding that, for low SES 12-year-olds, sex differences in attributions after the Thailand lesson followed the more common pattern might raise some doubt about the validity of the above speculations. The Thailand lesson, however, may have not appeared like a normal school reading task to the children. In this lesson, children were required to make inferences from photographs of Thailand. Very little reading and no writing was required. It is, therefore, conceivable that this activity elicited quite different reactions to those normally elicited in low SES boys and girls by school reading activities. Thus, the hypothesis that lower socioeconomic status boys attach little positive value to ability in reading remains plausible and could account for the reversal of the usual sex difference in ability attributions (even if teacher feedback was similar in low and high SES classrooms). It hardly needs to be said that this hypothesis required direct examination.

Logical Inferences About the Causes of Achievement: Sex Differences and Consequences of Mature Ability Inferences

Sex Differences

Further findings from two of the studies discussed above and one other suggest that there are differences between boys' and girls' reasoning about success and failure in addition to differences in the extent to which they use the different causal factors to explain their outcomes. These differences can be interpreted as indicating that girls are less inclined than boys to make logical inferences about ability from relevant cues.

In the first study suggesting such an effect, children explained the performance of two children they saw gain equal outcomes using different amounts of effort. Only by 12 and 13

Table 2.
Levels of Reasoning About Effort and Ability:
Numbers of Girls and Boys Reasoning at Each Level of Maturity

Sex	Level of Reasoning			
	1	2	3	4
Female	17	27	20	8
Male	16	19	12	25

Note: Level 4 is the most mature, where effort and ability are clearly distinguished as interdependent causes of level of performance.

years did most children clearly distinguish the concepts of ability and effort. The concept of ability, as capacity that is manifest only given optimum effort and which limits the effectiveness of effort, was not fully formed before this time. This mature concept of ability is the fourth of four levels of understanding of ability (Figure 1). In a 5 to 12-year-old sample, girls were significantly less likely than boys to display this mature conception of ability and effort (Nicholls, 1978, see Table 2).

Logical inferences based on own performance history and present performance would make children who generally achieve well relative to their peers more likely than low achievers to attribute any specific success to ability and failure to unstable factors. Similarly, those who generally perform poorly should attribute failure to poor ability and success to unstable factors more than would high-achieving students. With 12-year-olds, results consistent with this logical information processing model have been obtained in two studies (Nicholls, 1976a; Nicholls, 1979). No sex differences in relations between attributions and perceived and actual attainment were found at this age. That is, boys' and girls' attributions conform to the logical model to the same degree.

Two studies with 8 and 10-year-olds, however, show closer approximations to these predictions for boys than girls and higher correlations between actual attainment and perception of own attainment for boys than girls (Tables 3, 4, & 5). The sex

Level 1.—Effort and outcome are not distinguished as cause and effect. Explanations are tautological, and ability, effort, and outcome are not distinguished as separate dimensions. Children center on effort (people who try harder are smarter even if they get a lower score) or, less commonly, on outcome (people who get a higher score are said to work harder, even if they do not, and to be smarter).

Level 2.—Effort and outcome are distinguished as cause and effect. Effort is the prime cause of outcomes: Equal effort is expected to lead to equal outcomes. Ability, in the sense of capacity which can increase or limit the effectiveness of effort, is not suggested as a cause.

Where people get the same score but differ in effort, this is explained in terms of compensatory effort by the student who tried less (e.g., she/he worked really hard for while, worked at the end, might have started earlier, must have been thinking while fiddling) or in terms of misapplied effort by the person who tried harder (e.g., she/he tried too hard, went too quickly and made mistakes).

Limits to the effectiveness of effort are tacitly acknowledged, however, when one person tries harder than another but gets a lower score. It is acknowledged that the person getting the higher score with less effort must be, in some sense, more able. Though the concept of ability is not used, the reality of this situation is acknowledged: People are correctly classified and ordered in terms of effort and outcome.

Level 3.—The concept of ability is used intermittently. Effort is not the only cause of outcomes. Explanations of equal outcomes following different effort involve suggestions such as: The person trying less is faster, brighter, has a better understanding, or is naturally good at the activity. These explanations imply that high ability can compensate for lack of effort and that low ability limits the effect of effort. These implications are not, however, systematically applied. Alongside these explanations children assert that the students are as clever as each other or the hard worker is cleverer and that they would get the same scores if they worked as hard as each other.

Less frequently, explanations are of the level 2 type, but it is predicted that the student who tried less will improve more, if both try hard.

Level 4.—The concept of ability, in the sense of capacity which, if low, may limit or, if high, may increase the effectiveness of effort, is used systematically in explanations of observed behavior and outcomes and in predictions of outcomes when both students try hard. Ability is correctly inferred from effort and outcome, and outcomes are seen as determined jointly by effort and ability.

Figure 1. Levels of reasoning about ability and effort.

Table 3
Correlations of Self Perception of Reading Attainment with Grades in Reading and Causal Attributions for Success and Failure in Reading (Middle SES sample: Nicholls, 1979)

Age	Sex	N	Grades	Attribution to Success to				Attribution of Failure to			
				Ability	Effort	Difficulty	Luck	Ability	Effort	Difficulty	Luck
6	F	63	.07	.09	.01	-.12	.02	.03	-.02	.05	-.06
	M	84	.19	.03	.24*	-.03	-.26*	-.20	.04	.00	.15
8	F	54	.30*	.19	-.16	.10	-.15	.09	.08	-.18	.01
	M	71	.44***	.28**	-.12	-.09	-.08	-.37**	.11	.04	.22
10	F	58	.37**	.17	.10	-.07	-.16	-.14	.23	-.16	.07
	M	76	.55***	.40***	-.35**	-.04	-.07	-.07	.19	.16	.14
12	F	57	.76***	.59***	-.29*	-.12	-.49***	-.32*	-.24	-.25	.34**
	M	77	.67***	.57***	-.37**	-.05	-.49***	-.41***	.11	.04	.26*

Note: A posivive correlation indicates perception of high attainment associated with high grades or perception of cause as important.

*$p < .05$
**$p < .01$
***$p < .001$.

Reprinted by permission of author and publisher of *Journal of Educational Psychology*, 1979, 71, 94-99.

Table 4
Correlations of Causal Attributions Reading Grades (Middle SES sample: Nicholls, 1979)

Age	Sex	N	Attribution of Success to				Attribution of Failure to			
			Ability	Effort	Difficulty	Luck	Ability	Effort	Difficulty	Luck
6	F	63	.00	-.12	.11	.13	.01	-.03	-.04	.09
	M	84	.01	-.01	.14	-.15	-.14	-.07	.13	.09
8	F	54	.16	-.06	.12	-.24	.06	.15	-.12	-.10
	M	71	.24*	-.17	.00	-.09	-.27*	-.03	.11	-.22
10	F	58	.10	.08	-.09	-.07	-.11	.27*	-.05	-.14
	M	76	.31**	-.11	.05	-.28*	.02	-.18	.12	.07
12	F	57	.45***	-.43**	.13	-.38**	-.24	-.22	.10	.35**
	M	77	.34**	-.17	.08	-.39***	-.23*	.18	.01	.02

* $p < .05$
** $p < .01$
*** $p < .001$

Reprinted by permission of author and publisher of *Journal of Educational Psychology*, 1979, 71, 94, 99.

Table 5
Correlations of Causal Attributions with Perception of Own Attainment and Reading Vocabulary
(Lower SES sample: Sutton & Nicholls, 1977)

	Perception of Own Attainment				Reading Vocabulary			
	8-year-olds		10-year-olds		8-year-olds		10-year-olds	
	Girls	Boys	Girls	Boys	Girls	Boys	Girls	Boys
Perceived Attainment	—	—	—	—	.24*	.33*	.33*	.57*
Attributions of Success to								
Ability	.13	.23*	.32**	.60***	.30**	.36***	.47***	.55***
Unstable Effort	.05	-.01	-.11	-.10	.04	-.06	-.07	-.12
Stable Effort	.13	.26*	.23*	.07	.14	.19	-.09	-.07
Luck	-.08	-.28**	-.33**	-.23*	-.16	-.26*	-.08	-.12
Assistance	.03	-.10	-.12	-.34**	-.05	-.33**	-.23*	-.35***
Attribution of Failure to								
Ability	-.15	-.14	-.53***	-.66***	-.11	-.05	-.30**	-.44***
Unstable Effort	.03	.16	.10	.13	.27*	.19	-.03	.11
Stable Effort	-.01	-.07	-.23	-.14	-.08	.06	-.01	-.15
Luck	-.02	-.12	.27*	.38***	.12	.01	.09	.37***
Task Difficulty	.07	.03	.27*	.30**	-.16	-.01	.13	.19
Assistance	.05	.12	.32**	.20	-.06	-.17	.12	.03
N	78	81	74	82	78	81	74	82

*p<.05   **p<.01   ***p<.001

differences between correlations in these studies are not massive, but the trend is consistent. Eight and 10-year-old boys perceive their own attainment more accurately than girls and analyze the causes of their successes and failures in a fashion that is more logically related to their general performance level and their perception of that level. This parallels the sex difference in reasoning about effort and ability of observed actors.

It is interesting that this effect occurred in one study where girls' ability attributions were more self-derogatory than boys' (Nicholls, 1979) and in one where the reverse occurred (Sutton & Nicholls, 1977). Thus, this sex difference appears both conceptually and empirically distinguishable from the more widely noted type of sex difference in attributional bias.

My tentative explanation for this effect rests on the assumptions that girls are more generally on-task and obedient in the classroom (both confirmed here: Sutton & Nicholls, 1977), and value these characteristics in themselves more than boys do. The suggestion that girls may be more oriented to teacher approval and self-reinforce for effort, neatness, and good behavior more than boys is not a new one (Crandall, V. J., 1963) and is consistent with more recent evidence (Maehr & Nicholls, in press; Witryol, 1971). If this is so, girls' attention may often be focused on effort, neatness and satisfaction of teacher demands, so they may confuse these dimensions with cues indicating intellectual adequacy of performance and the contribution of ability to this performance. This would make them less likely than boys to perceive the intellectual quality of their work accurately to use history of intellectual outcomes as a basis for making logical inferences about their ability.

This explanation leaves the absence of sex differences in the extent to which attributions of 12-year-olds reflect the logical information processing model to be explained. Children's causal inferences and perception of own attainment may be more easily influenced by extraneous non-outcome related cues when the cognitive processes involved are not well developed. By 12, children's ability to compare their performance with others' and their causal attributional schemes may be sufficiently mature (Nicholls, 1978, 1979) to prevent such "distractors" having major effects on these processes.

Consequences of Mature Ability Inferences

The development of more accurate perception of attainment and of more logical attributional schemes, especially for inferring ability, might appear likely to reduce the sex difference in ability attributions. This, however, need not be the case. The mere fact that boys and girls have similar levels of academic attainment need not lead to similar levels of perceived ability. In fact, the development of a clear distinction between the concepts of ability and effort and of accurate perception of own attainment *appears likely to amplify both the sex difference in ability attributions and the behavioral consequences of this difference.*

When the concepts of effort and ability are clearly distinguished, one has the possibility of evading the negative implications of poor performance for one's ability by claiming low effort. Similarly, high performance can be seen as reflecting more favorably on one's ability if one can say that this performance was achieved with little effort. Given that girls apply greater effort in school than do boys, they thereby give themselves less chance than do boys to infer that they have high ability. The fact that girls' effort may often be directed at nonintellectual aspects of work may mean that much of this effort does relatively little to develop intellectual ability. It may also serve to make it harder for them to infer high ability from good performance and easier for them to infer low ability from poor performance. Therefore, girls' greater diligence in academic work may be of dubious value for their long-term academic and intellectual development and performance. Though they may be rewarded by teachers for this diligence, this very diligence may place them in the invidious position where it is difficult to infer that their ability is high and easy to infer that it is low.

High school children are more likely than elementary school children to take effort into account in this fashion when inferring their ability. Thus, the difference between the sexes in ability attributions should become more marked over the high school years. A recent study confirms this hypothesis (Kessel, 1979).

This trend occurs at the same time as perception of low ability probably comes to have more serious negative consequences for achievement behavior. The distinction between effort and ability implies that the effects of effort on performance are limited by capacity. The consequences of low perceived ability would probably be especially serious when ability is conceived of as capacity which is relatively independent of effort. In younger children who do not see ability as capacity, but who see ability as more related to and influenced by effort, low perceived ability would not have such serious negative consequences. That is, low perceived ability should produce a stronger sense of hopelessness in high school than in elementary school children. Thus, the tendency of girls to perceive their ability as lower than do boys, and to expect lower levels of achievement (Crandall, 1969; 1978), should have more negative consequences in the high school years than in the elementary school years.

It seems, therefore, that the development of the concept of ability as capacity and of the distinction between effort and ability would amplify both the sex difference in ability attributions and the behavioral consequences of this difference. This development could, therefore, be a contributor to the tendency of female achievement to fall behind male achievement during and after the high school years (Maccoby & Jacklin, 1974).

In a similar vein, if some school subjects appear to adolescents to require more ability (however they define it) than others, one would expect a sharper decline in female achievement in these subjects than in others. Success in mathematics could well be perceived as more dependent on ability than success in other subjects. If so, the differentiation of the concept of ability in early adolescence could be implicated in the lag in female mathematics achievement that emerges after the elementary school years (Maccoby & Jacklin, 1974).

Problems and Unfinished Business

So far in this paper, achievement motivation and achievement behavior have not been explicitly defined. But, a definition has been implied in the suggestions that males are more likely to attribute high ability to themselves and to seek situations allowing demonstration of ability and can, thereby, be described as more achievement-oriented than females. The implicit definition is that achievement behavior is behavior directed at maximizing the probability of attributing low ability to oneself. Though this is a worthy definition of achievement behavior (Kukla, 1978; Maehr & Nicholls, in press), if we use it as a reference or guide for assessing behavior that is directed at other goals, we will inevitably conclude that such behavior is deficient achievement behavior. This conclusion would, of course, be justified in the sense that the behavior would not be achievement behavior as defined. However, if we take our examination of this behavior no further, we will soundly fail to characterize it adequately. For example, girls may be concerned to establish, to themselves or others, that they are obedient and diligent members of their school class. If so, analysis of their classroom behavior as behavior directed at demonstrating and developing ability must obscure the essential features of that behavior. Boys may attempt to demonstrate ability and may feel especially good about themselves when they believe they have demonstrated ability. This, however, may be less true for girls. In this view, an adequate analysis of behavior requires that we describe it in terms of the psychological purposes it serves for the person, rather than in terms of the goal we decide it should have or the goals we happen to consider important.

For the most part, the discussion in this paper and the literature on sex differences in causal attributions has failed to do this. It has implicitly adopted the definition of achievement behavior as behavior directed at optimizing ability attributions and examined male and female behavior according to this criterion. As argued in the previous section, the goal of girls' academic behavior is often to "achieve" a sense of virtue through diligence and conformity. Girls presumably seek teacher approval to confirm this perception of themselves as "good girls." Middle class boys' behavior, on the other hand, appears more purely governed by the desire to demonstrate ability and avoid demonstrating lack of it (Maehr & Nicholls, in press).

This difference between the goals of boys' and girls' behavior in school was presented in explaining the sex difference in logical reasoning about the causes of success and failure. However, if it is now agreed that boys and girls may have different goals, it must also be agreed that success and failure (achieving or not achieving one's goals) may mean different things for boys and girls. In other words, when asked to explain why they do well in schoolwork (when this occurs) boys and girls may think of different constellations of events. For girls, doing well may mean getting the intellectual aspects of the work right as well as doing it in the approved manner (quietly, carefully, and neatly, for example) and at the approved time. For boys doing well may depend more largely on intellectual aspects of schoolwork. Thus, the apparently less logical reasoning of girls may only be less logical when judged from the point of view of the implicit definition of achievement motivation which does not adequately describe girls' motivation. If success for girls means getting schoolwork correct as well as trying hard, it is hardly surprising that their explanations for success and failure show a less clear distinction between effort and ability and that they perceive their intellectual attainments less accurately than boys.

A study of the structure of self-esteem of children in the USA provides evidence that compliments this analysis nicely (Entwistle & Hayduck, 1978, pp. 16-17). Factor analysis of a self-esteem scale produced different results with boys and girls. For boys, one factor obviously reflects academic-intellectual competence. Items loading on this factor were: arithmetic, good student, and learns new things quickly. For girls, there was no equivalent factor and the closest approximation had the following items loading on it: can look after others, can take care of herself, polite, honest, good student, learns new things quickly, writing and arithmetic. This factor represents moral or social as well as intellectual excellence, whereas for boys the social-moral items loaded on one almost purely social-moral factor. That is, boys distinguish the intellectual from the social-moral aspects of their behavior more clearly than girls do.

As Markus (1977) points out, individuals are necessarily selective in the information they process about themmselves.

She proposes that this selection process reflects self-schemata: cognitive generalizations about the self that organize the processing of self-related information. The implication of the above discussion is that boys and girls have different self-schemata that are relevant in academic situations. They, therefore, selectively attend to different information and process this in different ways that are consistent with their different self-schemata. These differences presumably reflect the fact that girls and boys have different personal goals in the classroom. In other words, *perception of success and failure and high or low self-esteem are consequent on perception of different personal qualities in boys and girls.*

Previous attribution theory research treated both males and females from a perspective that was more adequate for males than for females. Man has been the measure of all things, including woman. I have attempted to rectify this by outlining different analyses for boys and girls. With the advantage of hindsight it is possible to see that this exercise has been done the wrong way round. Ideally, research on causal attributions for success and failure should begin by establishing the goals of behavior in the situation of interest (Maehr & Nicholls, in press). Only then is it possible to specify the meaning or nature of success and failure for subjects and thus to ask them relevant questions about their perceptions of the causes of success and failure. Instead of doing this, most of us have started with unexamined assumptions about subjects' goals and about what they perceive as success and failure. Our analyses of their perceptions of the causes of success and failure have therefore reflected these assumptions. If we go back to square one and start with a systematic examination of goals and the subjective nature of success and failure, we should be able to shape the details of the above reconceptualization of boys' and girls' achievement motivation more adequately.

REFERENCES

Crandall, V.C. *Expecting sex differences and sex differences in expectancies: A developmental analysis.* Paper presented at 86th Annual Convention of the American Psychological Association, Toronto, Canada, August 31, 1978.

Crandall, V.C. Reinforcement effects of adult reactions and non-reactions on children's achievement expectations. *Child Development*, 1963, *34*, 335-354.

Crandall, V.C. Sex differences in expectancy of intellectual and academic reinforcement. In C.P. Smith (Ed.), *Achievement-related motives in children.* New York: Russell Sage, 1969.

Crandall, V.J. Achievement. In H.W. Stevenson (Ed.), *Child Psychology*, N.S.S.E., 1963.

Deaux, K. Sex: A perspective on the attribution process. In J.H. Harvey, W.J. Ickes, & R.F. Kidd (Eds.), *New directions in attribution research* (Vol 1). Hillsdale, N.J.: Lawrence Erlbaum Associates, 1976.

Dweck, C.S., Davidson, W., Nelson, S., & Enna, B. Sex differences in learned helplessness: (II) The contingencies of evaluative feedback in the classroom and (III) An experimental analysis. *Developmental Psychology*, 1978, *14*, 268-276.

Dweck, C.S. & Goetz, T.E. Attributions and learned helplessness. In J.H. Harvey, W.J. Ickes, & R.F. Kidd (Eds.), *New directions in attribution research* (Vol. 2). Hillsdale, N.J.: Lawrence Erlbaum Associates, 1978.

Edwards, D.W. Blacks versus whites: When is race a relevant variable? *Journal of Personality and Social Psychology*, 1974, *29*, 29-49.

Entwisle, D.R., & Hayduck, L.A. *Too great expectations: The academic outlook of young children.* Baltimore: Johns Hopkins University Press, 1978.

Frieze, I., Fisher, J., Hannsa, B.H., McHugh, M.C., & Valle, V.A. Attributions of the causes of success and failure as internal and external barriers to achievement in women. In J. Sherman & F. Denmark (Eds.), *Psychology of women: Future directions of research.* Psychological Dimensions, 1976.

Kessel, L.J. *Age and sex differences in causal attributions for mathematics and English achievement in adolescence.* Unpublished Ph.D. thesis, University of Illinois at Urbana-Champaign, 1979.

Kohlberg, L. A cognitive-developmental analysis of children's sex-role concepts and attitudes. In E.E. Maccoby (Ed.), *The development of sex differences.* Stanford: Stanford University Press, 1966.

Kukla, A. Foundations of an attributional theory of performance. *Psychological Review*, 1972, *79*, 434-470.

Kukla, A. An attributional theory of choice. In L. Berkowitz (Ed.), *Advances in Experimental Social Psychology* (Vol. 11). New York: Academic Press, 1978.

Maccoby, E.E. & Jacklin, C.N. *The psychology of sex differences.* Stanford: Stanford University Press, 1974.

Maehr, M.L., & Nicholls, J.G. Culture and achievement motivation: A second look. In N. Warren (Ed.), *Studies in cross-cultural psychology* (Vol. 3) New York: Academic Press, in press.

Markus, H. Self-schema and processing information about the self. *Journal of Personality and Social Psychology*, 1977, *35*, 63-78.

Neale, J. *An observational study of sex differences in pupil-teacher interactions.* Unpublished M.A. thesis, Victoria University of Wellington, 1978.

Nicholls, J.G. Causal attributions and other achievement-related cognitions: Effects of task outcome, attainment value and sex. *Journal of Personality and Social Psychology*, 1975, *31*, 379-389.

Nicholls, J.G. Effort is virtuous, but it's better to have ability: Evaluative responses to perceptions of effort and ability. *Journal of Research in Personality*, 1976, *10*, 306-315. (a)

Nicholls, J.G. When a scale measures more than its name denotes: The case of the Test Anxiety Scale for Children. *Journal of Consulting and Clinical Psychology*, 1976, *44*, 976-985. (b)

Nicholls, J.G. The development of the concepts of effort and ability, perception of academic attainment and the understanding that difficult tasks require more ability. *Child Development*, 1978, *49*, 800-814.

Nicholls, J.G. The development of perception of own attainment and causal attributions for success and failure in reading. *Journal of Educational Psychology*, 1979, *71*, 94-99.

Nicholls, J.G. *Quality and equality in intellectual development: The role of motivation in education.* Paper submitted for publication, 1979.

Parsons, J.E., Ruble, D.N., Hodges, K.L., & Small, A. Cognitive developmental factors in emerging sex differences in achievement-related expectancies. *Journal of Social Issues*, 1976, *32*.

Sohn, D. Affect-generating powers of effort and ability self-attributions of academic success and failure. *Journal of Educational Psychology*, 1977, *69*, 500-505.

Sutton, R., & Nicholls, J.G. *Classroom behavior and causal attributions for success and failure in ethnically mixed classrooms.* Paper presented at the meeting of the New Zealand Psychological Society, Auckland, 1977.

Valle, V.A., & Frieze, I.H. Stability of causal attributions as a mediator in changing expectations for success. *Journal of Personality and Social Psychology*, 1976, *33*, 579-587.

Veroff, J. Social comparison and the development of achievement motivation. In C.P. Smith (Ed.), *Achievement-related motives in children.* New York: Russell Sage Foundation, 1969.

Weiner, B. *Achievement motivation and attribution theory.* Morristown, N.J.: General Learning Press, 1974.

Weiner, B. A theory and motivation for some classroom experiences. *Journal of Educational Psychology*, 1979.

Witryol, S.L. Incentives and learning in children. In H.W. Reese (Ed.), *Advances in child development and behavior* (Vol 6). New York: Academic Press, 1971.

Witryol, S.L. Incentives and learning in children. In H.W. Reese (Ed.), *Advances in child development and behavior* (Vol 6). New York: Academic Press, 1971.

Chapter 13

Measuring Causal Attributions for Success and Failure*

TIMOTHY W. ELIG
The University of Pittsburgh

IRENE HANSON FRIEZE
The University of Pittsburgh

Although a great deal of research based on Weiner et al.'s (1971) model of causal attributions for success and failure events has been published during the 1970s, relatively little attention has been given to the question of how causal attributions should best be *measured* (Deaux & Farris, 1977; Smith, 1977). Although a few articles (e.g., Elig & Frieze, 1975; Frieze, 1976; Weiner, 1974; McHugh, 1975) have referred to the variety of measures used for assessing attributions, there has been no formal study of the implications of using one attribution measure over another, and researchers have tended to be unsystematic in their selection and use of the common techniques. This article explores the interrelationship of several measures of causal attributions to assess their validity and to make recommendations concerning the selection of instruments to be used in future studies of causal attributions for success and failure events.

As shown in Table 1, there are a number of commonly used techniques for assessing causal attributions: *open-ended*

---

*Reprinted by permission of authors and publisher of *Journal of Personality and Social Psychology*, 1979, 37, 621-634. Copyright (1979) by the American Psychological Association.

The authors would like to thank Grace French-Lazovik, Maureen McHugh, Diane Ruble, Eliot R. Smith, and Bernard Weiner for comments on earlier drafts of this article.

Portions of this article were presented at the meeting of the American Psychological Association in 1978 and served as part of the requirements for a master's degree for the first author.

responses, independent ratings, ipsative ratings, choice of one major cause, and *bipolar ratings.* Each of these methods has its own advantages and disadvantages in terms of practical considerations.

Most attribution studies use structured ratings rather than open-ended data. An open-ended procedure involves asking subjects to state in their own words why a particular event has occurred. These verbal responses can then be classified by a skilled rater into any of a set of previously defined attributional categories. Such a coding procedure is described more fully by Elig and Frieze (1975). Other systems have been developed by Bar-Tal and Darom (in press) and by Cooper and Burger (1978). The necessity for training coders and the time-consuming nature of this type of causal assessment contributes to its rare use in attribution research. However, there are also problems with structured response measures, which confine subjects to a limited set of factors defined in advance by the experimenter as important for the situation. This set may not include the factors of importance for some subjects. Open-ended response measures avoid this problem as well as the cuing of subjects toward considering causal possibilities that they may not have spontaneously considered (Frieze, 1976; Smith, 1977).

Weiner et al. (1971) postulated that individuals attribute the causes of success and failure primarily to four of the causal elements discussed by Heider (1958): ability, effort, task difficulty, and luck. Several studies using structured measures supported the belief that these factors are used by subjects in systematic ways to explain achievement outcomes (e.g., Frieze & Weiner, 1971; Weiner, Heckhausen, Meyer, & Cook, 1972; Weiner & Kukla, 1970). Frieze (1976) employed open-ended questionnaires to ascertain what causes college students naturally used to explain success or failure in two achievement tasks (an exam and an unspecified game). The results indicated that the four causal factors postulated by the Weiner et al. (1971) model were used by subjects for these situations and accounted for the large majority of causal attributions. However, two additional causal factors, mood and other people, were also indicated. Other recent studies (Elig & Frieze, 1975;

Table 1
Methods of Assessing Causes of Success and Failure

Method	Method used by	Example
Unstructured		
Open-ended	Elig & Frieze (1974, 1975) Frieze (1976) McHugh (1978)	Why do you think you *succeeded* on this task?
Structured		
1. Independent		
Unipolar ratings	Feather & Simon (1971b, 1972) Valle & Frieze (1976)	Rate the extent to which these factors caused your *success:* 1. If the factor to no extent caused the outcome. 9. If the factor to an extremely high extent caused the outcome. Your high general intelligence ____ How easy this type of task is ____
2. Ipsative		
Percentage assessment	Meyer (1970)	To what extent was your *success* caused by your high general intelligence ____ how easy this type of task is ____
Choice of one cause	Bailey, Helm, & Gladstone (1975)	Which of the following contributed the most to the outcome? Ability Effort
Bipolar ratings	Feather (1969) Feather & Simon (1971a) Weiner, Nierenberg, & Goldstein (1976)	My outcome was mainly due to: Ability—Luck
Paired comparisons	McMahan (1973)	Of each pair, circle which is more responsible for your outcome. Ability, luck Ability, effort

[a] Measured on a 5-point scale.

McHugh, 1978) that used open-ended questions for a variety of of social and achievement situations found still other causal factors that are frequently used to explain success and failure. Among those are personality, interest in the task, and physical appearance. This research supports the importance of the four causes proposed originally but suggests the importance of other factors not previously considered, especially for dealing with nonacademic situations. These results raise serious questions about studies relying only on the four causal factors of ability, effort, luck, and task difficulty in a structured format.

An additional advantage of open response assessment is that subjects may find open response questions easier and more natural to respond to. However, even though it is hypothesized in this study that subjects will prefer open-ended questions, it is also hypothesized that open-ended questions will be psychometrically inferior to more structured responses. First, the added step of coding the unstructured responses should lead to lower reliability for open-ended questions than for structured measures. Second, structured measures provide a closer approximation to interval or ratio measurement. Third, structured measures allow for degrees of attribution on various dimensions rather than the simple presence-absence or frequency of appearance measures typical for open-ended response coding.

A major distinction between various structured attribution measures is whether they involve ipsative or independent judgments. Ipsative measures are measures in which the score of one attribution must influence the score of other attributions, thus inducing negative correlations. Negative correlations are not forced by measures using independent judgments. With independence of ratings comes ease of analysis, since each attribution can be tested separately. However, independent ratings do not give as direct an assessment of the relative importance of attribution factors as is given by ipsative judgments. This ease of comparison is the major attraction of ipsative measurement.

Among the ipsative measures, the assignment of percentages to various causes is perhaps the best developed. Percentage ratings (e.g., Meyer, 1970) make explicit the basic assumption of interdependent judgments that the causes being rated account for the totality of cause for the outcome and that the total cause of an event can be parceled out to various particular causes. McHugh (1975) pointed out that percentage ratings give the clearest indication of the relative importance of various causes (e.g., luck rated relative to other causes seems unimportant.)

Another widely used method of causal assessment, the single

bipolar scale anchored by ability and luck used by Feather (1969) and Feather and Simon (1971a), does not allow for even the four causal factors identified by Weiner et al. (1971). It measures only two of the possible causes of events. This measure also confounds two dimensions that have been shown to be important for attributions: stability and locus of control (Weiner et al., 1971). However, multiple bipolar scales could be used in such a way as to overcome the particular problems of using a single bipolar scale (Weiner, Nierenberg, & Goldstein, 1976). Weiner et al. (1976) limited comparisons on each scale to a difference within only one dimension (e.g., luck and ability are not on a scale since they differ on both the stability and the location of cause dimensions). Such a limitation solves the theoretical problem of which dimension the causes were judged on as well as easing the practical problem of the very large number of comparisons needed in a paired comparison of many attributions. Within-dimension bipolar scales (Weiner et al., 1976) yield more sensitive measures than paired comparisons used by McMahan (1973), since the range of possible scores is increased by the use of scales.

Among these structured response measures, separate rating scales and percentage measures appeared to be the most interesting measures. These were selected for further study along with an unstructured, open-ended measure.

A few researchers have begun to question the indiscriminate use of the various attribution measures. (Deaux & Farris, 1977; Smith, 1977). Deaux and Farris analyzed a variety of measures by looking at differentiation across success and failure and at correlations across measures. They found some convergence, but they also found significant difference between what the various scales measure. This study extends their analysis by using a multitrait-multimethod test of the reliability of three of the most commonly used measures: open-ended measures, scale ratings, and percentage judgments. In addition, a formal assessment is made about subjects' perceptions of the face validity of the various measures.

Hypotheses

Two specific hypotheses that can be made concerning

various attribution measures involve a comparison of the open-ended response measure with the two structured response measures: (1) Convergent and discriminant validities will be lower for the open-ended response measure than for either of the two structured response measures and (2) the face validity to subjects of the open-ended response question will be better than that obtained by either structured response measure. That is, subjects will rank the open-ended measure as better and easier to answer. Another hypothesis is made based on the apparent superiority of the independent scale judgments: (3) the independent ratings will be superior to the percentage ratings in terms of convergent and discriminant validities.

Method

Subjects

The subjects were 252 students in introductory social psychology and personality psychology classes. Students were requested to participate in a study that would serve as a focal point for class discussions of experimental and measurement procedures in psychology. Subjects were free to participate or not as they chose. Two of a total of 254 students elected not to participate because of their dislike of anagrams.

Attribution Measures

Three types of attribution measures were used. The measures were worded separately for success and failure events, and subjects were directed to answer only the questions for their own outcome. The first attribution measure was always an open-ended question asking the subject, "Why do you think you *succeeded (failed)* on this task?" The other attribution measures were structured response measures.

Eight causal attributions were assessed in the structured response measures. These included the four basic Weiner et al. (1971) causes: high/low general intelligence, task ease/difficulty, good/bad luck, and high/low unusual effort. These were selected for theoretical reasons and because of their

Table 2
Measured Causal Attributions Varying According to Stability, Locus of Control, and Intentionality of the Attribution

Stability	Locus of control	
	Internal	External
Stable		
Unintentional	1. Your high (low) general intelligence.	2. How easy (difficult) (difficult) this type of task is
Intentional	5. The high (low) level of effort which you consistently demonstrate in whatever you do 8. Your high (low) desire to do well in everything you do	
Unstable		
Unintentional	7. Your good (bad) mood	3. Good (bad) luck
Intentional	4. How high (low) your interest in this task was 6. The unusually high (low) effort which you put forth in doing this task	

Note. Numbers indicate the order in which the attributions were presented. Failure wordings are in parentheses.

frequent use in the literature. The other four attributions were high/low stable effort (stable = consistent), high/low task interest, good/bad mood, and high/low motivation. These have been suggested as important causal factors for this type of situation by open-ended studies (Elig & Frieze, 1975; Frieze, 1976). The attributions were worded in the format suggested by Elig and Frieze (1975) and by Valle and Frieze (1976).[1] Table 2 presents these causes in a three-dimensional taxonomy (cf. Elig & Frieze, 1975; Rosenbaum, 1972).

Two structured response measures were used: unipolar 9-point rating scales and percentage ratings (see Table 1). Both of these were administered with two alternate forms (1) *determination* (e.g., "Please rate (indicate) how important you think each of the following factors was in determining your success or failure on the anagram task") and (2) *cause* (e.g., "Please rate (indicate) the extent to which each of the following factors caused your success or failure on the anagram task"). Each of these was then followed by instructions to respond to a set of

---
[1] Another issue, not dealt with in this article, is the way in which attributions are worded. This also needs to be systematically analyzed.

success or failure causes. Exact wording of the causes is shown in Table 2 for the success and failure conditions. Structured response measures were (a) rating scale of determinance, (b) percentage scale of determinance, (c) rating scale of causality, and (d) percentage scale of causality. With the open-ended measure, this resulted in a total of five different attribution measures.

The two forms were needed for alternate-form reliability estimates for the scale and percentage methods of assessing causal attributions. Reliability estimates for the open-ended measure were based on intercoder reliability. Since these causal attributions represent a causal analysis of a particular success-failure event, other forms of reliability estimates such as test-retest or coefficient alpha were not appropriate.

Subjects were given the four structured attribution measures and one unstructured attribution measure in a within-subject design. Sixteen orders of the four structured measures were used in a $2 \times 2 \times 2 \times 2$ variation of (a) whether the two measures using the same method were together or separate, (b) whether the scale or percentage method was asked first, and (c and d) whether the determination form or the causality form was asked first for the percentage method and for the scale method. To an unkown extent, this within-subject design will lead to an inflation or intermethod correlation estimates. Subjects may strive for consistency in their responses even when they are instructed to start each measure afresh and not to look forward or back in the test sheets. Such a problem confronts any assessment of a momentary state rather than of a stable trait. However, since our interest is in the pattern of correlations rather than in the overall absolute levels, this problem was not too troublesome. Varying order also allowed the assessment of order effects, but since no consistent order effects were found, the results for various orders are collapsed in the discussion of results.

Procedure

The experiment was administered during class time by a male experimenter in five classes. Subjects were told the following:

In a few minutes you will be asked to do a task consisting of 15 anagrams. An anagram is a group of scrambled letters that can be unscrambled to form a word...The brightest 25% of college students can solve at least eight of these anagrams. If you solve eight or more of these anagrams, you will have succeeded at the task; if you solve seven or less of the anagrams, you will have failed at this task...

Subjects were given 30 seconds to answer each of the 15 anagrams. The difficulty of the anagrams was manipulated so that half of the subjects received a set of mostly easy anagrams that led them to succeed (e.g., Mnegaa, Wadnet, Bolwe), and the other half of the subjects was given a set of mostly difficult anagrams that led them to fail (e.g., Sealgt, Iumsc, Oolruc) (Bar-Tal & Frieze, 1977).

After completing the anagram task, subjects were asked to total the number of anagrams solved and to rate their subjective feelings of success and failure on a 9-point scale. Subjects were then given as much time as they needed to answer the attribution questions previously discussed and a final set of questions that assessed the face validity of the attribution measures. Subjects were asked for their general impressions of the methods and then ranked them for difficulty and for global impressions of best and worst.

Subjects answered the questions at their own pace; no subject required more than 45 minutes. In each class, when all students were finished, the purpose and general nature of the study was discussed. All subjects were debriefed on the deceptions involved in the presentation of the anagram task.

Results and Discussion

Although the majority of subjects responded to the success-failure manipulation of anagram difficulty, there were a few subjects who did not. Thirty-six subjects who were given easy anagrams failed to solve eight of them and therefore failed the success criterion. Seventeen other subjects made up words for enough of the unsolvable anagrams to score a "success." This raised a question about whether these 53 subjects should be included in the analyses. A preliminary analysis indicated that there were no systematic differences in the responses of these "erring" subjects as compared to the responses of the other subjects. Therefore, all subjects were included in the analyses listed below. Similar data were obtained when these 53 subjects were eliminated from the analysis.

Open-Ended Attributions

The responses to the open-ended attribution measure were coded according to the scheme developed by Elig and Frieze (1975). In this coding scheme, each subject's response is segmented into phrases containing individual causes, each of which can be scored for the specific attribution category and for the dimensions of stability, location of cause, and intentionality. The index of each attribution category (such as ability and task ease-task difficulty) used in this study is the number of times it was mentioned by the subject. Two coders coded all responses; the percentage of agreement for the codings ranged between 82.8% and 93.5%. Percentage agreement, although a useful concept, is not a reliability estimate. Reliability was estimated by using the correlations reported in Table 3.

Table 3 also lists the attributions found for the open-ended question and it shows that all but a small proportion of these attributions were also measured by the structured response measures. Two types of responses are not shown in Table 3. Sixty-one responses (16.4% of total responses) simply repeated the outcome or deemed the outcome. Almost all of these responses were from the subjects who had failed to follow instructions. Another 65 responses (17.5% of total responses) were uncodable as attributions. Most of these responses were from subjects who misunderstood the question as asking for the reasons why they labeled the outcome as a success or failure rather than asking for the causes of the outcome.

Interest Correlation Validity

Interest correlation validities (convergent and discriminant) were assessed in multivariable—multimethod matrices

Table 3
Open-Ended Responses

Attribution category	Equivalent structured measure	Intercoder reliability a	No. subjects mentioning	% codable responses b
Ability	Intelligence	.91	51	20.8
Task difficulty	Task difficulty	.99	65	26.5
Luck	Luck	1.00	2	.8
Intrinsic motives	Interest	.97	17	6.9
Stable effort	Stable effort	.57	38	15.5
Unstable effort	Unstable effort	.51	11	4.5
Mood	Mood	.91	34	13.9
Personality	—	.70	8	3.3
Task × Ability interaction	—	.74	19	7.8

[a] $n = 252$. These are correlations of the frequency of the attribution being coded for the subject.

[b] A total of 247 attributions were made that were codable; that is, they did not simply repeat the outcome nor explain the process of labeling the outcome a success or failure. One codable attribution was given by 102 subjects, whereas 64 subjects made more than one codable attribution.

(Campbell & Fiske, 1959; Magnusson, 1967). Since alternate forms (cause and determination) of the structured response methods were used, separate multivariable—multimethod matrices were prepared for each. Table 4 presents the causal forms of the structured method. The matrix for the determination forms was highly similar[2].

Factor analysis procedures were used in the interpretation of the validities. The factor-analytic solution chosen was principal components analysis with varimax rotation. This solution accounts for reliable true variance with independent factors. We were most interested in the variance common to two or more attribution measures but had to allow for specific factors to emerge. Table 5 gives the factor loadings of the factor analysis of the multivariable—multimethod matrix presented in Table 4.[3] This analysis accounts for 91% of the reliable variance.

---

[2] For purposes of brevity, much of the specific comparison data obtained has been omitted from this article. These data are available on request from the authors.

[3] Because the percentage measures are ipsative, one percentage measure had to be dropped from the analysis. Interest was dropped in the analysts reported here. Though

Table 4
Multivariable-Multimethod Matrix

	Percent								Scale								Open response						
Attributions	I	T	L	I	S	U	M	M	I	T	L	I	S	U	M	M	A	T	L	M	S	U	M
Percent																							
Intelligence	.91																						
Task E-D[a]	-.27	.91																					
Luck	-.14	-.17	.97																				
Interest	-.31	-.17	-.24	.90																			
St. effort[b]	.10	-.43	-.17	-.18	.70																		
Us. effort[c]	-.16	-.33	-.10	.12	.19	.79																	
Mood	-.17	-.29	.00	-.12	-.06	.04	.88																
Motivation	.04	-.36	-.14	-.24	.43	-.02	-.19	.90															
Scale																							
Intelligence	.58	-.38	-.08	-.35	.43	-.03	-.11	.45	.85														
Task E-D[a]	-.10	.49	-.18	-.19	-.04	-.15	-.10	-.10	.03	.74													
Luck	-.03	-.15	.63	-.27	-.07	-.05	.08	-.03	.07	.07	-.08												
Interest	-.07	-.31	-.19	.43	.17	.22	.03	.02	.07	.07	-.08	.77											
St. effort[b]	.22	-.42	-.03	-.24	.61	.03	-.04	.55	.66	.07	.08	.29	.74										
Us. effort[c]	.08	-.38	-.08	.06	.28	.43	.01	.15	.26	.06	.03	.52	.85	.82									
Mood	.10	-.34	.07	-.22	.13	.07	.53	.07	.26	.04	.25	.27	.52	.35	.89								
Motivation	.30	-.40	-.08	-.27	.54	-.01	-.08	.60	.63	-.00	.02	.24	.36	.87	.48	.36	.88						
Open response																							
Ability	.06	-.00	-.14	-.08	.08	.05	.06	.05	.10	-.05	-.12	-.03	.11	-.00	-.01	.06	.91						
Task E-D[a]	-.09	.25	-.00	-.13	-.02	-.03	-.01	-.14	-.17	.09	-.11	-.14	-.14	-.11	-.15	-.16	-.15	.99					
Luck	.00	-.07	.16	-.02	-.02	-.01	.02	-.02	.04	-.12	.10	-.04	-.03	-.08	.12	-.01	-.04	-.05	1.0				
Motives[d]	.04	-.04	-.09	.06	.06	-.01	-.03	.05	.13	.09	-.06	.17	.14	.09	.07	.13	.06	-.11	.14	.97			
St. effort[b]	-.05	-.01	-.08	-.01	.03	.06	-.08	.23	.00	.01	-.05	.07	.08	.11	-.06	.07	-.05	-.10	-.04	.21	.57		
Us. effort[c]	.01	.05	.05	-.01	.02	.00	-.09	-.03	.06	.10	.02	.03	-.01	.03	-.06	-.03	-.00	-.07	-.02	.02	.02	.51	
Mood	-.12	-.08	-.04	.00	-.07	.10	.37	-.06	-.16	-.10	-.03	.12	-.14	.03	.22	-.14	-.03	-.09	.08	.06	.00	-.02	.91

Note. Reliability estimates appear in the main diagonal. Italics indicate convergent validities. The cause alternate form is presented here for the percent and scale methods. (The determination alternate form is available from the authors.) The main diagonal is alternate form reliability for the structured methods and intercoder reliability for the open response method. $n = 252$.

[a] Task ease-difficulty
[b] Stable effort.
[c] Unstable effort.

Table 5
Factor Loadings of Cause and Open-Response Measures

Attribution/method	Factor								Commun-ality
	1	2	3	4	5	6	7	8	
Stable effort/scale	.89								.87
Stable effort/percentage	.69								.57
Motivation/scale	.88								.84
Motivation/percentage	.79				-.27				.77
Intelligence/scale	.70				.50				.80
Mood/percentage		.87							.79
Mood/scale	.29	.71							.76
Unstable effort/percentage			.67	-.40					.67
Mood/open		.61							.50
Unstable effort/percentage			.67	-.40					.67
Unstable effort/scale	.36	.78							.76
Interest/scale			.74					.63	
Luck/scale				.86					.76
Luck/percentage				.87					.80
Task/ease/difficulty/scale					.88				.78
Task ease/difficulty/percentage	-.44	-.35			.62			.83	
Intelligence/percentage	.39					.80			.78
Stable effort/open						-.53	.44		.56
Intrinsic motives/open							.73		.60
Luck/open							.57		.49
Task ease/difficulty/open								-.70	.60
Ability/open					-.28			.61	
Ability/open					-.28			.61	.55
Unstable effort/open								.48	.40

The main diagonal of the matrix (Table 4) gives the reliability estimates for attribution-method unit. For the structured methods, these are alternate form estimates; for the open response method, they are intercoder reliabilities. With few exceptions, these are satisfactorily high. The open response measure shows low reliabilities for the effort attributions and shows generally low communalities indicating a need for further refinement of coding instructions to achieve more uniform, acceptable reliabilities for this measure.

The monomethod triangles (under the reliability diagonals) show the intercorrelations of the attributions measured the same way. It is obvious that the open response measure shows

---

this is not a totally satisfactory solution, it does eliminate the singularity program. Interest is included in other analyses, which are available from the authors.

the smallest (in absolute value) intercorrelations, indicating either relatively good independence for the attributions measured or a poor representation of the true interrelationships of the attributions. The true correlations may be attenuated by the small range of frequencies of the open response attributions.

The monomethod triangle for the open response method is more similar to the triangle for the percentage than for the scale method. In part, this seems due to the quasi-ipsative nature of the method. No subject in this study gave more than four responses to the open-ended question; thus, once an attribution was made, there was less chance for any other attribution to be made also, inducing small negative correlations. Negative correlations were directly induced by the ipsative percentage measures. The relatively high positive intercorrelations of the attributions measured by the scale method may in part have been due to a response bias of some subjects to use only one part of the 9-point scale. This was a response bias that cannot influence the percentage method or the open response method. Method variance can thus be seen to influence the monomethod intercorrelation pattern. The percentage method and the open response method seem to present a pattern different from the one evident in the scale measures.

Factor analyses can aid us in clarifying these patterns. Returning our attention to Table 5, we find the major evidence for method variance in the factor loadings of the open response measures, which form distinct method factors (Factors 7 and 8). However, with the exception of the intelligence attribution, the structured measures of a specific attribution both load on the same factor (see Footnote 3). Also, the monomethod blocks of the structured methods can be factor analyzed separately to clarify their patterns of correlations and the role of method variance. (The open response measure cannot be factor analyzed within method, since there is no alternate form for it as there are for the structured measures.) Tables 6 and 7 present the factor analyses of percentage and scale measures, respectively. The factors in these analyses account for 97% and 100% of the reliable variance of the percentage and scale measures, respectively. The strongest similarity between the percentage

Table 6
Factor Loadings of Percentage Measures

Attribution-alternate form	1	2	3	4	5	Commun-ality
Motivation-cause	.88					.81
Motivation-determination	.87					.79
Stable effort-determination	.65					.49
Stable effort-cause	.72					.61
Task ease/difficulty-cause	-.55	-.39	-.40	-.41	-.33	.89
Task ease/difficulty-determination	-.55	-.39				
	-.55	-.41	-.37	-.36	-.36	.86
Intelligence-cause		.95				.93
Intelligence-determination		.95				.93
Unstable effort-cause			.91			.84
Unstable effort-determination			.91			.85
Mood-determination				.95		.92
Mood-cause				.95		.92
Luck-determination					.97	.96
Luck-cause					.97	.96

Note. Interest was dropped from this analysis because of the ipsative nature of percentage measures.

Table 7
Factor Loadings of Scale Measures

Attribution-alternate form	1	2	3	4	5	Commun-ality
Motivation-cause	.880	.25				.86
Motivation-determination	.870	.26				.84
Stable effort-cause	.886	.29				.85
Stable effort-determination	.840	.31				.82
Intelligence-determination	.840					.73
Intelligence-cause	.840					.73
Interest-determination		.82				.73
Interest-cause		.85				.71
Unstable effort-cause	.360	.77				.74
Unstable effort-determination	.390	.71				.69
Mood-cause			.91			.94
Mood-determination			.91			.94
Luck-cause			.91			.89
Luck-determination				.92		.88
Task ease/difficulty-cause					.93	.87
Task ease/difficulty-determination					.93	.88

and scale monomethod triangles of the multivariable—multimethod matrix was that in each triangle, the largest correlation was for stable effort with motivation (motives in the open measure). These two attributions form the nucleus of the strongest factor (which appears to be a stable achievement factor) in all analyses. The analyses also agree to having mood and luck constituting independent factors.

There are some disagreements between the scale and percentage measures. Factor 1 includes task ease-difficulty for percentage measures and intelligence for scale measures. The factor analysis of the scale measures is more easily interpreted than that of the percentage measures. Factor 2 of the scale analysis can be called *unstable achievement*. Interest in a task should generate unusual effort leading to achievement. Interest and unstable effort correlate when measured by scale cause and scale determination, .52 and .43, respectively, indicating only 27% and 18% overlap in variance, respectively.

Tables 6 and 7 exhibit remarkably clear structures. The alternate forms of the structured methods have similar loadings, substantiating their interchangeability.

Returning our attention to Table 4, we note the convergent validities. The convergence of the two structured response methods seems high enough to warrant our further attention. The range of these validities in .43 to .63 for cause measures and .30 to .69 for determination measures. For attributions measured by cause, there is a mean overlap of 31.1% in variation of the structured methods, whereas the mean is 30.4% for the determination measures.

The convergence of the two structured methods can be clearly seen in Table 5. In this analysis, the scale and percentage measures of an attribution have their highest loading on the same factor except for intelligence. The two structured measures of intelligence converge to the point of loading on the same two factors. (Interest measured by percentages was not in the analysis; see Footnote 3.)

The convergent validities of the open response measure with the structured measures are woefully low (ranging from .00 to .37). These figures indicate that two studies claiming to measure the same attribution, one with a structured measure and one with the open response measure, will only have an average of 4.7% of the variance in common (and a maximum of 14%), with the remainder of the variance being due to noncommon factors (method and error). Mood is the only open response measure to show any amount of convergence. It correlated .37 with the percentage measure and .22 with the scale measure. Mood is the only open response measure to load on the same factor as its structured equivalent (see Table 5).

On either side of the convergent validity diagonals lie heterovariable—heteromethod triangles. Correlations in these triangles are of variables having neither trait nor method variance in common. From this consideration follows the second type of validity, discriminant validity. Discriminant validity is shown when a validity diagonal value is higher than the values lying in its column and row in the heterovariable—heteromethod triangles. For example, attributions to intelligence measured in percentages should correlate more highly with intelligence measured by a scale than they correlate with any other attribution measured by a scale.

The structured responses discriminate fairly well between variables. For the cause form, no convergent validity estimate is exceeded by a heterovariable—heteromethod value. In the determination form, only task ease—difficulty is problematic. Its convergent validity is exceeded by four other values. This relative lack of discrimination showed up in the factor analysis of the determination form (Table 6) by task ease—difficulty loading somewhat on every factor.

The discrimination shown by the open response method is not nearly as good. Only mood always discriminated. The other attributions failed to show discrimination an average of 4.7 times out of 13 comparisons. It seems reasonable to blame the open response rather than the structured response measures for this lack of discrimination, since good discrimination was shown between the structured measures themselves.

As predicted, results indicate some superiority of structured measures. Structured response reliabilities are higher and are substantiated by communalities, whereas the open response measure intercoder reliabilities are not substantiated by communalities. Convergent and discriminant validities for structured measures are satisfactory, whereas open response convergent validities are low (except for mood). Except for mood, other open response convergence values are exceeded by an average of one third of the values in their row and column of the heterovariable—heteromethod triangles, indicating poor discrimination.

Face Validation

The second hypothesis concerns the reactions of those subjects whose attributions were being measured. This hypothesis states that to subjects, the face appearance of the open response measure will be more positive than that of either structured response measure. Table 8 presents the frequency with which the subjects termed each method easiest, hardest, best, and worst. As can be seen, subjects differed in ranking the methods on all criteria. Contrary to the hypothesis, however, the differences are not between the open response method compared with structured methods. There were no significant differences between the open response and scale methods; even the largest difference between these methods on the evaluation of best is not significant, $\chi^2(1) = 2.4$. Differences arose from a general disliking of the percentage method. Subjects said they felt that the percentage measure was hard to compute and was not the best reflection of what they felt were the reasons for the outcome.

Table 8
Frequency of Attribution Measurement Methods Selected as Easiest, Hardest, Best, and Worst

	Method			
Evaluation	Open response	Scale	Percent	X^2
Easiest	76	79	47	9.282**
Hardest	52	49	97	21.909***
Best	77	58	47	7.595*
Worst	45	51	89	18.462***

*$p < .05$.
**$p < .01$.
***$p < .001$.

Outcome Effects

The final analysis concerned the variations across the methods for success as compared with failure conditions. Results of the analysis of variance for each attribution are presented in Table 9. As can be seen, success was more attributed than failure to ability, stable effort, and motivation for all four structured measures. These outcome effects were not found for the open response measures. There was convergence of the open response measure and both forms of the

Table 9
Mean Attributions for Success and Failure

	Structured response measure				
	Percentage		9-point scale		Open response measure
Outcome and attribution	Cause	Determination	Cause	Determination	
Success					
High intelligence	22.10***	21.41***	5.42***	5.54***	.19
Task ease	17.02***	16.55***	5.14	5.27	.15**
Good luck	8.47	8.04	2.97	3.06**	.02
High interest	11.71***	11.24***	4.86	5.08	.11*
High stable effort	7.50	7.76	4.57***	4.91III	.05
Good mood	6.60	6.80	4.16III	4.32III	.09
High motivation	14.35***	14.90***	6.21***	6.02***	—
Personality	—	—	—	—	.03
Task-ability match	—	—	—	—	.05
Failure					
Low intelligence	8.89	9.72	2.31	3.38	.21
Task difficulty	32.27	32.63	5.02	4.90	.34
Bad luck	7.64	7.33	2.33	2.30	.00
Low interest	22.50	21.77	4.51	4.96	.03
Low stable effort	4.85	4.18	2.08	2.36	.15
Bad mood	9.13	9.37	2.65	2.74	.16
Low motivation	5.38	5.49	2.11		
Low motivation	5.38	5.49	2.11	2.37	—
Personality	—	—	—	—	.03
Task-ability mismatch	—	—	—	—	.10

Note. Significant difference between success and failure attributions are indicated on the success means.
*$p < .05$.
**$p < .01$.
***$p < .001$.

percentage method (but not the scale method) in that both had greater attributions to low interest and task difficulty for failure than to high interest or task ease for success. Both scale measures also yielded an outcome effect such that success was more attributed to luck, unstable effort, and mood.

Conclusions

As predicted, open-ended response measures of causal attributions have poorer interest validity and reliability than structured response measures. However, contrary to prediction, structured scale measures were seen by subjects as easy to respond to, like the open-ended measures. Percentage ratings were seen by these subjects as having the least face validity. Thus, one of the postulated advantages of the percentage measure was not confirmed. And a supposed weakness of scale measures was not found. Thus, at least for college students solving anagrams or doing similar tasks in which the basic causal categories are well understood, *the scale method is clearly a superior technique.* This strong support for the scale method of independent assessment of the contribution of various causal factors also confirms findings of Deaux and Farris (1977), who used other criteria for their conclusions. They also found little support for the percentage method.

Although open-ended procedures are weaker on the basis of psychometric criteria than scale ratings, they have utility for the researcher who is asking for causal attributions in a new situation. Clearly, people use different categories of causal explanations in different settings (e.g., Elig & Frieze, 1975), and without open-ended pretesting, the experimenter cannot know which causal factors to include in structured measures. Using an open-ended question in later stages of research can serve as a continuing validation of the attribution scales that subjects are asked to rate.

Data from the factor analyses and the success—failure comparisons suggest that open-ended and scale and percentage ratings not only vary in their psychometric properties but also yield different types of data. Hypotheses that concern particular causal attributions can be supported by one type of measurement and disconfirmed by another; this was demonstrated in the outcome effects reported here.

The attributional measures discussed in this article have been the measures typically used and are representative of three general methods of attribution measurement. Other methods and other within-method variations are possible (e.g., "How much ability did he have?" vs. "How much was ability a cause of his success?"). Many of these possible variations have been used in attribution research. As indicated in this article, these measurement variations may account for some of the discrepancies in the attribution literature.

Fiske (1971) has suggested that in cases in which multiple measures yield conflicting results, the construct be operationally defined by a single measure. These are multiple criteria for selection of a measure, but many of them would lead to the use of the scale measure for attribution research. Results reported here indicate that scale measures have moderately good intermethod correlations with percentage measures, do not force intercorrelations among attributions, and have good face validity. Another important criterion for selecting the specific measurement technique that defines the construct is the construct validity of various procedures. Other data not reported here (see Elig & Frieze, 1978) indicate that the scale method used in this study provides generally better support for some of the basic theoretical relationships between causal attributions and future expectancies and affect than do either the percentage or open response methods. Even if attribution researchers do not agree on a single measurement technique, these reliability and validity issues should be carefully considered before selecting the measures to be used in future investigations of causal attributions.

REFERENCES

Bailey, R. C., Helm, B., & Gladstone, R. The effects of success and failure in a real-life setting: Performance, attribution, affect, and expectancy. *Journal of Personality*, 1975, *89*, 137-147.

Bar-Tal, D., & Darom, E. Pupils' attributions for success and failure. *Child Development*, in press.

Bar-Tal, D., & Frieze, I. H. Achievement motivation for males and females as a determinant of attributions for success and failure. *Sex Roles*, 1977, *3*, 301-313.

Campbell, D. T., & Fiske, D. W. Convergent and discriminant validation by the multitrait—multimethod matrix. *Psychological Bulletin*, 1959, *56*, 81-105.

Cooper, H. M., & Burger, J. M. *Categorizing open ended academic attributions: Replication of earlier findings.* Paper presented at the annual meeting of the American Psychological Association, Toronto, August, 1978.

Deaux, K., & Farris, E. *Causal attributions for performance: Approaching substance via method.* Unpublished manuscript, Purdue University, 1977.

Elig, T. W., & Frieze, I. H. A multi-dimensional coding scheme of causal attributions in social and academic situations. *Personality and Social Psychology Bulletin*, 1974, *1*, 94-96.

Elig, T. W., & Frieze, I. H. A multi-dimensional scheme for coding and interpreting perceived causality for success and failure events. JSAS *Catalog of Selected Documents in Psychology*, 1975, *5*, 313. (Ms. No. 1069)

Elig, T. W., & Frieze, I. H. *Construct validity of causal attributions for success and failure.* Manuscript in preparation, University of Pittsburgh, 1978.

Feather, N. T. Attributions of responsibility and valence of success and failure in relation to initial confidence as task performance. *Journal of Personality and Social Psychology*, 1969, *13*, 129-144.

Feather, N. T., & Simon, J. G. Attribution of responsibility and valence of outcome in relation to initial confidence and success and failure of self and other. *Journal of Personality and Social Psychology*, 1971, *18*, 173-188. (a)

Feather, N. T., & Simon, J. G. Causal attributions for success and failure in relation to expectations of success based upon selective or manipulative control. *Journal of Personality*, 1971, *39*, 527-541. (b)

Feather, N. T., & Simon, J. G. Luck and the unexpected outcome: A field replication of laboratory findings. *Australian Journal of Psychology*, 1972, *24*, 113-117.

Fiske, D. W. *Measuring the concepts of personality.* Chicago: Aldine, 1971.

Frieze, I. H. Causal attributions and information seeking to explain success and failure. *Journal of Research in Personality*, 1976, *10*, 293-305.

Frieze, I. H., & Weiner, B. Cue utilization and attributional judgments for success and failure. *Journal of Personality*, 1971, *39*, 591-606.

Heider, F. *The psychology of interpersonal relations.* New York: Wiley, 1958.

Magnusson, D. [*Test theory*] (H. Mabon, trans.). Reading, Mass.: Addison-Wesley, 1967.

McHugh, M. C. *Sex differences in causal attributions: A critical review.*

Paper presented at the annual meeting of the Eastern Psychological Association, New York, April 1975.

McHugh, M. C. *Attributions of causality made by alcoholics and nonalcoholics for success and failure in selected situations.* Unpublished manuscript, University of Pittsburgh, 1978.

McMahan, I. D. Relationships between causal attributions and the expectancy of success. *Journal of Personality and Social Psychology,* 1973, *28,* 108-114.

Meyer, W. U. *Selbstverantwortlichkeit und Leistungsmotivation.* Unpublished doctoral dissertation, Ruher Universitat, Bochum, Germany, 1970.

Rosenbaum, R. M. *A dimensional analysis of the perceived causes of success and failure* (Doctoral dissertation, University of California, Los Angeles, 1972). *Dissertation Abstracts International,* 1973, *33,* 5040B. (University Microfilms No. 73-10,475)

Smith, E. R. *On the concept and measurement of cause in studies of causal attribution.* Paper presented at the annual meeting of the American Psychological Association, San Francisco, August 1977.

Valle, V. A., & Frieze, I. H. Stability of causal attributions as a mediator in changing expectations for success. *Journal of Personality and Social Psychology,* 1976, *33,* 579-587.

Weiner, B. Achievement motivation as conceptualized by an attribution theorist. In B. Weiner (Ed.), *Achievement motivation and attribution theory.* New York: General Learning Press, 1974.

Weiner, B., et al. *Perceiving the causes of success and failure.* New York: General Learning Press, 1971.

Weiner, B., Heckhausen, H., Meyer, W., & Cook, R. E. Causal ascriptions and achievement behavior: A conceptual analysis of effort and reanalysis of locus of control. *Journal of Personality and Social Psychology,* 1972, *21,* 239-248.

Weiner, B., & Kukla, A. An attributional analysis of achievement motivation. *Journal of Personality and Social Psychology,* 1970, *15,* 1-20.

Weiner, B., Nierenberg, R., & Goldstein, M. Social learning (locus of control) versus attributional (causal stability) interpretations of expectancy of success. *Journal of Personality,* 1976, *44,* 52-68.

Chapter 14

Alleviating Learned Helplessness in a Wilderness Setting: An Application of Attribution Theory to Outward Bound

RICHARD S. NEWMAN
The University of Michigan

Introduction

A therapeutic intervention for the alleviation of learned helplessness is discussed in terms of theories of causal attributions, self-concept development, and environmental psychology. Outward Bound, a wilderness training program, is presented as an example of this intervention.

It is proposed that there are three ways in which deficits of learned helplessness might be reduced in a wilderness setting. First, there are opportunities for acquiring skills, mastery, and a sense of competence or controllability in specific areas. Second, there is environmental facilitation of clear and realistic patterns of causal attributions and expectations. Third, there are opportunities for perceptions of competence to positively affect one's self-concept and self-esteem. An attempt is made to provide a rationale, or a framework, that explains how the "ideal" Outward Bound process might serve as a model for the alleviation of learned helplessness.

Learned Helplessness and Causal Attributions

The phenomenon of learned helplessness was first studied systematically in animals by Seligman and Maier (1967). They found that animals who were pretreated with unavoidable shock later failed to escape that outcome, even when the trauma could have been easily avoided by their performing some simple response that was originally in their repertoire. The authors proposed that these animals had learned that *their responding had no effect* on whether there would be termination or continuation of the shock, i.e., there was an absence of contingency between response and outcome. A review of the many animal studies and an account of the transition from these to a formulation of learned helplessness in humans are found in Seligman (1975).

Seligman (1975) claimed that learning response-outcome independence leads to a generalized sense of futility in responding. He found that there are cognitive, motivational, and emotional deficits associated with this phenomenon and that often fear concerning the uncontrollability of an aversive outcome gives way to psychopathological depression.

A less clinical and more school-related orientation to learned helplessness has been adopted by Dweck and her associates (see Dweck and Goetz, 1978). Specifically, they have found that learned helplessness can develop when children perceive the termination of failure in achievement situations to be independent of their responding. Dweck and Reppucci (1973) have contrasted learned helplessness with *mastery-orientation*. Learned helpless children, upon encountering academic failure, become incapable of solving the same problems they easily solved only shortly before the exposure to failure. The mastery-oriented students, on the other hand, persist in the face of prolonged failure. These two groups of children can be distinguished according to their attributional patterns, i.e., how they explain their successes and failures.

Most of the recent work on learned helplessness in academic settings has stressed attributional mediation as explaining why actions and outcomes are perceived to be independent. Interest in a learned helplessness model of depression has recently given rise to an attributional framework and reformulation of Seligman's original work (for a review, see Abramson, Seligman, and Teasdale, 1978). For the purpose of this chapter, it is not necessary to distinguish the clinical orientation of learned helplessness from the academic or achievement orientation.

Most developmental work from an attributional perspective has been focused on intellective functioning and causal schemata (for a review, see Weiner and Kun, 1976). However,

Weiner and his associates are now considering aspects of motivational and emotional development in achievement-related contexts from the same perspective (see Weiner, Kun, and Benesh-Weiner, 1979; Weiner, Russell, and Lerman, in press; and Weiner, 1979).

An attributional link between cognition and emotion seems well-suited for an investigation of the deficits of learned helplessness, i.e., deficits related to cognition, motivation, emotion, and possibly self-esteem. It is these deficits and their causes, according to an attributional framework, that need to be addressed in attempts at therapeutic and preventative intervention (see Abramson et al., 1978).

Because of the cognitive basis of causal schemata and the attribution process (see Kelley, 1972), it is natural to first look in the "cognitive direction" for clues to the alleviation of learned helplessness.

According to an Expectancy-Value theory of motivation (see Atkinson, 1964), the intensity of motivation is determined jointly by the expectation for success and by the anticipated affect or incentive value of the goal. In turn, causal attributions for success and failure have been shown to influence both the expectancy of success and the affective consequences of achievement performance. And so, it has been reasoned and shown, that because of their influence on expectancy and affect, causal attributions influence various motivational indices, e.g., persistence of behavior (see Weiner, 1974).

According to Weiner (1979), it is essential to create a classification scheme of causes, in order to identify their underlying properties. These underlying properties or dimensions have various psychological functions or consequences. Weiner's new scheme of causal attributions includes three dimensions: *stability*, *locus of causality*, and *controllability*. He also acknowledges a fourth possibility, *globality*, which is included in the attributional scheme of Abramson et al.'s (1978) learned helplessness model of depression.

These attributional dimensions will now be discussed in terms of the psychological consequences related to the deficits of learned helplessness—deficits in cognition, motivation, emotion, and possibly self-esteem.

It is the negative cognitive set which accounts for the "learned" in "learned helplessness." In other words, learning response-outcome independence interferes with later perception of response-outcome dependence. Seligman (1975) has noted that this is simply a case of proactive interference. Since cognitive deficits are a "given" in learned helplessness, they are not differentially linked to one dimension of attribution or another.

Attributional Dimension of Stability

The primary conceptual linkage of the stability dimension is with expectancy for success. Attributions of an outcome to stable factors, e.g., ability or task difficulty, produce greater increments in expectancy after success and greater decrements after failure than do attributions to unstable causes, e.g., mood or effort.

Persistence behavior has been widely used to support the Expectancy-Value theory of achievement motivation (see Feather, 1962). Referring to persistence and expectancy of success, Weiner (1979) stated:

> It is suggested that resistance to extinction is a function of attributions to the causal dimension of stability during the period of nonreinforcement.

Weiner and his associates (see Weiner, 1979) have demonstrated clearly that expectancy changes are related to the dimension of stability, and not to locus of causality, as was once thought to be the case.

It is persistence, or motivation in the face of failure, that is considered the major behavioral contrast between mastery-orientation and learned helplessness in achievement situations (Dweck and Goetz, 1978). Dweck and Reppucci (1973) and Andrews and Debus (1978) have demonstrated the self-defeating attributional schemata of the learned helpless: failure being ascribed to stable causes, e.g., lack of ability. This is in contrast to the pattern of unstable causes, e.g., lack of effort, on the part of mastery-oriented children.

It is this difference in effort versus ability attributions for failure that has been the conceptual basis for attempts to retrain attributions among the learned helpless in achievement-situations. For example, Dweck (1975) showed the efficacy of training failure-prone children to attribute failure to lack of effort, instead of lack of ability.

Abramson et al. (1978) have formulated expectancy into their model and refer to it as chronicity, also governed by the stability dimension of attributions. "Helplessness is called *chronic* when it is either long-lived or recurrent and *transient* when short-lived and nonrecurrent." In summary, motivational deficits which characterize learned helplessness (Seligman, 1975) are influenced by the stability of one's causal attributions.

In addition to this primary linkage of the stability dimension, Weiner, Russell, and Lerman (in press) have speculated on its secondary linkage to esteem-related emotions. The attributional dimension of stability seems to have affective implications of hopelessness versus optimism. This "suggests that only attributions conveying that events will not change in the future beget feelings of helplessness, giving up, and depression" (Weiner, 1979).

There is often a cyclical or self-reinforcing relationship between self-concept and one's pattern of attributions along this dimension of stability. A person who feels competent will often ascribe failure to unstable causes, and not to lack of ability. This will not reduce subsequent expectancy of success and the person will sustain a high ability self-concept. On the other hand, success is expected and will probably be ascribed to stable ability: this increases subsequent expectancy of success and also reinforces one's self-concept.

The converse type of cycle occurs with an individual with a low self-concept of ability. Failure will be ascribed to a stable factor, e.g., low ability, and success will be a "surprise" and will be ascribed to an unstable factor, e.g., luck or unexpected good will on the part of an instructor. This latter cycle is characteristic of learned helplessness.

Attributional Dimension of Locus of Causality

Rotter's (1966) construct of *locus of control,* measuring

internality versus externality of responsibility for an outcome, has recently been re-examined because of confusion over its meaning (see Weisz and Stipek, in press). Embedded within this construct are two others which should be distinguished from each other and both considered as distinct dimensions of causal attributions: they are (a) *perceived contingency* or *causality* and (b) *perceived competence* or *personal control*.

Locus of causality refers to perceptions of responsiveness of an outcome to variations in actors' attributes or behavior. The actors referred to are all one's peers (including oneself) or "relevant others" (Weisz and Stipek, in press).

Weiner et al. (1971) initially postulated a simple dimensional link: locus of causality was related to affective or emotional consequences of success and failure. Affective reactions were thought to be maximized given internal attributions for success, e.g., feelings of pride, and for failure, e.g., feelings of shame. These reactions were thought to be minimized given external attributions.

Recently though, Weiner (1979) and Weiner, Russell, and Lerman (in press) have described a more complicated linkage between attributions and emotional reaction. The dimension of locus of causality still is seen as having the primary association with affect, but there are other linkages also.

One level of affective response is due simply to the outcome, per se. Success elicits certain affects, e.g., happiness; failure elicits others, e.g., sadness. This type of emotional response is supposedly the most short-lived, yet the most intense, of all affects in achievement contexts.

The second level of response is due to specific causal attributions, some of which are delineated in Weiner (1979).

The third level is linked to dimensions of causal attributions, e.g., stability and locus of causality. This type of response is suspected of having greater longevity than the other two, because of implications for how one views oneself, i.e., these are self-esteem related emotions.

Locus of causality enters the picture here as a major determinant of these responses. The affects for achievement success that are most intimately associated with self-esteem, e.g., competence, confidence, and pride, are associated with internal attributions (ability, effort, and personality). Internal attributions for failure tend to lead to feelings of shame and guilt and a sense of resignation and aimlessness.

On the other hand, external attributions do not seem to influence affective response as strongly. External attributions for success, e.g., ease of task difficulty or use of a crutch of some sort, seem to only occasionally lead to feelings of guilt. External ascriptions for failure, e.g., excessive difficulty or some condition in nature, minimize shame.

Emotional or affective deficits, and sometimes poor self-esteem, are characteristic of learned helplessness. Abramson et al. (1978) speculate that internality of attributions for failure tends to make affective and self-esteem consequences more severe or intense. They also specify that for depressed affect to emerge from learned helplessness, there necessarily must be expectation of response-outcome independence about either the loss of a highly desired outcome or about the occurrence of a highly aversive outcome.

They do differentiate between affective and self-esteem consequences of attributions, which Weiner (1979) does not do. In the Abramson et al. model, self-esteem consequences are not explained by (primarily) the locus of causality and (secondarily) the stability dimension of attributions, which is Weiner's contention. Instead, self-esteem deficits are explained by the *personal versus universal* nature of the learned helplessness, and this distinction is intimately associated with the interplay or contrast between the dimensions of locus of causality and controllability, which is discussed next.

Attributional Dimension of Controllability

Controllability is the second of the two components of Rotter's (1966) original construct of locus of control. Weisz and Stipek (in press) refer to this dimension as *perceived competence*. It is personal control, concerning the individual actor (excluding his peers and relevant others) that is referred to as "perceived effectiveness in controlling one's environment—i.e., the capacity to produce desirable effects and avoid or reverse undesirable ones" (Weisz and Stipek, in press).

There is an important distinction between perceived competence and perceived contingency. One might perceive a task as contingent on a certain action, but personally perceive himself as unable to perform that action. The perception of contingency is based on what any relevant other is thought to be capable of. The perception of competence is strictly personal; it refers to one's own ability. This illustrates a case of *personal helplessness*.

On the other hand, if a person feels not only like he is incompetent at a task, but also that none of his peers could perform it either, then that person may see himself as *universally helpless*. Abramson et al. (1978) give examples of a father's efforts to find a cure for his child's leukemia, or an unemployed man looking for work in the midst of a nationwide economic crisis. Persons who are in this universal helplessness situation may suffer the same behavioral deficits as in personal helplessness, i.e., cognitive, motivational, and affective, but do not suffer the loss of self-esteem that accompanies personal helplessness.

Trying to distinguish between two common unstable and internal causes for failure, namely lack of effort and boredom, points to the need for this dimension of controllability. To most people, effort seems much more controllable than boredom. In this case, Abramson et al. speculate that the lack-of-effort attribution produces more self-blame, self-criticism, and guilt than the other because it pertains to a factor that is seemingly controllable.

Weiner (1979) links the controllability dimension with various experiential states, that have been explained with a variety of theories. According to attribution theory, there is pleasure gained in perceiving control over the factors affecting a successful outcome. Competence motivation theory (see Harter, 1978, and White, 1959) postulates pleasure coming from successful mastery attempts; the pleasure is in the form of feelings of efficacy or competence. Self-determination theory (see Deci, 1975 and de Charms, 1968) speaks of pleasure in perceiving control over one's behavior or over the context of that behavior, e.g., selection of tasks. According to de Charms (1968), there is pleasure in feeling like an *origin* (versus a *pawn*), i.e., feeling that what one is doing is the result of his own free choice.

Attributional Dimension of Globality

The last causal dimension to be considered here, globality, has been acknowledged by Weiner (1979) as a viable fourth. It has been formulated into Abramson et al.'s learned helplessness model of depression, but its "goodness of fit" in the model has not yet been well tested.

As an analog to the stability dimension's implications for temporal generalization of learned helplessness deficits, the globality dimension supposedly influences *stimulus generalization of deficits*. Global factors perceived as causes of failure tend to influence expectations of future helplessness over a broad, general range of situations. Specific attributions imply helplessness only in the original situation. Attributing failure to very specific and very unstable factors should predict little transfer of helplessness.

Tennen and Eller (1977) tried to manipulate attributions by labelling unsolvable problems as progressively "easier" or "harder." They reasoned that failure on "easy" problems would be attributed to lack of ability (internal, stable, and relatively global) whereas failure on the "hard" problems would be attributed to task difficulty (external, unstable, and more specific). Subjects then went on to solvable problems. Attribution to low ability (from the "easy" problems) was associated with cognitive deficits on these later tasks; attribution to task difficulty (from the "hard" problems) was associated with facilitation of later cognitive functioning.

Abramson et al. (1978) conclude from this that lack of performance deficits in the "difficult-task" group is explained by the specificity of failure attributions, that did not generalize to the later tasks. This is a speculative conclusion because of the failure to isolate the globality dimension from the others. More research needs to be done concerning the validity of using this intuitively-sound dimension in an attributional model. For the purposes here, it will be included as a possible explanation for the generality of learned helplessness deficits.

Alleviation of Deficits of Learned Helplessness

Abramson et al. (1978) conclude their review with this:

> Taken together, the studies examining depressives' attributions for success and failure suggest that depressives often make internal, global, and stable attributions for failure and may make external, specific, and perhaps less stable attributions for their success.

Depressive deficits in learned helplessness can be characterized as general, depending on the globality of the attributions for helplessness; chronic, depending on the stability of the attributions; and involving loss of self-esteem, depending on the locus of causality and the controllability.

Four general therapeutic strategies have emerged from the learned helplessness model of depression of Abramson et al. They are:

(1) *Change the estimated probability of the outcome.* Reduce the likelihood of aversive outcomes and increase the likelihood of desired outcomes.

(2) *Reduce the aversiveness of un-relievable outcomes and the desirability of unattainable outcomes.*

(3) *Change the expectation from uncontrollability to controllability* when the outcomes are attainable. This might entail training appropriate skills when the responses are not yet in the individual's repertoire, but can be.

(4) *Change unrealistic attributions for failure toward external, unstable, specific factors*; change unrealistic attributions for success toward internal, stable, global factors.

The four strategies above are included within the wilderness intervention that is proposed in this paper. Basically, it seems that in order to reduce the deficits of learned helplessness, it is necessary for one to acquire skills, mastery, and a sense of competence or controllability; to develop clear and realistic patterns of causal attributions and expectations; and to have perceptions of competence impact positively upon one's self-concept and self-esteem.

Of course, it is not so simple for one to follow these lines of attack. For instance, skills and mastery are not acquired easily once learned helplessness patterns of attribution and behavior are in play. Moreover, every individual has limitations of some sort that preclude various skills from being learned and mastered. In many cases, it is important to intervene at the point of mediation, where there characteristically are unrealistic, distorted, failure-prone attributions and overgeneralization of expectancy of failure.

For devising an intervention model, it is necessary to first examine the cognitive basis of causal schemata and the attribution process.

Cognitive Clarity

Causal Schemata

Cognitive deficits in learned helplessness are of a cyclical nature in that the perception of uncontrollability leads to subsequently affected cognitions that tend to generalize across settings and strengthen or validate the original association of helplessness. Causal schemata and the attribution process can be viewed as mediating these cognitions:

> The mature individual undoubtedly has acquired a repertoire of abstract ideas about the operation and interaction of causal factors. These conceptions afford him a solution to the need for economical and fast attributional analysis, by providing a framework within which bits and pieces of relevant information can be fitted in order to draw reasonably good causal inferences. (Kelley, 1972, p. 152)

In contrast to this mature individual whom Kelley describes, it is possible to view those with learned helplessness as being "developmentally immature" in their understanding of covariation of cause and effect. They might be functionally viewed as

> young children who have limited information-processing capacities and weakly formed causal schemata; they may conceive of only a small number of causes; they may not differentiate stable from unstable causes.(Weiner and Kun, 1976, p. 9)

An important type of cause-effect covariation, that will be considered here, is *effort-outcome*. Feelings of mastery are produced by perceptions of effort-outcome covariation (Weiner, Kun, & Benesh-Weiner, 1979), and it is effort-outcome covariation that is weakly perceived by individuals exhibiting "helpless" behavior (Weiner & Kun, 1976). The authors were referring here to the children in the Dweck and Reppucci (1973) study who were less likely to attribute success to high effort or failure to a lack of effort than to a presence or absence of ability.

Causal schemata enable an individual to integrate and use information gleaned from spatially and temporally distinct occasions. These schemata can be seen as heuristics for problem-solving or perception, in particular social perception. According to Bruner (1957, p. 130), "perception is a process of categorization in which organisms move inferentially from cues to categorical identity." The cues of this type of social perception are the outcomes that are experienced as a result of one's actions, and the categorical identity can refer to the causes that are attributed to having produced those outcomes. Kelley (1972) has likened the attribution process to a statistical analysis of variance (ANOVA) in which effects are attributed to causal factors with which they covary uniquely, rather than to the factors of which they are relatively independent.

The ANOVA is an idealized model, in that data collection and careful interpretation of results are clearly limited by time constraints: decisions or attributions often have to be made quickly. And this is how causal schemata provide an invaluable element of economy to our decision-making process.

But in the case of learned helplessness, it might be argued that the individual's attributional system has become maladaptive: too much economy has led to overshadowing of individual data points and cue subtleties. Overgeneralization of failure or uncontrollability from one setting to all, or from the present to always, can be metaphorically likened to a causal schema that treats all data points in an ANOVA as their mean, i.e., the null model is predicted as the best fit for all the individual data. Economy is an adaptive human information-processing feature, but it is also potentially maladaptive (see Gregory, 1969) and can lead to misperceptions of helplessness, as a result of various subjective probability heuristics, e.g., representativeness, primacy, or conservatism (see Huesmann, 1978).

According to the recent reformulation of learned helplessness discussed earlier (Abramson et al., 1978), there tend to be generalized attributions for failure along various dimensions, and these generalizations may be reflected in Kelley's (1972) statement:

> Once it is learned, a schema may be activated by any number of appropriate sets of data or cues, and it thereby has "mobility" in that it is applicable to a broad range of objects and situations. (p. 153)

Referring to this statement, it should be added that inappropriate sets of data or cues might also activate a schema and lead to attributions that are wrong or damaging or both (Valins and Nisbett, 1972).

Cognitive Maps: Confusion and Clarity

The above discussion has briefly touched on causal schemata and their attributions as the cognitive processes mediating learned helplessness. A different conceptualization of how information may be represented and stored, as a basis for knowledge about one's environment is the *cognitive map* (see Tolman, 1948).

There are many kinds of cognitive maps: some deal literally with the spatial domain of one's environment; others concern temporal relationships or affective bonds (see Downs and Stea, 1973). The common feature is that maps consist of a network of connections or relationships that exist among representations of "places" or events in the world. The map is a model of the environment that allows one to follow along its paths of associations so as to make predictions, inferences, or perceptions about the world, based on past experiences.

Hebb (1949) laid the groundwork for this theorizing about the structure of thought with his neural descriptions of cell assembly formation. Kaplan (1978b) has discussed the relevance of cognitive map theory to any simplification of reality. Of particular interest here is its relevance to the representation of one's experiences with success and failure and the causes of those outcomes.

With the notion of cognitive mapping, there is a basis for speculating about a link between the human though process and one's environment. This link and its implications for individuals accustomed to maladaptive attributional processing will be explored. Kaplan and Kaplan (1978) provide a partial rationale for this exploration:

> The way an individual experiences and reacts to a given environment begins to be understandable in the context of an experience-based internal structure that corresponds, at least in certain respects, to the environment in question. A cognitive map so generated provides the basis for hopes, expectancies, and plans. Gaps in the map might account for ignorance, apparent irrationality, and misguided despair. (p. 43)

Cognitive confusion, or metaphorically "being lost" in one's wayfinding in his/her environment, and its motivational and emotional consequences are illustrated in Bruner and Postman's (1949) experiment of *induced incongruity*. Recognition of playing became a very distressing experience for people because of changes in the color of the cards, i.e., changes in the way one was accustomed to seeing them. A learned helplessness-type of depression emerged with some subjects; they could not persist in the task because of the uncontrollable confusion.

Without a clear state of mind, an individual has difficulty perceiving subtleties in his environment, sensibly anticipating or predicting the future, and accurately assigning causal attributions for behavioral outcomes. Cognitive confusion can make it difficult for an individual to perceptually differentiate stimuli, e.g., in keeping straight in one's mind the difference between specific and global causes of outcomes, between stable and unstable causes, between internal and external causes, or between controllable and uncontrollable causes.

In contrast to cognitive confusion and lack of perceptual differentiation, *cognitive clarity* is the outcome of successful selective attention. This state of mind is characterized by rapt attention, freedom from distraction, and clear perception of subtleties in the environment.

James (1892) distinguished between voluntary and involuntary attention. Effort provided by an inhibitory mechanism of voluntary attention is involved in the suppression of distractions, and it is the contention of Kaplan (1978a) that "the stresses of modern life could lead to the fatigue of this mechanism." Moreover, recovery of voluntary attention should be facilitated by the availability of environments providing opportunity for involuntary attention.

Natural settings, in particular the wilderness, can be described as conducive to recovery from information-processing stress and overload due to the inherent fascination in the environment. Both the content and the process factors of information-processing contribute to this fascination and successful involuntary attention, and hence to cognitive clarity. Informational content and processes, in an Outward Bound setting, might provide the cognitive clarity to indirectly help alleviate the deficits exhibited in learned helplessness, by facilitating new insights into one's attributional patterns of stability, internality, controllability, and globality.

Informational Content in the Wilderness

A seemingly large attraction of a wilderness setting is its low level of stress and noise, human density, and stimulus ambiguity (e.g., see Wohlwill, 1970).

Contending with noise requires inhibition from the human distraction-suppression mechanism. Without a drain on cognitive activity from this source, less of a person's limited mental capacity has to be devoted to this problem (Kaplan, 1978a). Not only is there an aversiveness from stressful noise per se, but also from anxiety connected with feelings of helplessness, i.e., uncontrollability or unpredictability with respect to the onset and offset of that noise (Glass and Singer, 1973). Wilderness noise is of low level, and is usually perceived as under the control of oneself (or of someone else in one's group). Uncontrollable sounds of animals and the sound of rushing water typically are perceived as softness, as blending into the background, and are not perceived aversively or stressfully (at least after initial familiarization and comfort with them have been achieved).

Sherrod and Cohen (1978) discuss the behavioral effects of population density. It seems that pathological effects of high density come not from the density per se, but from the uncontrollability that may be perceived in the environment. According to Cohen (in press), unpredictable and uncontrollable environments require close monitoring for protection from potential surprise or threat. This type of monitoring requires attentional capacity that would be otherwise available for various information-processing needs. Crowding, assuming it implies perceptions of uncontrollability and unpredictability, has been shown to create a sense of learned helplessness—influencing expectancies of non-contingency between response and outcome in post-crowding problem solving tasks, that were, in fact, controllable (Rodin, 1976).

It would seem that a sense of controllability in a low density setting "may not only reduce the effects of uncontrollable high density, but also may serve to alter an individual's expectation about the value of voluntary responses and to influence one's self perceptions as a competent human being" (Sherrod and Cohen, 1978).

Whatever perceptual uncertainty and ambiguity that do exist in the wilderness can usually be dealt with in a wide range of "mental health producing" ways, due to few societal sanctions. The sanctions and limits that exist in the wilderness are set by nature, and are usually perceived, accepted, and respected by the "trespassers" as impartial and neutral, even in their occasional arbitrariness, e.g., with weather conditions.

Although a person never before exposed to the wilderness would find a baffling array of unfamiliar stimuli, initial confusion and anxiety resulting from unfamiliarity would soon be sorted out. In fact, that initial confusion is sometimes quite worthwhile, as will be noted later. The wilderness setting for an Outward Bound program is frequently very alien to the participants and provides a certain "shock value" (Schultze, 1971). It is important that the area be physically dominating, in terms of providing threat, since danger is important for its value of fascination in the attentional process. However, the danger is clearcut, locatable, and by-in-large "copable" or controllable. Mountains, canyons, caves, and coastlines are sought out because of their striking and fascinating topographies.

Finally, Thomas (1977) has discussed conditions conducive to memory and learning in the wilderness:

> If one catalogues the list of independent variables that laboratory studies have indicated to be favorable to learning, they are largely those that one naturally encounters when learning in the wilderness: concreteness, cue salience, using imagery, contextual and spatial cues, and varying the stimulus properties. (p. 11)

As an example of this conduciveness, gathering of wild plants, for many people, is a fascinating task which initially may seem overwhelming, but soon becomes easier because of multiple environmental cues that are not in conflict with one another, but instead lead to cross-confirmation. Features of information-processing in the wilderness, in particular the concreteness, cue salience, and lack of distraction and ambiguity, seem to facilitate clearness of perception.

Informational Processes in the Wilderness

Various informational processed in the wilderness also can be seen as involving fascination and as providing rapt attention and cognitive clarity. Many Outward Bound activities, which will be discussed in more detail in the next section, are governed by these processes, and hence might facilitate realistic attributions and self-perceptions.

Processes most compelling of attention seem to be those that make sense and also demand involvement. Making sense and comprehending seem to be innate motivators for people. In contrast to the comfort in finding one's way through the environment, is the discomfort and panic of being lost, as exemplified in the perceptual incongruity of the Bruner and Postman (1949) study. Orienteering with compass and map are obvious examples of activities in the wilderness that involve a striving for making sense. Contours that are kinesthetically experienced and simultaneously associated with the visual representations on topographical maps provide real concreteness in perceptual processing.

Involvement is the second key aspect of processes that compel one's attention. Curiosity, exploration, purpose, challenge, coping with risk and danger, and struggling to attain basic needs all provide involvement and hence fascination.

Involvement in dealing with conflict is seen as an important aspect of attention that can lead to human creativity. There often is a need to go to primitive and less reasonably structured aspects of functioning to find new "insight," according to Farber (1966). Anxiety, disorganization, and even some derangement, given in "copable doses," can be beneficial: these conditions allow a person to learn that unexpected situations can often be dealt with. To understand one's mental and physical limitations, i.e., to be able to differentiate that point at which there is panic (and not earlier) is an important aspect of self-awareness.

Houston (1968) discusses physical challenge and human involvement, specifically in mountain climbing:

> We must differentiate risk from danger. Experienced climbers understand, enjoy, and seek risk because it presents a difficulty to overcome and can be estimated and controlled. He equally abhors danger because it is beyond his control. Injury while crossing a slope swept by avalanche is unpredictable and dangerous, whereas the risk of an exposed climb up an overhanging face lies within his capability. (p. 52)

To a climber, one of the beauties of the mountains is the clearness and vividness of perceptions of goals, stress, success, and failure. Climbing is an exercise in perceiving control: courting and controlling risk, and avoiding uncontrollable dangers. Involvement here is an exercise in fine-grained differentiation of perceptions—an opportunity to sort out and make accurate attributions for success and failure. Houston (1968) quotes John Harlin, a famous climber, shortly before he was killed on a difficult route; he indirectly talks of attribution theory and self-awareness:

> I have used climbing as a medium for introspection into my own mind and have tried to understand my reactions to stimuli, particularly

between emotion and muscle coordination. Before training, the coordination of mind and body is not stable when one is on a two-thousand-foot ice wall with a tenuous belay. After training, the personal understanding of oneself that occurs in the intricate alpine experience can be developed and used outside of this experience. In other words, it can be borrowed and projected. This ultimately leads to a physical and emotional control of one's self. I believe that this control is an important prerequisite to creativity. (p. 58)

Catton (1969) has discussed coping with uncertainty as a process that people find naturally fascinating. On mountain climbing expeditions, he observed that for optimal fascination and motivation there needs to be a balance between involvement and making sense. Kaplan (1978a) has also expressed this necessity for "the complementary facets of a person's experience with the environment: ... making sense without involvement characterizes the boredom with the familiar; involvement without making sense is the essence of being lost" (p. . 89).

Thus far, it has been argued that elements conducive to cognitive clarity are present in a wilderness setting. This clarity would seem to facilitate a re-orientation of causal schemata or maps that have given rise to inaccurate attributions along any of the four dimensions outlined by Weiner (1979). Theoretically, a clear network of representations of the self should provide accurate causal attributions and ease of wayfinding and dealing with uncertainty in everyday life.

Outward Bound now needs to be examined, and in order to speculate about its efficacy in alleviating learned helplessness, it is necessary to refer back to attribution theory and also to discuss the formation of self-concept.

Outward Bound

Outward Bound is a wilderness training program that has been adapted for therapeutic purposes. Many schools and institutions throughout the country have applied the technique with adolescents who might be clinically described as "learned helpless."

Very briefly, the program typically consists of a four-week experience, with ten participants and one or two instructors in some wilderness setting, e.g., the desert, the ocean, or high elevation mountains. The group is instructed in all necessary survival skills and is then faced with various high-risk challenges in a non-artificial situation, in which both the individual and the group as a whole need to function smoothly in order to attain their goals. During the course, there are individual challenges (e.g., rock climbing, rappelling, or foraging for food) and group challenges (e.g., team ascent of mountains, ocean sailing, or group initiative problems).

The term "Outward Bound" is used here in a generic sense. In other words, Outward Bound and other similar programs are referred to synonymously because theoretically they are all based on a similar process. This consists of the interaction among the physical and social setting, the participants and instructors, and a number of problem-solving activities or challenges (see Walsh and Golins, 1976). It is this "ideal" process which is proposed as a model for the alleviation of learned helplessness.

Physical and Social Setting

The Outward Bound setting provides contrast to the participants, i.e., it helps them "to see generality which tends to be overlooked by human beings in a familiar environment or to gain a new perspective on the old, familiar environment from which (they) come" (Walsh and Golins, 1976). This could have the effect of facilitating discussion and realizations about outcome likelihoods and causal attributions that are differential across settings.

Seligman (1975) talked about discriminative control as a limitation on the generality of helplessness. Striking environmental differences between a mountain terrain and a classroom, known for previous failure, might help in the recognition of a need to be discriminative. On the other hand, there are also commonalities between settings, and these too have to be explored in terms of implications for differentiation or generalization.

Because of the group size there is easy accessibility to the decision-making process, and hence there is personal involvement and little sense of arbitrary hierarchy. This involvement in decision-making should increase people's perceptions of controllability in this setting and maybe elsewhere (Sherrod and Cohen, 1978).

Theoretically at least, an individual will be aware of an added dimension in his consideration of success and failure, and that is the performance of the group. An external attribution for success made to teamwork might take on some degree of internality in the individual's eyes. This might have a positive effect on the expectancy of his own future success if his contribution to the team is perceived by him as characteristic of a general and stable ability that is easily reproducible in the future, in other group settings. Also, group attribution for failure may sometimes by less debilitating for an individual than some personal, internal cause.

Sharing of experiences within the group tends to encourage effective social comparison for the purpose of checking one's self-evaluations (Schachter, 1959). The principle of social comparison is fundamental to the formation of self-concept (Rosenberg, 1979) since people learn about themselves by comparing themselves to other *referrent individuals* (Pettigrew, 1967). Valins and Nisbett (1972) note the value in using social consensus as a check on debilitating self-ascriptions of personal inadequacy. Group discussions are facilitated by the instructor, often for this purpose of questioning the legitimacy of causal attributions by individuals.

The Instructor

The instructor is an important source of appraisal for the participants. Theoretically, both direct reflections of the instructor's appraisal and the participant's perceptions of the instructor's attitudes, i.e., one's *perceived self*, are critical for self-concept development (Rosenberg, 1979). The function of appraisal from the instructor is similar to that of social comparison from the other members of the group: to provide accurate self-information that will facilitate accurate expectancies of outcomes and accurate causal attributions. Usually, there is a high degree of value placed on this appraisal because of the significance perceived in the instructor, who is in the position of a role model for the participants.

Coopersmith (1967) isolated three general antecedents of self-esteem, that can be paraphrased into an instructor-participant framework:

> total or nearly total acceptance of the participant by the instructor; clearly defined and enforced limits; and the respect and latitude for individual actions that exist within the defined limits (p. 236).

Instructors ideally strive to provide these three conditions, with limit setting being greatly aided by the various constraints of nature.

Problem-Solving

Incremental and manageable nature of tasks. Outward Bound activities are engineered in terms of complexity and consequence so as to lead to an incremental learning of skills and a gradual decline of fear. Careful needs assessment is made by the instructor, who must be sensitive to varying levels of ability among the patrol members. The instructor helps in the setting of realistic self-goals which

> strengthen the proper attributions regarding the causes of success and failure. When an individual works on a task within his reach, degree of effort becomes the overriding determinant of success (and failure)... and reinforcing the causal link between personal effort and outcome helps combat a sense of learned helplessness. (Covington and Beery, 1976 p. 109-110)

On the other hand, participants are encouraged to seek out and challenge their limits. Crucial to the affective reactions to success at a task is the perceived level of difficulty (Weiner and Kukla, 1970). If the task is completely mastered or is perceived as too easy, then attributions for success tend to shift from the self (e.g., effort or ability) to the ease of the task, lowering the positive affective reaction (Weiner, Kun, and Benesh-Weiner, 1979). Also, White (1959) would argue that individuals have a natural inclination to actively develop new competencies.

Success at moderately difficult tasks usually indicates this development and thus produces greater positive affect than success at easy tasks.

Recognition and discussion of manageability should theoretically help people appreciate the typically rapid transition from a feeling of total and overwhelming incompetence in a new situation to a sense of newly-acquired knowledge and skills being automatic. The novelty feature of the wilderness helps ensure the sense of being overwhelmed, and manageable problems help ensure the contrast of competence. Ideally, this kind of recognition should convince people to be careful about making attributions for failure to stable and global causes.

The incremental nature of tasks might also help people who deny their successes by attributing them to luck or who even sabotage their own effort because of the threat posed by success, namely expectations by others for further or greater success. Mette (1971) has speculated that accepting only partial or incremental credit may allow confidence to build gradually, without arousing threat.

Incremental challenges are fairly easily mappable, one-to-one, with variations in the terrain. Technical rock climbing provides a good example of an activity in which just about everyone can succeed at some level, and in which gradual increases in difficulty can be provided with very slight variations of the climbing route. The complexity of a coastline and its system of currents also provide clear gradations of difficulty.

Concrete and consequential nature of tasks. A second characteristic of Outward Bound activities is their concreteness; they are often clearly perceived as problems discrete in time and space. Having to set up shelter at a high altitude at night, as a storm sets in, is unmistakably concrete. When climbing a peak, there is little ambiguity concerning the goal and the attainment of that goal.

Absolute standards of achievement clarify objectives and tend to stimulate achievement motivation (Covington and Beery, 1976). With concreteness, it would seem that there would be a greater likelihood of failure feedback implying simply falling short of a specific goal, rather than (inappropriately) implying more, e.g., about the worth of the individual.

Concreteness of a task would seem to encourage task analysis and the discovery of what specific component of the task is difficult. Constructive interpretation of failure might be easier here, where one can clearly differentiate what he can do from what he cannot do. An individual might perceive himself as an adequate navigator, given visible landmarks and buoys, but realize that for an extended trip requiring charts and compass he would probably fail. He is reminded of the distinction between a temporary skill deficit, that can be remediated, and a permanent lack of ability. Such a self-cognition implies failure attributions being unstable and specific, thereby reducing the chronicity, generality, and intensity in any feelings of helplessness the individual might have (Abramson et al., 1978).

Therapeutic Interaction

The desired effect of the Outward Bound process should theoretically take place as the physical and social setting, the instructor and participants, and the problem-solving activities interact. A discussion of this therapeutic interaction follows.

Mastery and competence. The problems just mentioned, along with skill training and peer and instructor support, account for a high probability of increased self-rewards, success, and mastery.

As a consequence of mastery, people are less likely to defensively withdraw out of a *self-consistency motive*, and instead are more likely to be governed by a *self-esteem motive* (Rosenberg, 1979). In other words, in a success-oriented environment, there is more likelihood of positive change of self-concept via the component of self-confidence, i.e., the belief that one can make things happen in accord with inner wishes. This belief is closely related to a sense of competence or efficacy.

Clarity of causal attributions and expectations. Individuals are encouraged to critically analyze their attributions so as to minimize the likelihood of misperceiving a case of universal helplessness as the more deleterious personal helplessness. They are encouraged to critically analyze and use social comparison information so as to appreciate individual differences among themselves. Accurate causal attributions, along both the dimensions of locus of causality and controllability, cannot be made without accurate information about oneself in relation to relevant others.

Klein, Fencil-Morse, and Seligman (1976) note the clinical importance for a depressed individual to accurately differentiate causal attributions along these two dimensions. "It makes clinical sense that showing the patient that his performance is not inferior to that of others should reduce his feelings of unworthiness." In this study, performance deficits associated with unsolvability were eliminated by instructions to blame failure "on the harshness of the environment," i.e, on the difficulty of the task. This difficulty was experienced similarly by all relevant others.

In nature, there is no upper limit to physical challenge, so universal failure is often inevitable, e.g., given a veritable blank wall to climb. This should be marvelled out, for the sake of respect for nature with its occasional non-contingencies: there are some tasks that are impossible, and they should be accepted as such.

However, the loss of self-esteem comes when the individual cannot hike for 15 minutes without stopping, whereas everyone else can. He has personally failed, whereas "relevant others" have succeeded. At this point, the person should be encouraged to question his social comparison information as to what constitutes a "relevant other." For example, possibly there is an important difference between himself and the others that he should keep in mind, e.g., age or a medical condition that he has to resign himself to.

Attributions for failure to lack of effort are most motivating for an individual, given the assumption that this failure can be turned around to success by a greater output of effort (Dweck, 1975). This ideal assumes some base level of ability that does exist, but in order to be brought to bear, must be "energized" by some amount of effort, initiative, and spontaneity. However, if the ability is lacking, then trying to increase attributions for failure to lack of effort will just lead, it seems, to exacerbated frustration and failure because such a person is going to fail regardless of his effort.

It seems that it is important to learn to be realistic and differentiating with one's attributions, perhaps occasionally resigning oneself to a lack of a specific ability. Also, it is very important for a person to be able to differentiate a temporary skill deficit from a more stable and global incompetence. Because of individual differences in people, and a relative lack of ability in some specific areas for everyone, attribution retraining focusing on individual differences should be strived for. This differentiation is what Valins and Nisbett (1972) refer to in their emphasis on one's understanding *situational* aspects of an outcome, in contrast to making personal, *dispositional* interpretations. It should be a responsibility of the "ideal" instructor to help in the reduction of the desireability of outcomes which are unattainable, and in the reduction of the aversiveness of outcomes which are unrelievable. Assistance in one's resignation to limitations should be accompanied with encouragement of alternative and more realistic goals, in the context of emphasizing existing strengths.

There should also be encouragement of failure attributions being treated differentially, as *strategy-specific*, so that alternative and more realistic strategies might be employed in future tasks. Diener and Dweck (1978) found that in addition to attribution retraining, learned helpless children might also benefit from being trained to focus on self-instructions and self-monitoring following failure, as do mastery-oriented children. In other words, training of self-monitoring or *executive functioning* (see Butterfield and Belmont, 1977), which involve an important element of cognitive self-awareness, might be a relevant goal of Outward Bound. This type of post-failure cognition, in addition to accurate attributions, might increase the likelihood of persistence leading to future success, and not simply to unproductive perseveration.

Occasional failure engineered into Dweck's (1975) *Attribution Retraining Treatment* had a definite value. Similarly, occasional failure, e.g., due to bad weather conditions, is naturally engineered into wilderness living and Outward Bound tasks and serves a useful purpose. Kliman (1978) observed the psychological value that can come from the acknowledgement and acceptance of situational crises, which provide an opportunity for people to seek out adaptive ways of survival and mastery. There is value in:

> the opportunity to ingest small, tolerable doses of sadness and anxiety, doses which could strengthen them for possibly larger, more unexpected and uncontrollable assaults to come. (p. 8)

Centrality and self-esteem. The preceding ways in which the elements of Outward Bound may interact have an overall purpose of producing a certain therapeutic effect. Providing opportunities for the development of a sense of competence and for the questioning and altering of unrealistic attributions and expectations should aid a learned helpless individual to see himself/herself differently. Perceptions of self-confidence, competence, or personal control should be clarified. These terms can be used synonomously, according to Rosenberg, (1979):

> The theme common to all these ideas is that of feeling oneself to be an active agent in one's own life (rather than the object of external forces), of believing generally that one can work one's will on a more or less recalcitrant world. (p. 31)

The terms refer to a single component of the total picture of the self (i.e., self-concept), and they also refer to a single component of the totality of the feelings of worth one has about oneself (i.e., self-esteem).

Rosenberg (1979) has outlined four principles involved in self-concept formation: *reflected appraisals, social comparisons, self-attributions,* and *psychological centrality.* Thus far, the first three of these principles have been discussed, although in a structure slightly different from his, i.e., reflected appraisals and social comparisons seem to be antecedent to the attribution process in the therapeutic model discussed here. The interactional strategies that have been described are governed by these three principles, and impact upon one's self-concept formation, or re-formation, as a result of the component of self-confidence. Rosenberg's fourth principle, psychological centrality, might be viewed as part of the theoretical rationale for the "solo" experience at the end of an Outward Bound course, an activity not yet discussed.

Solo is markedly different from the rest of the activities; it can be described as a time for reflection on the whole Outward Bound process. The experience traditionally lasts for three days and nights, with minimal equipment and food. Afterward, many people report feelings of acceptance of their limitations within the complexity of nature. They see themselves as adaptable, flexible, appreciative of their strengths and capabilities—both physical and mental. Often there are reports of gaining awareness of oneself as an "evolutionary animal," as fitting into a larger, more complex scheme of things. People question egocentric thought and egocentric values. Reflection concerns priorities and values clarification, and personal accounts often sound quite spiritual in nature.

It is obviously difficult to analyze and classify the significance of such reports because of their personal and individualistic nature. However, the theoretical value of this quietude and introspection lies in the clarity of perception of the content and structure of one's self-concept. The various components of self-concept are complexly related and synthesized into a global self-concept, which gives rise to feelings of global self-esteem. Moreover,

> one cannot appreciate the significance of a specific self-concept component for global self-esteem if one fails to recognize the importance or centrality of that component to the individual. (Rosenberg, 1979, p. 73)

The elements of one's self-concept do not simply add up to the whole; there is a complex hierarchical structure, based on the centrality of each element. Perhaps reflection allows this heirarchical structure to be better understood. Values clarification, about what is important and what isn't, and about what one thinks should be important and not important, may take place. In other words, one may perceive his/her *extant self-concept* in comparison to the *desired self-concept* (Rosenberg, 1979).

Possibly the reflection and values clarification during solo allow the hierarchical structure to be re-arranged. Following this theory, the acceptance of unavoidable weaknesses, e.g., being a poor organizer, and the relegation of those components of self-concept to a status of less importance or centrality than components perceived competently, e.g., being a good facilitator in interpersonal conflicts, might positively affect one's global self-concept and global self-esteem. It should be noted that speculation about this impact on global self-concept and global self-esteem is predicated on the argument that the self-confidence or competence emerging from mastery of fairly specific tasks in the wilderness is, or will be, perceived as psychologically central to the individual.

It was noted earlier that positive affective responses, including competence, confidence, and pride, which have been linked to dimensions of causal attributions for success (e.g., to internality) are supposedly associated with self-esteem (see Weiner, Russell, & Lerman, in press and Weiner, Kun, & Benesh-Weiner, 1979). Cognitive clarity in the wilderness may facilitate a re-orientation of one's attributions, which in turn may facilitate a change in self-concept and self-esteem.

The hope is that reorganization of self-concept and self-esteem, along with new habits or patterns of causal attributions, would direct the course of one's subsequent experiences, within environments that are both cognitively and emotionally supportive.

Naturally, community, family, and school support are all needed to sustain improvements in self-esteem and to reinforce the realistic attributional patterns learned in the wilderness. Although not directly in the realm of learned helplessness, the work of Kelly and Baer (1971) and Kelly (1974) on adolescent delinquency provides an illustration of the need for environmental changes to sustain positive effects from Outward Bound.

Conclusion

A therapeutic intervention for the alleviation of learned helplessness has been discussed in terms of theories of causal attributions, self-concept development, and environmental psychology. From this framework, the ideal Outward Bound process emerges as a therapeutic model.

Elements of a wilderness environment provide a general clarity of perception, and this may facilitate a re-orientation of one's perceptions of competence. Unrealistic expectations and causal attributions may be questioned and altered. Specifically, differentiation of attributions along the dimensions of stability, internality, controllability, and globality should help reduce the cognitive, motivational, emotional, and self-esteem-related deficits of learned helplessness. Outward Bound activities may have an effect on one's self-concept and self-esteem if changes in perceptions of competence in specific areas affected during the program are considered psychologically central to the individual.

It should be noted that a conglomeration of different ages is represented in the studies cited in this paper, although most have been in the range of upper elementary school to adolescent (which has been the implicit focus). However, no attempt has been made to deal with developmental issues of self-concept formation or attribution theory.

The therapeutic model that has been proposed is idealistic, yet grounded in the reality of actual Outward Bound programs providing actual psychological changes. Most empirical attempts at evaluating Outward Bound's impact have been elusive though (see Smith, Gabriel, Schott, and Padia, 1976; Godfrey, 1974; and Harmon, 1974). Kelly (1974) speculated on the need for a theoretical basis for Outward Bound evaluation:

> Perhaps we strive to quantify an unquantifiable effect since cognition and emotion are two distinct spheres of experience and efforts to relate one to the other suffer in translation... This is not to imply that it cannot be measured, but that present techniques are of questionable validity. (p. 11)

Possibly an attributional framework, associating cognition with emotion, can serve as a springboard for future research on learned helplessness intervention in the wilderness, and elsewhere.

REFERENCES

Abramson, L., Seligman, M., & Teasdale, J. Learned helplessness in humans: critique and reformulation. *Journal of Abnormal Psychology*, 1978, *87*, 49-74.

Andrews, G., & Debus, R. Persistence and the causal perception of failure: modifying cognitive attributions. *Journal of Educational Psychology*, 1978, *70*, 154-166.

Atkinson, J.W. *An introduction to motivation*. Princeton, N.J.: Van Nostrand, 1964.

Bruner, J.S. On perceptual readiness. *Psychological Review*, 1957, *64*, 123-152.

Bruner, J.S., & Postman, L. On the perception of incongruity: a paradigm. *Journal of Personality*, 1949, *18*, 206-223.

Butterfield, E.C., & Belmont, J.M. Assessing and improving the executive cognitive functions of mentally retarded people. In I. Bialer & M. Sternlicht (Eds.), *Psychological issues in mental retardation*. New York: Psychological Dimensions, Inc., 1977.

Catton, W.R., Jr. Motivations of wilderness users. *Pulp & Paper Magazine of Canada*. Quebec: National Business Publications, Dec. 19, 1969, 121-122.

Cohen, S. Environmental load and the allocation of attention. In A. Baum & S. Valins (Eds.), *Advances in environmental research*. Norwood, N.J.: Erlbaum, in press.

Coopersmith, S. *The antecedents of self-esteem*. San Francisco: Freeman, 1967.

Covington, M., & Beery, R. *Self-worth and school learning*. New York: Holt, Rinehart & Winston, 1976.

deCharms, R. *Personal causation*. New York: Academic Press, 1968.

Deci, E.L. *Intrinsic motivation*. New York: Plenum, 1975.

Diener, C., & Dweck, C. An analysis of learned helplessness: Continuous changes in performance, strategy and achievement cognitions following failure. *Journal of Personality and Social Psychology*, 1978, *36*, 451-462.

Downs, R.M., & Stea, D. (Eds.) *Image and environment: Cognitive mapping and spatial behavior*. Chicago: Aldine, 1973.

Dweck, C.S. The role of expectations and attributions in the alleviation of learned helplessness. *Journal of Personality and Social Psychology*, 1975, *31*, 674-685.

Dweck, C.S., & Goetz, T.E. Attributions and learned helplessness. In J.H. Harvey, W. Ickes, & R.F. Kidd (Eds.), *New directions in attribution research* (Vol. 2). Hillsdale, N.J.: Erlbaum, 1978.

Dweck, C.S., & Reppucci, N.D. Learned helplessness and reinforcement responsibility in children. *Journal of Personality and Social Psychology*, 1973, *25*, 109-116.

Farber, S.M. Quality of living—Stress and creativity. In F. Darling &

J. Milton (Eds.), *Future environments of North America*. Garden City, N.Y.: The Natural History Press, 1966.

Feather, N.T. The study of persistence. *Psychological Bulletin*, 1962, *59*, 94-115.

Glass, D.C., & Singer, J.E. Experimental studies of uncontrollable and unpredictable noise. *Representative Research in Social Psychology*, 1973, *4*, 165-180.

Godfrey, R. *A review of research & evaluation literature on Outward Bound and related educational programs*. Paper presented at the Conference on Experiential Education, Estes Park, CO, 1974.

Gregory, R.L. On how so little information controls so much behavior. In C. Waddington (Ed.), *Towards a theoretical biology: Two sketches*. Edinburgh: Edinburgh University Press, 1969.

Harmon, P. *The measurement of affective education*. Unpublished report, Harmon Associates, San Francisco, 1974.

Harter, S. Effectance motivation reconsidered: Toward a developmental model. *Human Development*, 1978, *21*, 34-64.

Hebb, D.O. *The organization of behavior*. New York: Wiley, 1949.

Houston, C.S. The last blue mountain. In S.Z. Klausner (Ed.), *Why man takes chances*. N.Y.: Doubleday & Co., 1968.

Huesmann, L.R. Cognitive processes and models of depression. *Journal of Abnormal Psychology*, 1978, *87*, 194-198.

James, W. *Psychology: The briefer course*. (1892). N.Y.: Collier Books, 1962.

Kaplan, S. Attention and fascination: The search for cognitive clarity. In S. Kaplan & R. Kaplan (Eds.), *Humanscape: Environments for people*. North Scituate, Ma.: Duxbury Press, 1978. (a)

Kaplan, S. On knowing the environment. In S. Kaplan & R. Kaplan (Eds.), *Humanscape: Environments for people*. North Scituate, Ma.: Duxbury Press, 1978. (b)

Kaplan, S. & Kaplan, R. (Eds.) *Humanscape: Environments for people*. North Scituate, Ma.: Duxbury Press, 1978.

Kelley, H.H. Causal schemata and the attribution process. In E. Jones, D. Kanouse, H. Kelley, R. Nisbett, S. Valins, & B. Weiner (Eds.), *Attribution: Perceiving the causes of behavior*. Morristown, N.J.: General Learning Press, 1972.

Kelly, F.J., & Baer, D.J. Physical challenge as a treatment for delinquency. *Crime and Delinquency*, Oct., 1971, 437-445.

Kelly, F.J. *Outward Bound and delinquency: A ten year experience*. Paper presented at Conference on Experiential Education, Estes Park, CO, 1974.

Klein, D., Fencil-Morse, E., & Seligman, M. Learned helplessness, depression, and the attribution of failure. *Journal of Personality and Social Psychology*, 1976, *33*, 508-516.

Kliman, A.S. *Crisis: Psychological first-aid for recovery and growth*. N.Y.: Holt, Rinehart & Winston, 1978.

Mettee, D.R. Rejection of unexpected success as a function of the negative consequences of accepting success. *Journal of Personality and Social*

Psychology, 1971, *17*, 332-341.

Pettigrew, T.F. Social evaluation theory: Convergences and applications. In D. Levine (Ed.), *Nebraska Symposium on Motivation, 1967*. Lincoln, Nebraska: University of Nebraska Press, 1967.

Rodin, J. Density, perceived choice, and response to controllable and uncontrollable outcomes. *Journal of Experimental Social Psychology*, 1976, *12*, 564-578.

Rosenberg, M. *Conceiving the self*. N.Y.: Basic Books, 1979.

Rotter, J.B. Generalized expectancies for internal versus external control of reinforcement. *Psychological Monographs*, 1966, *1*, (whole no. 609).

Schachter, S. *The psychology of affiliation*. Stanford, CA: Stanford University Press, 1959.

Schulze, J.R. *An analysis of the impact of Outward Bound on twelve high schools*. Unpublished report, Outward Bound, Inc., Reston, Va., 1971.

Seligman, M.E.P. *Helplessness: On depression, development, and death*. San Francisco: Freeman, 1975.

Seligman, M.E.P., & Maier, S.F. Failure to escape traumatic shock. *Journal of Experimental Psychology*, 1967, *74*, 1-9.

Sherrod, D.R., & Cohen, S. Density, personal control, and design. In S. Kaplan & R. Kaplan (Eds.), *Humanscape: Environments for people*. North Scituate, Mass.: Duxbury Press, 1978.

Smith, M.L., Gabriel, R., Schott, J., & Padia, W.L. Evaluation of the effects of Outward Bound. In G. Glass (Ed.), *Evaluation studies review annual* (Vol. I). Beverly Hills, CA: Sage Publications, 1976.

Tennen, H. & Eller, S.J. Attributional components of learned helplessness and facilitation. *Journal of Personality and Social Psychology*, 1977, *35*, 265-271.

Thomas, J.C. *Cognitive psychology from the perspective of wilderness survival*. Research report #28063, IBM Thomas J. Watson Research Laboratory, Yorktown Heights, N.Y., 1977.

Tolman, E.C. Cognitive maps in rats and men. *Psychological Review*, 1948, *55*, 189-203.

Valins, S., & Nisbett, R. Attribution processes in the development and treatment of emotional disorders. In E. Jones, D. Kanouse, H. Kelley, R. Nisbett, S. Valins, & B. Weiner (Eds.), *Attribution: Perceiving the causes of behavior*. Morristown, N.J.: General Learning Press, 1972.

Walsh, V., & Golins, G. *The exploration of the Outward Bound Process*. Unpublished report, Colorado Outward Bound, Denver, CO, 1976.

Weiner, B. (Ed.) *Achievement motivation and attribution theory*. Morristown, N.J.: General Learning Press, 1974.

Weiner, B. A theory of motivation for some classroom experiences. *Journal of Educational Psychology*, 1979, *71*, 3-25.

Weiner, B., & Kun, A. The development of causal attributions and the growth of achievement and social motivation. In S. Feldman & D. Bush (Eds.), *Cognitive development and social development*. Hillsdale, N.J.: Erlbaum, 1976.

Weiner, B., Kun, A., & Benesh-Weiner, M. The development of mastery, emotions, and morality from an attributional perspective. *In Minnesota Symposium on Child Development* (Vol. 13). Hillsdale, N.J.: Erlbaum, 1979.

Weiner, B., Russell, D., & Lerman, D. The cognition-emotion process in achievement-related contexts. *Journal of Personality and Social Psychology*, in press.

Weiner, B., Frieze, I.H., Kukla, A., Reed, L., Rest, S., & Rosenbaum, R.M. *Perceiving the causes of success and failure.* Morristown, N.J.: General Learning Press, 1971.

Weisz, J.R., & Stipek, D. Developmental change in perceived control: A critical review of the locus of control research and a proposed reorientation. *Psychological Bulletin*, in press.

White, R. Motivation reconsidered: The concept of competence. *Psychological Review*, 1959, *66*, 297-323.

Wohlwill, J. The concept of sensory overload. In J. Archea & C. Eastman (Eds.), *EDRA II*, Proceedings of the Second Annual Environmental Design Research Association Conference. Stroudsburg, Pa.: Dowden, Hutchinson, & Ross, 1970.

V
Sex Differences in Achievement Motivation

Chapter 15

Achievement Motivation and Values: An Alternative Perspective

JACQUELYNNE E. PARSONS
The University of Michigan

SUSAN B. GOFF
The University of Michigan

Unequal participation of the sexes in the domain of employment has become increasingly difficult to ignore. Although increasing numbers of women are working, these women are still concentrated in the lower levels of the professional hierarchy in spite of attempts in recent years to decrease discrimination in hiring and salaries of women. For example, the percentage of women in professional and technical occupations has increased from 39% in 1968 to only 42% in 1976 while during the same time period the percentage of women clerical workers increased from 73% to 80%. It is interesting to note in 1970, when women occupied 40% of the professional and technical positions, more than 62% of these women were nurses, physical therapists, dieticians and elementary and secondary school teachers. In comparison, only 12.6% of men occupied these "female" professional occupations (Carnegie Commission on Higher Education, 1973). Even within the professional domain, women cluster in the lower realm of the status hierarchy. The underemployment of women implied by these figures is widespread.

Although highly important, institutional barriers and sex typing of jobs are not entirely responsible for this phenomenon. There is evidence that other factors might also contribute to the fact that women are underrepresented in professional careers. Psychological investigations have highlighted several such factors which could affect female professional participation by influencing career aspirations in such a way as to predetermine the training young women seek and the skills they acquire. Motivational factors are undoubtedly one of the key contributors.

Over the past several years we have become interested in the motivational factors influencing long range social goals such as career or occupational choice, major selection in college, decisions to have children, return to work, etc. Our interest in this area initially grew out of our concern over the underrepresentation of women in professional careers discussed above and the decisions of many college women not to pursue traditional masculine achievement paths. Like many of the contemporary researchers in this field, we started out trying to identify those characteristics of non-traditional women or role innovators which distinguished them from more traditional women and those factors which have constrained many women's efforts to attain non-traditional goals. We were particularly interested in high need achievement women who were capable of achieving these goals. But in the past few years, we have redirected our focus. It seemed to us that models which assumed that choosing a non-traditional career reflected maturity and enlightenment while choosing a traditional career reflected immaturity and sex-role rigidity were inherently biased. How could a model predict an individual's achievement behavior if it assumed that all non-competitive achievement behavior, for example, was pathological? The question better suited to answering our inquiries was *"Why do women make the choices they make?"* and not "Why don't they act more like men?".

In an attempt to answer this question, we have returned to basic motivational models and have chosen to treat long range life-defining choices as analogous to task choices. Assuming that life choices can be conceptualized as task choices, we have focused most of our thinking around the expectancy × value models of Lewin and Tolman. Further, given our basic training in motivational structures, we began our thinking with Atkinson's original suggestion that the tendency to approach a given goal (T_s) would be a function of a motivational component, an expectancy component, and an incentive value component. Finally, we have come to believe, as have many, that the likelihood of the selection of any particular goal will be a function of its Ts in addition to the T_s's of the alternative choices and the tendency to avoid each of these choices.

This paper shall focus upon three topics basic to a more complete understanding of achievement behavior: (1) an analysis of the traditional achievement model and its problems handling long range life goals; (2) a discussion of the role of values in motivation and the difficulty we had in distinguishing between values and motives; and (3) a discussion of Bakan's agency/communion dichotomy and its utility in aiding our understanding of men's and women's achievement oriented behavior.

Need Achievement Model and its Problems

Among the many factors that undoubtedly affect career aspirations (see Parsons, Ruble, and Frieze, 1978, for full discussion), the motive to achieve has received considerable attention. McClelland and Atkinson have been key figures in this research arena. Within the McClelland/Atkinson tradition, achievement behavior and career choice have been linked to two basic motives: the hope for success motive (M_s) and the motive to avoid failure (M_{af}). These motives are assumed to be latent, stable characteristics acquired early in life. They are aroused in situations in which the standard of performance is evaluated against some measure of excellence: M_s being aroused at the prospect of doing well and by the anticipated feeling of pride that accompanies success, and M_{af} being aroused by the prospect of failure and the shame that accompanies it. The strength of the positive tendency to approach success, T_s, and the negative tendency to avoid failure, T_{-f}, are multiplicative functions of their respective motives, incentive values, and the perceived chance of success or failure: $T_s = M_s \times I_s \times P_s$ and $T_{-f} = M_{af} \times I_f \times P_f$, respectively (Atkinson, 1958). The tendency to respond to an achievement situation is predicted by the tendency to approach success minus the tendency to avoid failure plus tendencies aroused by extrinsic factors (e.g., peer pressure): $T_a - M_s \times I_s \times P_s - (M_{af} \times I_f \times P_f) +$ extrinsic factors (Ext.)

To simplify the model, Atkinson made the following assumptions about the relations between the various components: (1) the sum of the probability of success and the probability of failure is equal to one; (2) the sum of the incentive for success and the incentive for avoiding failure is equal to one; (3) the incentive for success is equivalent to one minus the probability for success. Notable among these assumptions is the inverse relationship between incentive value and the likelihood of or expectancy for success (i.e., easy tasks would have a low incentive value and difficult tasks would have a high incentive value). Incentive value is most clearly defined as an extension of one's expectancies, containing no independent significance. This model marked the beginning of the dominance of expectancies or calculated probability of success and failure in the investigation of the achievement motive.

While Atkinson's mathematical model predicted men's career aspirations and related adult achievement behaviors reasonably well, it did not predict reliably either women's adult achievement behavior or career aspirations. Inspection of this model suggests three basic problems that could account for its limited utility in explaining the long range life goals of women: *elimination of incentive value* as a key determinant of approach behavior, *oversimplification of the concept of probability of success* and *limitations on the range of achievement goals* considered in designing the formula. Each of these is discussed below.

First, by equating incentive value with one minus the probability of success ($1 - P_s$), Atkinson had severely limited the explanatory power of the one factor that is most clearly linked to gender-role socialization, and thus, may account for the differing life goals of men and women. Atkinson's formulation had the effect of emphasizing the causal importance of expectancies and individual differences in motive strength at the expense of incentive value. Additionally, since Atkinson did not originally define expectancies as a subjective variable, individual differences in expectancies for particular tasks have not been investigated systematically until quite recently. As a result subsequent work in traditional need achievement theory and its offshoots (e.g., Horner, 1968) have focused primarily on motives. Unfortunately, few consistent sex differences in motive strengths have emerged. Since sex-roles socialization is primarily aimed at creating a sex-differentiated perception of valued life goals, it seems that attention to the mediating role of task incentive values would move us closer to an understanding of the differences in achievement patterns of men and women.

Second, it is unlikely that people attach equal probabilities of success to the same life choices. Atkinson attached a stable probability of success estimate to different tasks according to his perception of the ease of accomplishment. In a ring toss task, his definition of the probability of success was based on the average ability of a group to succeed. For simple and highly controlled tasks, this interpretation of "probability of success" might be valid. However, in a more complex situation, such as deciding what type of job or career to pursue, the degrees of freedom increase to the point that a simple determination of probability of success according to task ease is a dramatic understatement of the reality of the situation. It is much more likely that evaluations of probability of success will vary greatly among people and between the sexes.

Third, at an even more basic level, it is not clear that the model can predict future oriented, life-defining choices. The model predicts that high need achievement individuals will select tasks with a 50% chance of success. While this prediction seems plausible for some tasks (primarily short term or recreational activities, M. Maehr, personal communication) it seems untenable for other tasks. In particular, it seems unlikely that we select long term goals at which we are as likely to succeed as to fail. The cost of failure is too great in relation to the amount of time and energy invested. Instead it seems more reasonable that we select life goals that are both challenging, reasonably probable, and important to us. Thus, it seems that Atkinson's derivation does not yield a model that intuitively would predict life choices behavior.

It is apparent that there are a variety of reasons why Atkinson's model would not be very appropriate for an analysis of life goals in general and for an analysis of the difference between men's and women's life goals in particular. But few researchers have attempted to look at any of the aforementioned problems. Their investigations have centered, instead, on the motivational components and recently on the expectancy component; they have asked the question "How might men and women differ in their motivational structures and expectancies and what additional motives are needed to explain the differences in men's and women's life choices?"

As a case in point, Horner (1968) accounted for the failure of Atkinson's model to predict female adult achievement patterns by noting the exclusion of an additional avoidance motive: the fear of success motives (M_{-s}). She proposed that fear of success was higher in females than in males and that the inclusion of M_{-s} into Atkinson's model would increase its accuracy as a predictor of female achievement behavior. More specifically, Horner suggested that women are less likely to approach achievement situations not because of a weakness in the achievement motive but rather because potential success aroused their fear of success motive, which, in turn, created enough anxiety to impede the tendency to approach achievement situations. Responses to verbal leads such as "after first term finals, Ann finds herself at the top of her medical school class," were used to measure the "fear of success" motive. A response was considered high in fear of success if it contained negative imagery reflecting anxiety about the success. The negative imagery most frequently took the form of Ann's physical unattractiveness and lonely Friday and Saturday nights with her books. Horner interpreted these responses as a projection of the writer's fear of success.

Based on the differential response patterns of college women to the female cues and college men to the male cues and the relationships of these patterns to a measure of achievement behavior, Horner concluded that women have higher fear of success than men and that fear of success does interfere with achievement behavior in some settings. She generalized these results to the broader domain of adult achievement patterns including career aspirations. Subsequent research and theoretical analysis have not substantiated these conclusions (Condry & Dyer, 1976). Thus, despite initial support and widespread enthusiasm, the fear of success model as conceptualized by Horner has not clarified our understanding of female achievement behavior or more general life choices.

The focus of the criticism of Horner's work has ranged from methodological inadequacies to conceptual disagreement. The criticisms have been adequately reviewed elsewhere (e.g., Condry & Dyer, 1976) and will not be reviewed here. We will focus, instead, on our evaluation of the two major shortcomings inherent in both the Atkinson model and in Horner's refinement: limiting the definition of achievement to the *agentic domain* (Bakan, 1966) and *ignoring the importance of incentive value* as a causal determinant in life goal selection.

Agentic Achievement

At a fundamental level, we believe that the failure of both of these achievement models to account for female adult achievement behavior lies in their narrow operationalization of success and achievement goals. The achievement tasks selected and the criterion of success were usually individual tasks on which success results from one's own actions and attributes rather than cooperative or group achievement tasks on which success results from *joint efforts*; the adult achievement tasks were generally careers with success being defined in terms of the status of the professions aspired to rather than familiar or social roles, or the intrinsic value of the task. Viewed from this perspective, it is apparent that the assumptions underlying both the traditional Atkinsonian achievement model and Horner's subsequent refinement reflected what Bakan (1966) has labelled an "agentic" perspective on achievement: a perspective of achievement in which success is defined by a personal achievement with a view of self as separate from others and success as a result of one's own actions and attributes.

Defining achievement in the agentic mode has had three

major consequences for this field of research, which have reduced the explanatory power of the existing models and related research for long range life choices of both men and women. First, because career status is so highly valued in our society, the research question asked by Horner and others led to the inclusion of a negatively valued, avoidance motive to explain why women were not attaining high career status. She neglected to identify the careers that women desired to pursue or the characteristics of their chosen careers that mediated these choices.

Second, the agentically biased models have, in the past, ignored motive systems other than the one stemming from a simple need to achieve. Douvan and Adelson (1966) and Hoffman (1972) noted the part that the motive to affiliate plays in the development of self-esteem. Hoffman proposed that affiliation was seen as success by women, and furthermore, was an affirmation of the self. But neither of these papers connected the woman's perceived value of social behavior with achievement motivation. When the female's affiliative orientation was applied to achievement motivation in Horner's (1968) thesis, it was negatively valued as "fear of success" instead of being labeled as a need for affiliation within the achievement setting. As suggested by Stein and Bailey (1973), we see affiliative factors as integral to female achievement motives.

In their recent work, Atkinson and Raynor (1974) have demonstrated the need to consider both more than one motive and the interaction of the individual's motive type with situational cues in predicting achievement behavior. Veroff has also argued that more than one social motivational force operates on any given choice. Finally, in a recent review of the motivational influence literature, Denmark, Tangri, and McCandless (1975) discuss the importance of yet another motive: need power. They point to the work of Winter (1973) as clear support of a relationship between a need for power and career choice. Denmark et al. conclude their review with a plea for research on the interaction of these three motives: need achievement, need affiliation, and need power on life choice.

Third, the focus of achievement was directed to the goal itself, instead of the process involved in gaining the final reward. Veroff (1977) has pointed out the importance of dissecting the more global concept of achievement into two levels: the process of achieving and the impact of the accomplishment. Traditional models of achievement have ignored the process of achievement, focusing instead on the outcome and its meaning for the achiever. For example, Horner directed her study to the projected effect of final success upon one's available social rewards. Fear of success was the fear of the negative consequences of success; but a careful look at some of the responses classified as evidencing fear of success actually suggest that the subjects were concerned with the costs of success in terms of the process of seeking success rather than the costs of success in terms of the social rejection following success. The lonely Saturday nights studying could reflect concern over the cost of succeeding in medical school rather than fear of being rejected once one had become a successful doctor.

Veroff (1977) discussed evidence that women and men may well differ in their orientation to process versus impact achievement orientation. For example, Zander, Fuller and Armstrong (1973) found that women's pride and shame about themselves were more influenced by their teams' efforts while men's pride and shame were more influenced by their competitive competence. Similarly, Veroff, McClelland and Ruhland (1975) have found that power achievement or the need for your achievement to have an impact on someone else was higher in males than females.

The concern over cost of becoming a doctor could also reflect an assessment of the relative worth of the ultimate goal. People have many goals for their lives. The likelihood of the selection of any one goal is dependent, to some extent, upon its impact on the whole constellation of goals an individual holds. For example, a woman might well desire both a professional career and a family. But if she sees these goals as conflicting, then her choice should reflect her relative priorities. The findings of both Parsons, Frieze, and Ruble (1978) and Poloma and Garland (1971) suggest that women's attitudes regarding the demands inherent in the wife/mother role influence occupational aspirations. Professional career aspirations are more probable when career obligations are not perceived as interfering with the fulfillment of the wife/mother role. If women believe that facilitative institutions and spouse support are available which can lessen the burden of childcare without harming the child, they may choose a nontraditional life style. However, if these institutions and support are not available or if existing childcare facilities are believed to be inadequate in quality, selection of a professional career is unlikely. Additionally, if a woman feels it is important to be the major socializer of her children during the preschool period and is committed to being available to her children throughout their childhood years, then she is unlikely to select a professional career that allows for little flexibility in both career commitment and time scheduling across those years.

In a study which tested these suggestions, we had women rate the importance of various careers including mothering. We found that male stereotyped occupations were seen as relatively more difficult than comparable female stereotyped occupations but were not seen as of any more importance to the women themselves. Further, success at mothering was rated as more important than success at any of the other occupations. In line with this finding, the women reported that they would be willing to exert the most effort to be "successful" mothers, would feel the best about this success, and the worst about failing to meet this goal. Finally, being a "successful" mother was seen as difficult but highly probable. It seems likely, given this pattern, that any occupation that seriously threatened these women's ability to become "successful" mothers would not be seen as very appealing.

In an attempt to explore this issue of the integration of family and career further, we interviewed 15 male undergraduates and 15 female undergraduates in the Spring of 1977. The open-ended interview schedule focused on the following topics: (a) marriage plans, (b) career plans, (c) plans to have children, and (d) plans for child-rearing. Given the data reported above, we felt it was especially important to assess career oriented college women's attitudes toward child bearing and rearing. We felt that high level professional commitment was dependent on either endorsement of and faith in daycare arrangements or expectation of major involvement of the father in childrearing. Further, we felt that future societal change would be reflected in these students' current plans. To our surprise, these college students did not endorse either of these alternatives. All 30 students planned to have a career after college. All of the men and 13 out of 15 women planned both to marry and to have children. Of these, 11 of the females and 8 of the males did not want their children in daycare centers. Females cited the desire to raise their own children as the primary reason for their reluctance to use daycare. Males also stressed the importance of the family in raising the children. In response to the question "If both you and your mate have full time careers and you both want to have children, how would you handle this situation?" Only 2 males and 6 females expected the father to share the child rearing role. Nine of the females and 11 of the males expected the mother to assume the bulk of childrearing responsibilities. Taken together, these results suggest that college students today expect that either they (if female) or their spouse (if male) will take time out of a career to raise the family. Further, given today's job market, this goal, in essence, precludes a high level professional career commitment during the early family years. Those women who desire a professional career will be forced either to (1) lower these aspirations, or (2) opt for a non-traditional (non-male) career path—entering into the profession late or establishing themselves in their career early and then taking time out for a family. Both of these options will structure the career choices available to these women.

Omission of Incentive Value

The most serious problem, in our estimation, with the traditional achievement model is the omission of incentive value. The question of underlying values, either personal or those inherent in the task, was not handled in Atkinson's original need achievement model. Viewed from within the traditional achievement model, the failure of a highly able individual to aspire to a high level occupation is incomprehensible. For example, striving towards and achieving the top position in one's medical school class is prophetic of future high-status, power, and financial security. Both Atkinson and Horner decided that any person with a strong need to achieve would view this type of success as highly desirable, more desirable, certainly, than affiliative success. A person designated as having high need achievement by their measures, but for whom these goals are not particularly salient is seen as an anomoly. However, the motive to approach a task is undoubtedly influenced by the *underlying value structures* of the individual or the task. In the past, need achievement models have failed to take into account variability on these values. Task values has been treated as a constant. In our egocentric way, we have assumed that "success" in life-defining roles like careers is simply fulfilling the requirements of the task according to the experimenter's perception of a common standard of achievement. The consequence of assigning a stable, agentic character to the motivational and incentive value components of an achievement task is instability within the model and a lack of predictive power for those individuals with a different value system than the experimenter. It is important to realize that the intrinsic value of a task may vary as a result of differences in underlying value constructs; and therefore that it must be considered as variable within the motivational construct, rather than constant.

But, why was task value overlooked in need achievement models? As was discussed earlier, it was algebraically eliminated. Largely as a consequence of this reduction, little research has been done on the impact of values on achievement choices. Career counseling researchers have devoted some attention to this issue but have done little more than identify a relationship between the global values one holds and the profession one is in. In one of the most comprehensive attempts to develop a typology of values and to relate individual differences on these global values to behaviors including occupation, Rokeach (1973) demonstrated that members of different occupations are discriminable in terms of their pattern of value endorsements. But it seems these studies are not really capturing incentive value as it is conceptualized within an expectancy × value model of task choice. None studied the relationship between the value of various tasks and goals to the individual and that individual's task choice.

Work within the field of developmental study of achievement orientation, primarily that of the Crandalls (1962, 1969), has measured attainment value and has related it to task choice. Others have related attainment value to persistence on tasks (Battle, 1965, 1966; Stein & Bailey, 1973). In general these studies have demonstrated the relationship between attainment value and task choice in the experimental setting. But, they have not investigated the relationship between values and long range goal choices.

It is our belief that the failure of the need achievement model to illuminate women's achievement behaviors and life choices reflects these shortcomings. Most importantly, the failure of the need achievement model to consider individual differences in task perception and value system has severely limited the utility of the model. It is to this omission that we will direct the remainder of this paper.

Values and Life Choices

The first step in our investigation of values and long range goal choice is to define more clearly task value. Task value has been conceptualized broadly as *a quality of the task which contributes to the increasing or declining probability that a given individual will approach the task.* This quality of the task can be further defined in terms of three primary components: (1) the *utility value* of a given task in aiding the achievement of some long range goals, for example, the value of taking a high school math course in terms of its importance for becoming an engineer; (2) the *incentive value* of engaging in the task, for example, the value of taking a high school math course in terms of the enjoyment one gets from solving math problems; and (3) the *incentive value of successfully achieving one's goal*, for example, the value of taking a high school math course in terms of the enjoyment one gets from getting an A in a math course.

Incentive value itself can be further divided into several components, two of which are particularly important to us.

Incentive value can be conceptualized in terms of the immediate rewards, intrinsic or extrinsic, that performance of a task will provide for the individual. For example, tennis could be intrinsically rewarding because it makes one feel healthy or extrinsically rewarding because one is paid for the performance. Incentive value can also be conceptualized in terms of the global values that an activity fulfills. Tasks can be perceived as related to certain global values such as competition, altruism, nurturance, power, status, or intellectual quality. If one holds one of these values as very important, then one may select activities that are related to that value. For example, tennis could have a high incentive value because one values competitive competence and tennis allows one to demonstrate to oneself one's competitive competence. In order to determine the incentive value of a particular task, the researcher must first discover the individual's perception of the values inherent in the task, or in other words, which kinds of needs the individual believes that the task will fulfill. Then the researcher must determine whether the individual believes that participation in the task will lead to a fulfillment of needs or will reaffirm the individual's self-concept.

The fact that incentive value can be conceptualized in this latter way indicates the importance of the phenomenological study of the relation between global values and task choice. If the individual's task selection is influenced by the incentive values of the task, and if the incentive value can be related to the individual's basic value structure, then it is important to measure both the individual's value structure and the individual's perception of the relationship of various tasks to these values that should predict the incentive value of the task. If the incentive value of a task is influenced by the congruity of one's value structures and one's perception of the relationship of the task to these structures, then one can reasonably assume that the resounding effect of values upon task choice and task persistence can be felt on multiple levels.

The implications of this analysis for our understanding of the differences in life choices between men and women are clear. Since sex role socialization has a major impact on individuals' goals and values (Frieze et al., 1978), it is reasonable to expect that women's value structure will differ from that of men.

In terms of task value, sex difference in value structure can manifest itself in several ways. For one, women and men could attach different incentive values for engaging in and successfully completing various activities. For example, women may place more importance on spending time with friends than men and, thus, be more likely to approach this activity.

Alternatively, women's hierarchy of values might differ from that of men's. That is, when asked to rank order the importance of various activities and adult life goals, women may display a different pattern than men. For example, as discussed earlier, if women see the parenting role as more important than a professional career role while men rate these roles as equally important, then it is to be expected that women would be more likely to resolve life decisions in favor of their parenting goals. This differential would be especially marked if women see the career options as not only of lower importance but also as detrimental to the successful completion of their parenting goals.

At a more fundamental level, sex role socialization could create a sex differentiated hierarchy of global values. That is, women and men could order their central core values differently. Consequently, various life tasks satisfying different core values would have different incentive values for men and women. For example, if women see "helping others" as a more important core value than do men, occupations which allow one to help others would have a higher incentive value for women than men.

The above mentioned example illustrates the impact of a sex-differentiated core value structure on the importance of both processes and goals. A sex-differentiated core value structure could also influence the very definitions of success and failure on a whole variety of tasks and activities. Men and women may well differ in their conceptualization of the requirements for successful task participation and completion. Consequently, men and women would attach different incentive values to the various options and would approach and structure their task involvement differently. The parenting role provides an excellent example of this process. If males define success in the parenting role as an extension of their achievement roles, then they may respond to parenthood with increased commitment to their career goals and with emphasis on encouraging competitive achievement in their children. In contrast, if women define success in the parenting role as high levels of involvement in the children's lives, they may respond to parenthood with decreased commitment to their career goals.

Career roles can also be influenced by this process. If men and women differ in their definitions of career success they should structure their career activities quite differently. For example, if women are more likely to define the medical profession as an opportunity to help others and be involved with the patient's lives, while men are more likely to define the medical profession as an opportunity to achieve in a high status occupation, then women doctors should be more likely to structure their medical career around helping as many people as possible in as broad a capacity as possible while male doctors should be more likely to structure their career around high status specialities. This is, in fact, the case. Women are much more likely to become general practitioners and much less likely to seek out a speciality than men (Heins, 1978).

Given all of these influences on the incentive values of task choices, it is to be expected that the utility value of various activities will also be sex-differentiated. If the utility value is determined, partly, on the basis of the usefulness of an activity for reaching a future goal, then sex differences in future goals will result in differences in the utility values of various activities. This, in turn, will result in differential approach behaviors. For example, high ability girls are more likely to drop out of math in high school than are high ability boys. While a host of reasons may be responsible, differentially perceived utility value is undoubtedly important.

One final issue that struck us as we researched the value literature was the distinction between values and motives. If one is focusing on incentive value as the anticipated reward for engaging in or successfully completing a given task, then incentive value is conceptually distinct from motive. But if one is defining incentive value in terms of the congruence between one's basic value structure and one's perception of various tasks, then the distinction is less clear. M. Brewster Smith recognized this problem in 1969. He defined a value as a standard by which we judge our behavior. Because values act as a standard, they can influence effort, pride, shame and task choice. He concluded that *"In this sense a value may also be a motive, when one's values influence one's choices they do so by virtue of motivational force."* How, then, does a value differ from a motive? This remains an important question for those of us attempting to measure both and to develop a model relating each to task choice. But it is clear that incentive and utility values must be considered in any model of life choices.

Bakan's Model

We will turn now to a discussion of Bakan's model of agency and communion which we feel provides an alternative conceptualization of the interplay between motives and values. Bakan (1966) suggested that there are two basic modalities of life that encompass both values and motivation: "communion" and "agency"; and that men and women, as collectives, differ in their orientation to these modalities. Women, much more than men, have been noted as conforming to the *"communion"* disposition of existence: characterized by *openness, noncontractual cooperation* and the *sense of being at one with others*. Within the traditional motive theories, communion could, then, be described as an integration of achievement and affiliation motives and values. In contrast, men's orientation to life goals conforms more to the "agency" modality: characterized by isolation, self-protection, self-assertion, self-expansion, by the urge to master, and to remain separate from others. Again within a motivational framework, agency could be considered as the segregation of achievement and affiliation. If there is such a tendency for men and women to cluster around the separate modalities, then it should affect many facets of behavior, including life and career goals.

In an investigation of college students' considerations for

choosing a particular career, Astin (1975) found that women appeared to be more motivated by intrinsic considerations and men by extrinsic considerations. In particular, women were more likely than men to endorse the following job characteristics influential in their career choice: relevance of the field to their intrinsic interests, the opportunity to contribute to society, to work with ideas and people, to be helpful to others, and to express one's identity. In contrast, more men than women rated the following as important job characteristics: the opportunity for high salaries, high prestige, rapid advancement, and a stable future. These value structures are similar to Bakan's descriptions of the "communion" and "agency" modalities. The desire to be at one with others, and to gain rewards through one's interaction with others reflects communal characteristics, while the desire to be both separate from "the people" and to ascend to a position which few attain and many respect reflects a more agentic orientation.

The degree to which the traditional model of achievement and success reflects an agentic perspective becomes increasingly clear. The achievement motive has been validated in studies of individual, goal oriented tasks. A view of achievement and success from a communion perspective raises issues which have not been considered fully in the past. Agentic rewards alone offer little reinforcement to a communally oriented person. Agency-type rewards can be defined as the fulfillment of long term goals. Attention is paid to the ultimate fulfillment of the goal rather than the characteristics of the task and the process of achieving it. A communion oriented person is perceived to attend to the factors mediating the actual achievement of the goal. The worthiness of the goal, for example, cannot be estimated without regard to continued intrinsic interest in the task, possible affiliative loss or gain, possible benefit or harm to others, and/or the likelihood that the pursuit of the task will lead to self-growth and realization of certain intrapsychic goals. In view of this, the woman's tendency to focus on the process of achieving an end (Veroff, 1977) may be directly applicable to life goals and career aspirations. Women may weight the procedural costs in an achievement setting more than men.

A communal perspective also points to the need to consider more than one motive system. The procedural costs alluded to above often take the form of affiliative losses: an inability to interact socially because of a task's demand on one's time, an inability to directly assist people in one's career due to its nature, a lack of affiliation with one's co-workers due to the supervisory quality of the position or an incompatibility between career demands and parenting goals. When one considers the relation of intrapsychic variables to women's attitudes toward achievement (as representatives of a more communal mode), and their resultant behavior, the necessity of incorporating affiliative factors within the traditional achievement model becomes pre-eminent.

Finally, consideration of a communion model of achievement requires attention to much more specific definitions of success. While agentic success is defined, to some extent, in terms of one's distinctiveness from one's peers, communion-type success does not rely on the social comparison process. One can feel successful if one's group succeeds at a task they have defined as important or if one has contributed to harmony within one's social group. One can also feel successful if one is able to help someone succeed at his or her own goals, with minor regard to one's own welfare.

Thus, it can be seen that the consideration of people's communal needs is as essential as the consideration of their agentic needs in predicting their career aspirations and life goals. In addition to the predictive power it will lend the model with respect to women's achievement behavior and life choices, we expect that the inclusion of communion considerations in our achievement models will shed more light on intrapyschic factors (i.e., values) mediating men's achievement behavior. It cannot be assumed that men pursue goals without any regard to communion-type factors. However, in view of traditional societal demands upon men to adhere to an agentic-type path for career and life success, it is hardly surprising that internal values have not been considered as related to achievement. Perhaps a more realistic illustration of achievement behavior will aid in dispelling some of the more inhumane concepts and demands of "success" as we know it.

Measurement

As schemes for measuring agency and communion began to take shape, we perceived two distinct methodological approaches. The first approach engaged the use of the traditional Atkinsonian model with some major modifications. The second involved the development of a new scale which would reflect, more directly, agency and communion orientation in conjunction with the convergence of several other measures. What follows are our beginning attempts as conceptualizing measurement strategies within each of these approaches.

Modified motivational assessment. The reader will recall the many objections registered against the use of this model in attempting to predict long range achievement choices. In spite of its many deficiencies, however, we have constructed a formula using this paradigm which, by incorporating an additional motive structure, might provide a better estimate of the communion orientation to achievement.

Within a traditional motivational model, communal factors could be described as affiliative motives in that affiliation typifies the most distinct difference between agency and communion, that is, the attitude towards the importance of oneself apart from others versus the importance of others in conjunction with oneself. Take the need achievement motive: need affiliation can be divided into the motive to gain affiliative success (M_{as}) and the motive to avoid affiliative failure (M_{aaf}). These motives are assumed to be aroused whenever performance involves interpersonal interaction; and the degree of success or failure is measured against internal standards of the achievement of certain intimacy levels with a certain situation. M_{as} will be aroused at the prospect of gaining new friends, achieving deeper levels of intimacy, and maintaining that intimacy. M_{aaf} will be aroused by possible rejection or reduction of levels of intimacy with significant others, or of not gaining any new deep friendships. As in Atkinson's model, the strengths of the tendencies to approach affiliative success and avoid affiliative failure are multiplicative functions of their respective motives, incentive values and probability of success or failure ($T_{as} = M_{as} \times I_{as} \times P_{as}$; $T_{aaf} =$

$M_{aaf} = I_{aaf} = P_{aaf}$).

Combining the motive systems, the tendency to approach an achievement situation becomes the sum of the tendency to approach within task affiliative success minus the tendency to avoid within task affiliative failure ($T_{as} - T_{aaf} = T_a$ affiliation) and the tendency to approach goal oriented success minus the tendency to avoid goal oriented failure ($T_s - T_{af} = T_a$ achievement). To this function, the tendency to respond to extrinsic factors in an achievement situation could be added. The resulting formula appears as such:

$$T_A = (T_{as} - T_{aaf}) + (T_s - T_{af}) + T_{ext}$$ or
$$T_A = (T_a \text{ affiliation}) + (T_a \text{ achievement}) + T_{ext}.$$

Thus, one could calculate a single estimate of one's approach tendency in a given situation by considering both the achievement T_a and the affiliative T_a aroused by that situation. For individuals with high need achievement and high affiliation, the T_A for any given task would be high to the extent that the oportunity for both achievement and affiliative success were high; the T_A for any given task would be lowered to the extent that the probability of affiliative failure was high. In contrast, for individuals with low need for affiliation, the T_A for a given task would be high to the extent that T_a achievement was high; variations in the opportunity for affiliative success or failure would have little impact. Comparisons of the T_{AS} of a variety of individuals for a particular task would allow one to test hypotheses regarding individual differences in task selection. Comparisons of the T_{AS} of a variety of tasks for one individual would allow one to test hypotheses regarding within subject variations in task selection.

Alternatively, using a strategy similar to that used by Spence and Helmreich (1978) in scoring the PAQ for Androgyny, one could divide the population based on their scores on Mehrabian's need achievement and need affiliative scales into quadrants. This would yield four groups: high need affiliators and high need achievers (Type 1), high need affiliators and low need achievers (Type 2), low need affiliators and high need achievers (Type 3), and low need affiliators and low need achievers (Type 4). For our purposes the most interesting comparisons would be between Types 1, 2 and 3: Type 1 potentially reflecting the integrated individual Bakan alludes to, Type 2 the communion individual, and Type 3 the agentic individual. Intuitively, it seems that Type 1 individuals are caught in the kind of double bind that is common to bright, competent women. While on one level these individuals could be characterized like Androgynous individuals as having the best of both worlds, on the other hand, they will often be caught between conflicting highly valued alternatives. In-depth analyses then of the differences in the resolution of conflicting choices between Type 1 and Type 3 individuals might illuminate the differences in the life choices and goals of academically competent men and women.

Values Assessment. The procedures discussed above rely upon traditional notions of the relation of motivation to behavior as a vehicle for describing the interaction of achievement orientations. The following procedure reflects our attempts to engage personal values more directly in the measurement of agency and communion. We have developed a scale based on our perception of the ways in which values might be articulated in attitudes concerning sensible and desirable philosophies and behaviors in work, leisure, and social aspect of life.

The original *Achievement Orientation Scale* consisted of 39 forced choice items. Each item described two attitudes, beliefs, or behaviors which were thought to reflect agency or communion-typed values. For instance, the item "Which, for you, defines "peace" more . . . 1) quiet, serenity, solitude . . . a mind at ease and cleared of the fog of daily, petty anxieties; or 2) warmth between people, acceptance because of and in spite of one's faults . . . no quarrelling, jealousy, or haughtiness," attempted to determine whether the individual desired to rise above, or get away from people rather than be together, as one. Several items measured these types of values in different ways. An extension of this differential orientation is seen in items discriminating the desire to "achieve" in one's career from the desire to develop friendships or help other's in one's career e.g., "Who would you rather be? 1) Someone who is a good friend . . . would do anything for you . . . one of my favorite people; or 2) Some who will go far in their field . . . has gained a lot of people's respect . . . devotes a lot of time to work and loves it."

Based on several construct and content validity analyses,[1] the initial set of 39 items was reduced to 24 items divided into 3 subscales: (1) a group of items differentiating individuals with high agentic orientation from those with low agentic orientation (Subscale 1); (2) a group of items differentiating individuals with high communion orientation from those with low communion orientation (Subscale 2); (3) a group of items differentiating individuals with high agentic orientation from individuals with high communion orientation (Subscale 3). Scores on this last subscale reflect the resolution of the conflict between agentic and communion values. It is on this subscale that we have focused our validation research.

While we have just begun to test the strength and validity of this subscale, and have not assessed age and wider population appropriateness, our initial results are encouraging. In response to a question inquiring as to their probable career choice, our subjects (all college women) answered in ways that were easily categorized into either human service oriented, helping careers (doctor, nurse, social worker, and teacher) or self-oriented, non-helping careers (engineer, business executive, artist). Further the women within each career type were sorted into one of three groups according to their reasons for entering their career: (1) helping reasons (i.e., benefit others); (2) external, non-helping reasons (potential high salary, potential fame); (3) internal non-helping reasons (personal interest, intellectual challenge). A one-way analysis of variance disclosed no effects of career type per se on any of the AOS subscales. The reason for entering the career, however, was related to scores on Subscale 3. Women entering their career for helping or internal, non-helping reason scored significantly higher (more communion orientation) than women entering for non-helping reasons. In addition, low scores (agency orientation) on Subscale 3 were related to higher educational aspirations, plans to make more money and plans to work a larger percentage of working years.

---

[1]Details can be obtained from the authors.

Responses to open-ended questions investigating phenomenological perceptions of criteria for success in life and in one's career, or for failure in life or in one's career were coded for content, then categorized into several types of responses—ranging in degree of self-orientation from achievement of external rewards (money, power, or status) to benefiting others. Low scores on Subscale 3 (high agency) were associated with beliefs that making a lot of money and gaining fame would signal success in life. High scores on the same subscale were associated with the belief that to have helped others would be indicative of success.

REFERENCES

Astin, H. Women and work. Paper presented at New Directions for Research on Women. Madison, Wisconsin, 1975.
Atkinson, J.W. *Motives in fantasy action and society.* Princeton: Van Nostrand, 1958.
Atkinson, J.W., & Raynor, J.O. (Eds.) *Motivation and achievement.* Washington, D.C.: Winston (Halsted Press/Wiley), 1974.
Bakan, D. *The duality of human existence.* Chicago: Rand McNally, 1966.
Battle, E. Motivational determinants of academic task persistence. *Journal of Personality and Social Psychology,* 1965, *2,* 209-218.
Battle, E. Motivational determinants of academic competence. *Journal of Personality and Social Psychology,* 1966, *4,* 634-642.
Carnegie Commission on Higher Education. *College graduates and jobs.* New York: McGraw-Hill Co., 1973.
Condry, J., & Dyer S. Fear of success: Attribution of cause to the victim. *Journal of Social Issues,* 1976, *32*(3), 63-83.
Crandall, V.C. Sex differences in expectancy of intellectual and academic reinforcement. In C.P. Smith (Ed.), *Achievement-related behaviors in children.* New York: Russell Sage Foundation, 1969.
Crandall, V.J., Katkovsky, W., & Preston, A. Motivational and ability determinants of young children's intellectual achievement behavior. *Child Development,* 1962, *33,* 643-661.
Denmark, F., Tangri, S., & McCandless, S. *N-achievement, n-affiliation, n-power: Alternative perspectives.* Paper presented at New Directions for Research on Women, Madison, Wisconsin, 1975.
Douvan, E., & Adelson, J. *The adolescent experience.* New York: Wiley, 1966.
Frieze, I.H., Parsons, J.E., Johnson, P., Ruble, D.N., & Zellman, P. *Women and sex roles.* New York: Norton, 1978.
Heins, M. Practice and life patterns of women and men physicians. In S. Golden (Ed.), *Work, family roles and support systems.* Ann Arbor,

Michigan: Center for Continuing Education of Women, 1978.

Hoffman, L. Early childhood experiences and women's achievement motives. *Journal of Social Issues*, 1972, *28*, 129-156.

Horner, M. *Sex differences in achievement motivation and performance in competitive and non-competitive situations.* Unpublished doctoral dissertation, University of Michigan, Ann Arbor, 1968.

Parsons, J., & Ruble, D. Introduction. In Ruble, D., & Parsons, J. (Eds.) *Sex roles: Persistance in change. Journal of Social Issues*, 1976, *32*, 1-6.

Parsons, J., Frieze, I., & Ruble, D. Intrapsychic factors influencing career aspirations in college women. *Sex-Roles*, 1978, *4*, 337-348.

Poloma, M.M., & Garland, T.W. Jobs or careers? The case of the professionally employed married in Europe and America. *International Journal of Comparative Sociology*, Part II, 1971.

Rokeach, M. *The nature of human values.* New York: The Free Press, 1973.

Smith, M.B. *Social psychology and human values.* Chicago: Aldine, 1969.

Spence, J.T., & Helmreich, R.L. *Masculinity and feminimity.* Austin, Texas: University of Texas Press, 1978.

Stein, A.M., & Bailey, M.M. The socialization of achievement orientation in women. *Psychological Bulletin*, 1973, *80*, 345-364.

Veroff, J. Process vs. impact in men's and women's motivation. *Psychology of Women Quarterly*, 1977, *1*, 283-293.

Veroff, J., McClelland, L., & Ruhland, D. Varieties of achievement motivation. In M. Mednick, S. Tangri, & L. Hoffman (Eds.), *Women: Social psychological perspectives on achievement.* Washington: Hemisphere Publishing Co., 1975.

White, R.W. Motivation reconsidered: The concept of competence. *Psychological Review*, 1959, *66*, 297-333.

Winter, D. *The power motive.* New York: The Free Press, 1972.

Zander, A., Fuller, R., & Armstrong, W. Attributed pride or shame in group and self. *Journal of Personality and Social Psychology*, 1973, *23*, 346-352.

Chapter 16

Achievement and Vocational Behavior of Women in Iran: A Social and Psychological Study*

FARIDEH SALILI
The University of Hong Kong

Theories of achievement motivation and the existing body of data suggest that achievement behavior is a function of the strength of achievement motive. However, many of these theories apply only to men. Little is known about female achievement motivation. Available data about women are contradictory and inconsistent (cf., Bardwick 1971). For example, standard measures of achievement motivation correlate with male academic achievement, but not female (McClelland et al., 1953). Atkinson (1958) admits that sex difference in achievement motivation is "perhaps the most persistent unresolved problem in research on need achievement."

According to McClelland (1958) and Atkinson (1964) the achievement motive is a motive to compete in a situation where a standard of excellence is involved. A highly achievement-motivated person is one who has developed an internal standard of excellence, is independent, persistent, chooses to perform tasks of moderate difficulty, and has a clearly understood goal in mind. This definition implies that an achievement-motivated person does not rely on external support or social approval; he strives to achieve because he has an internal standard of excellence. It is then not surprising that research findings show that the model predicts only male achievement behavior and not female. Social approval and need to affiliate are important aspects of the feminine role which women have internalised in the process of socialization (cf., Bem 1974).

---

*Reprinted by permission of author and publisher of *Journal of Vocational Behavior*. It should be noted that the present study was completed prior to the "Islamic Revolution" in Iran.

Furthermore, while the achievement motive is assessed partly by behavior, its interaction with other motives and situational determinants of behavior has not been given enough thought (cf., Bardwick 1971).

As to the origin of the achievement motive, Maehr (1978) argues that the theory derived from the work of McClelland and his colleagues has defined the achievement motive in terms of culturally learned personality patterns, while the role of situational-contextual factors in achievement is largely neglected. Therefore, McClelland's approach is biased towards an *enthnocentric* conception of achievement because it is based on western standards and values. While the pervasive effects of being a member of a certain social or cultural group are hard to ignore, there is no clear understanding of how, to what extent, and under what conditions these factors can affect achievement motivation. Maehr (1978) proposes that the phenomenon should be studied, taking into consideration personality as well as situational-contextual factors that affect achievement motivation: situation or context is important when examining differences in achievement motivation "across social and cultural groups" while "consideration of personality is necessary when studying the individual achiever."

Maehr's proposal is particularly useful for studying achievement and career behavior of women. There are reasons to believe that achievement motivation of women is influenced not only by early environmental and learning experiences; situational-contextual factors are among the most important determinants for women. For example, due to social and economic pressures during World War II, women ventured into male dominated careers and became economically more active. Various other studies (Frazier, 1976; Pettigrew 1964; Horner 1968) have reported that black women show different motivation and aspiration patterns from those of white women. Forces in the social system cause black women to be more dominant, aggressive and have higher aspiration. Weston et al. (1975) report that black college women do not exhibit similar fear of success compared to that of white women in an intellectually competitive situation.

While social and situational factors seem to be important

determinants of female achievement, early socialization and cultural learning cannot be ignored. Sex role socialization has so great an impact that even with a change in the structure of society and women's economic independence there is still great pressure on both males and females for sex role differentiation. This is evidenced by Trey's (1972) analysis of women's role in World War II. When the crises were over, women returned to their original activity patterns. Mednick's (1975) analysis of the kibbutz also shows that despite all the efforts towards sex role change and male-female equality, sex role differentiation still strongly existed.

The purpose of this chapter is to present results of several studies conducted in Iran in which males are compared with females on various psychological measures. The results of these studies are compared with those reported in the U.S. Such comparisons will allow us to see to what extent achievement and vocational behavior of women are influenced by sociocultural and environmental factors. Samples in these studies are from two major cities in Iran, namely Mashhad and Tehran. Some relevant facts concerning the education and employment of women as well as an ethnographic description of women in urban areas of Iran will be presented first.

Urban Iranian Women

According to the census of 1972 (Plan Organization, Iran, 1977) 20.4% of urban women were literate. Of these the lower class consisted of labourers, servants, factory workers, small shopkeepers, and craftsmen. Many of them have migrated from villages into the cities in recent years. While they are now exposed to the material and cultural life of cities, their mentality and outlook are basically similar to those of their rural sisters. Most of them are strictly under the control of men and have not much freedom of choice with their lives. Lack of education, as well as adherence to traditional religious beliefs about the role of women in society are most prevailing characteristics of these women. Their most rewarded function is childrearing, especially of boys: male offspring provide a sense of personal achievement and security.

Middle- and upper-class females can enjoy better educational opportunities, but there is still more attention paid to sons than daughters. Due to traditional and religious attitudes, social interactions are generally more restricted for women. This is coupled with a de-emphasis of independent career achievement for women. Nearly always, males are encouraged to venture into the wider world, take up new jobs, given more opportunity for education or technical training, while women are encouraged to stay home and continue their traditional activities (cf., Callaway 1975). One widespread female goal is to marry the new technocrats or professionals with social prestige and economic security.

On the other hand, the passing of the old largely peasant economy means the transition of extended families, with economic burdens on the shoulders of heads of extended households, to nuclear families in the city, where each couple has to take care of its own finance. With rising costs of living, the need for women in nuclear families to work is one aspect of reality that cannot be suppressed by religious or traditional attitudes. Even so, such work is considered more an economic expedience than careers in themselves. And household responsibilities and rearing of children still remain largely if not solely the work for women.

Education

According to the census of 1968, 20% of women in the whole country were literate, and in 1973, 30% were literate (cf., Women's Organization, Iran, 1976). Although literacy is making some progress, female students still choose academic majors which are different from those of males. The census of 1974 (cf., Women's Organization, Iran, 1976) shows that only 10% of girls choose to study mathematics and related sciences in high school, in comparison to 35% of boys. This greatly reduces the chances of girls entering university, because of the heavy emphasis on these subjects in entrance exams, especially for fields such as medicine, engineering, and related areas.

The census of 1975 (Women's Organization, Iran, 1976)

shows that the percentage of girls graduating from universities is much lower in every field except the humanities (51.4%). In 1972, out of 721 graduates of medicine there were only 121 females. The census of 1975 also shows that out of 364 students of law there were only 70 females. Out of 5,963 students studying various branches of engineering there were only 698 females. The same pattern appears in technical schools. In 1975, only 17% of the total number of students entering technical schools were girls, mainly in the fields of nursing, secretarial jobs, elementary teaching, and fine arts. The numbers of girls in other technical fields are negligible or non-existent.

Employment of Women

The 1972 census (Plan Organization, Iran, 1977) reported that about 7.5% of urban women were economically active. Altogether women constituted 11% of the employed population. In the universities of Iran 3.2% of heads of departments or more senior persons were women. In all government ministries together, 4.4% of those at or above the level of director general were women. Of the people with doctorate degrees employed by government, 7.5% were female (Women's Organization, Iran, 1976). In comparison with developed nations, these figures are very low. In the U.S. for example, in 1974, 40% of professional and technical workers were women and 20% of managers and administrators were women (U.S. Women's Bureau, 1974).

In Iran, with an elitist class mainly consisting of male descendents of the former landlord class, and more restricted social interactions for women, opportunities for responsible positions for women are few and not necessarily based on individual merit. With development projects around the country and problems of coordination, geographical mobility is one necessary aspect of any responsible position, whether based in the capital of Tehran or in the provinces. But women have more binding commitments to family and therefore are less mobile. Together with necessary career interruptions such as childbirth, women are disadvantaged in the actual ability to be on the spot where and when the job requires, whether it be in social gatherings of the men of the former landed class or be around the country for official business.

Vocational Aspirations of Women

As we have just described above, various statistics show that there are considerable differences between the employment of men and women in Iran, with women mainly occupying lower paid and lower prestige jobs. The question raised here is whether there is actually any difference between the vocational aspirations of women and their actual employment situation. A second question is whether socioeconomic factors apart from sex might be responsible for variations in vocational aspiration among women.

Iranian students, 235 male and 258 female, aged 9 to 18 years old, were simply asked to indicate the occupation they would like when they completed their studies. Totally, some 92 responses were given by the subjects. The responses were then independently classified into four categories by two Iranian judges, according to skills required and social status (with the number 1 assigned to the highest category, 2 to the next, etc.). The judges agreed in their classifications in most cases and agreed on the others after discussion. Subjects' choice of vocational aspiration were then compared on the bases of age, sex and socioeconomic background. In order to compare age differences in vocational aspirations subjects were classified into three age groups (9-11, 12-14, and 15-plus). An analysis of variance showed that vocational aspiration did not differ significantly at different age levels.

In order to examine the effect of socioeconomic background, three Iranian judges were asked to classify subjects' socioeconomic background into 3 levels (upper, middle, and lower classes), according to the information given about each student's background, including parents' education, occupation, income, and housing condition. The judges reached complete agreement in 98% of cases and the decision of a majority of two judges was used to classify other cases. The ANOVA test indicated that there was a significant relation between socioeconomic background and aspired vocation, with a mean category choice of 1.25 for upper class subjects, 1.29 for middle class subjects, and 1.49 for lower class subjects (F = 7.76, $p<.001$). Although all the means were close to the highest choice classification of 1, there were still socioeconomic differences apparent. Sex differences were also significant, with female subjects having lower vocational aspiration, having a mean of 1.42 compared with 1.26 for males (F = 10.68, $p<.005$). It is interesting to note that similar sex differences have been found in the U.S. (e.g., Strong, 1955; Tyler, 1964; Campbell, 1974; Harmon, 1975; Farmer, 1976). The present study then shows that lower-class Iranian women are not only disadvantaged by sex, but also by their humble background.

Discriminating Attitudes

Having discovered that males and females in Iran differ significantly in vocational aspiration, a question which arises is: what types of sex role discriminations are being made by both sexes and are there agreements or disagreements between the sexes in some particular areas? Traditionally, in Iran as well as in other places (e.g., Farmer and Backer, 1976) women's role is considered to be at home taking care of husband and children, and housekeeping. Women are considered to be weak and not suitable to hold responsible positions at work. Employing women is considered by many Iranians to be not only a waste, because they get married and either quit work or do not take their jobs seriously, it is also considered a source of distraction for males (Women's Organization, Iran, 1976). To study the extent of sex role discrimination and to investigate the above questions, the following study was designed.

The subjects were 99 undergraduates and recent highschool graduates in Mashhad, Iran, including 44 males and 55 females from middle or upper socioeconomic backgrounds. It is worthy to note that subjects are among the better educated, younger generation of Iran. A thirteen-item questionnaire (see Table 1) based on Birk, Cooper and Tanney (1973) was used. This questionnaire was translated into Persian and back-translated for accuracy. It was then administered together with

Table 1
Questionnaire on Sex Role Discrimination

1. Women, rather than men, should have primary responsibility for ensuring that their children grow up healthy physically and psychologically.

   Agree M 65.1% F 60% *Disagree* M 18.6% F 21.8% *Neutral* M 16.3% F 18.2%

2. Women are absent from work more than men because of illness, therefore they cost their company more.

   Agree M 38.6% F 40.9% *Disagree* M 20.5% F 29.1% *Neutral* M 40.9% F 30.9%

3. Women are not seriously attached to the labour force; that is, they work only for extra "pin money".

   Agree M 47.7% F 30.9% *Disagree* M 27.3% F 54.6% *Neutral* M 25% F 14.6%

4. Women don't work as many years or as regularly as men; their training is costly, and largely wasted.

   Agree M 40.9% F 27.3% *Disagree* M 31.8% F 20% *Neutral* M 27.3% F 14.5%

5. When women work, they deprive men of job opportunities; therefore, women should quit those jobs they hold.

   Agree M 25 % F 3.5% *Disagree* M 59.1% F 92.9% *Neutral* M 15.9% F 3.6%

6. Women should stick to "women's jobs" and should not compete for "men's jobs".

   Agree M 54.6% F 10.9% *Disagree* M 31.8% F 81.8% *Neutral* M 13.6% F 7.3%

7. Women don't want responsibility on the job; they would prefer not to have promotion or job changes which add to their load.

   Agree M 22.7% F 14.5% *Disagree* M 38.6% F 74.6% *Neutral* M 38.7% F 10.9%

8. Children of working mothers are more likely to become juvenile delinquents than children of non-working women.

   Agree M 43.2% F 10.4% *Disagree* M 29.6% F 70.9% *Neutral* M 27.2% F 12.7%

9. Men don't like to work for women supervisors.

Agree M 72.9%   Disagree M 4.6%   Neutral M 22.7%
       F 70.9%   F 5.5%   F 23.6%

10. Women rather than men should have primary responsibility for housekeeping.

Agree M 93.1%   Disagree M 4.6%   Neutral M 2.3%
       F 89.1%   F 7.3%   F 3.6%

11. A girl doesn't have to support herself; her husband or family will support her.

Agree M 52.3%   Disagree M 34.1%   Neutral M 13.6%
       F 9.1%   F 74.5%   F 16.4%

12. Women have a higher turnover and absenteeism rate than men.

Agree M 27.3%   Disagree M 25%   Neutral M 47.7%
       F 25.5%   F 50.9%   F 23.6%

13. Women get married and then quit their work.

Agree M 34%   Disagree M 45.5%   Neutral M 20.5%
       F 23.6%   F 43.6%   F 32.8%

---

other tests, including the Bem sex role scale and a test for risk-taking behavior which will be discussed later. Subjects were tested in small groups of about 15 each, and the order of the different tests administered was randomized. Subjects were asked to indicate the extent of their agreement or disagreement with each item on a scale of 1 to 5, with a score of 1 for strongly agree and a score of 5 for strongly disagree.

Table 1 shows the actual items and the percentages of agreements, disagreements, and neutrals for male and female subjects. It is worthy to note that while male responses were sex discriminating, so were female responses, although less so. Female views on motherhood and household duties were particularly self-discriminatory. The majority of men (93.1%) and women (89.1%) believed that women, rather than men, should have primary responsibility for housekeeping. Both males (72.9%) and females (70.9%) agreed that men don't like to work for female supervisors. *About 48% of the men but also*

30.9% of the women believed that women don't take their jobs seriously. Similar findings have been reported in American studies (cf., Kievit, 1972; Darling, 1973; Farmer and Backer, 1976).

Sex Role Orientation

Apart from attitudes towards sex role accomplishment, studies have reported that males and females differ considerably in their personality and value orientation. Positive female traits are usually being warm, affectionate, sociable, understanding, and helpful towards others while positive masculine traits are being competitive, independent, and oriented towards mastery (cf., Bem 1974). Veroff (1969) argues that the importance of interpersonal gratification, constantly emphasized in the socialization of girls, means that few girls develop a strong need to achieve. Therefore, "girls tend to fuse the need to achieve with the need to affiliate, using achievement as a means of securing acceptance and love." Sears (1962) found that affiliation rather than achievement needs of girls correlated with academic success. Similar findings have been reported by Tyler, Rafferty, and Tyler (1962). In the present study an attempt was made to investigate sex differences in sex role orientation and to establish the relationship of these characteristics to one important aspect of vocational and achievement behavior, namely risk taking.

The same 99 subjects as above (44 male and 55 female) were tested. The Bem sex role inventory and a questionnaire testing risk-taking behavior, based on Kogan and Dorros (1975) were translated and back-translated to check for accuracy. The Bem inventory items were loaded either on female value characteristics, male value characteristics, or social desirability. Subjects were asked to rate on a 5-point scale how well each item described themselves. Scores computed from the Bem sex role inventory were: a masculinity score, a femininity score, a social desirability score, and a simple androgyny difference score which is the difference between the femininity and masculinity scores. The mean masculinity and femininity scores for males were 3.63 and 3.34 respectively, while the corresponding mean scores for females were 3.36 and 3.56. An analysis of variance with males versus female subjects as one factor and male versus female scores as the other factor indicated a significant interaction effect, with male subjects scoring higher on the masculinity items and female subjects scoring higher on the femininity items, significant at the .0001 level (F (1,79) = 41.3). Bem (1974) has reported similar findings in the U.S.

Social desirability scores of male and female subjects did not differ significantly. They all described themselves in mildly socially desirably ways, with a mean score of 3.77 for females and 3.78 for males. There was a significant difference between the androgyny scores of males and females, with a mean of -.27 for males and .22 for females (t (97) = 4.19, $p < .005$). This shows that male subjects were more sex-typed than female subjects.

Correlation analysis involving masculinity scores, femininity scores, and social desirability scores, for male as well as female subjects were made. Results showed that femininity scores were significantly correlated with social desirability scores ($r = .37$ for male subjects, $p < .01$; and $r = .48$ for female subjects, $p < .001$). The correlation between androgyny scores and social desirability scores was significant only for female subjects ($r = .24$, $p < .05$). This indicates that those female subjects who had higher androgyny scores, that is, those who were more female sex-typed, also had higher social desirability scores in their self description. Bem (1974) has also reported a similar finding in the U.S. These findings suggest that *females rely more on social reinforcement than males.*

Correlation analyses were made between risk-taking scores and respectively, masculinity, femininity, and androgyny scores, for males and females separately. Only two significant correlations were found, both for females. The correlation between the masculinity scores of female subjects and their risk-taking scores was .37 ($p < .01$). The correlation of their risk-taking scores and androgyny scores was -.34 ($p < .01$). Both correlations show that female subjects who had more masculine value characteristics also tended to take higher risks in test items.

Sex Differences in Achievement Orientation

In the U.S. and elsewhere a number of studies suggest that female attributional patterns differ from those of the male. Rotter (1966) reported that female college students are more externally oriented than male. Some other studies (Bar-Tal and Frieze, 1973; Simon and Feather, 1973) have reported that females attribute the cause of success and failure to luck more often than males. Women also tend to attribute cause of success less to their own ability (McMahan, 1971) and cause of failure more to lack of ability. On the other hand, attribution of success or failure to internal or external factors affects people's self-esteem: maximum self-esteem is related to internal attribution and success ascribed to internal factors results in greater pride (Weiner, 1970; 1972). Our finding of female sex role orientation reported above suggests that social reinforcement is the most important for females and hence women's reliance on external factors for reward. Therefore, a woman achieving in a male oriented task may not only feel less proud but also very anxious if rejected by important others (cf., Horner, 1968). In western studies, women with traditional female role orientation tend to score lower on achievement-motivation measures than those with non-traditional female orientation (Bem, 1974; Alper, 1974; Lipman-Blumen, 1972). The following study is an attempt to investigate sex differences in the internal-external attribution of cause of success or failure.

Subjects were 175 male and 201 female, from elementary and high schools and of all socioeconomic backgrounds. The Intellectual Achievement Responsibility Scale of Crandall, Katkovsky, and Crandall (1965) was adopted for use in Iran in a similar way as in the above studies. The test includes 34 items representing positive or negative experiences relating to school achievement. This is followed by two alternatives for subjects to attribute causes of events, to either internal (ability or effort) or external (environmental) factors. Subjects were asked to choose one of the two alternatives.

The same scoring procedure used by Dweck and Reppucci (1973) was adopted. Scores were computed for measures of responsibility for success, responsibility for failure, responsibility for effort, responsibility for success attributed to ability, success attributed to effort, failure attributed to ability, failure attributed to effort, as well as overall internal responsibility. ANOVAs were done to test for significant sex differences. The mean overall internal responsibility score for females was 48.4 and for males 52.0 (F (1,374) = 5.09, $p < .05$). Significant sex differences were also found for five other of the above subscores, and a marginal one for the remaining internal attribution of success to ability ($p < .07$), all in the same direction with male subjects attributing cause more to internal factors. As it is evidenced by the results, our findings strongly match those of the U.S. studies discussed earlier.

Conclusion

In this chapter I have discussed two types of determinants of achievement motivation; the first is social and the second is psychological. Needless to say the two are closely related. In the case of the social environment in Iran, we saw the lack of opportunities, both educational and otherwise for development of career achievement by women. This is coupled with discriminating sex role attitudes and conservative religious views about women's place in society. On the other hand, we noted some similarities in the psychological make-up of Iranian women and those of developing nations. Both Iranian women and western women are more externally oriented than men; they also agree that household activities and child rearing are their responsibility. They both have lower job aspirations and rely more on social reinforcement for achievement, as was indicated by the high correlation between femininity scores and social desirability. Our results also show that *those women who had higher masculinity scores also had higher risk-taking scores*. All these suggest that Iranian women, like their counterparts in developed nations, experience home-career conflict, depend on social approval, and feel anxiety and less proud than men when succeeding in male dominated tasks (cf., Horner, 1968).

If the basic psychological make-up of women everywhere appears to be similar, the greater difference between the condition of women in developing and women in developed physical handicaps for achievement of Iranian women. These barriers are lower in developed countries.

The actual greater physical handicaps for women in developing countries also mean that the chances for the sensitization of these women to their problems and role are smaller. In developed countries like the U.S. most women are aware of movements in society towards equality, etc. The media that brings this awareness, the means of communication, are also precisely those aspects of material life that women in developing countries are less accessible to. The problem of educating both men and women in developing countries towards sex equality is therefore compounded.

On the other hand, developing countries could possibly, at times, recognize an urgency in national purpose which might enhance equality of sex roles. In his analysis of women's role in World War II, Trey (1972) shows that under a situation of social and economic pressure, when all members of society have to be active and productive, a possibility of change in sex role exists. It is possible, then, that developing countries might consider the need to advance, to close the gap with developed countries as an unusual situation requiring departure from stereotypes of sex roles.

REFERENCES

Alper, T. Achievement motivation in women: Now-you-see-it-now-you-don't. *American Psychologist*, 1974, *29*, 194-203.

Atkinson, J.W. (Ed.). *Motives in fantasy, action and society*. Princeton, N.J.: Van Nostrand, 1958.

Atkinson, J.W. *An introduction to motivation*. Princeton, N.J.: Van Nostrand, 1964.

Bardwick, J.M. *Psychology of women: A study of bio-cultural conflict* New York: Harper & Row, 1971.

Bar-Tal, D., & Frieze, I. *Achievement motivation and gender as determinants of attributions for success and failure*. Unpublished manuscript, University of Pittsburg, 1973.

Bem, S. The measurement of psychological androgyny. *Journal of Consulting and Clinical Psychology*, 1974, *42*, 155-162.

Birk, J., Cooper, J., & Tanney, M. *Racial and sex role stereotyping in career information illustration*. Paper presented at the meeting of the American Psychological Association, Montreal, 1973.

Callaway, H. *The analysis of an educational setting: An anthropological approach.* Paper presented at the International Seminar sponsored by the Women's Organization, Tehran, Iran, 1975.

Campbell, D.F. *SVID-SVII manual.* Stanford University Press, 1974.

Crandall, V.C., Katkovsky, W., & Crandall, V. Children's belief in their own control of reinforcement in intellectual academic achievement situations. *Child Development,* 1965, *36,* 91-109.

Darling, M. *The role of women in the economy: Meeting of Experts.* Unpublished manuscript, Washington, D.C., 1973.

Dweck, C.S., & Reppucci, N.D. Learned helplessness and reinforcement responsibility in children. *Journal of Personality and Social Psychology,* 1973, *25,* 109-116.

Farmer, H. What inhibits achievement and career motivation in women. *The Counselling Psychologist,* 1976, *6,* 12-14.

Farmer, H., & Backer, T. *Women at work: A counsellor's sourcebook.* New York: Behavioral Publications, 1976.

Frazier, E.F. *Black bourgeoisie.* New York, Collier Books, 1976.

Harmon, L. Career counselling for women. In D. Carter, & E. Rawlings (Eds.), *Psychotherapy for women: Treatment toward equality.* Springfield, Il: Thomas & Sons, 1975.

Horner, M.S. *Sex differences in achievement motivation and performance in competitive situations.* Unpublished doctoral dissertation, University of Michigan, 1968.

Kievit, M. *Review and synthesis of research on women in the world of work.* U.S. Government Printing Office, Washington, D.C. 1972.

Kogan, N., & Dorros, K. *Sex differences in risk taking and its attribution.* Paper presented at the meeting of the American Psychological Association, Chicago, 1975.

Lipman-Blumen, J. How ideology shapes women's lives. *Scientific American,* 1972, *226,* 34-42.

Maehr, M.L. Sociocultural origins of achievement motivation. In Bar-Tal, D. & Saxe, L. *Social psychology of Education: Theory and research.* New York, Halsted Press, 1978.

McClelland, D.C., Atkinson, J., Clark, R., & Lowell, E. *The achievement motive.* Appleton, New York, 1953.

McClelland, D.C. The importance of early learning in the formation of motives. In J.W. Atkinson, (Ed.), *Motives in fantasy, action and society.* Princeton, N.J.: Van Nostrand, 1958.

McMahan, I.D. *Sex differences in causal attributions following success and failure.* Paper presented at the European Psychological Association, 1971.

Mednick, M.T.S. Social change and sex role inertia: The case of the Kibbutz. In Mednick, M.T.S., Tangri, S.S., & Hoffman, L.W. *Women and achievement: Social and motivational analyses.* New York: Hemisphere, 1975.

Pettigrew, L.A. *Profile of the negro Americans*. Princeton, N.J.: Van, Nostrand, 1964.
Plan Organization, Iran. *Yearly census report of the center for statistics*, Tehran, 1977.
Rotter, J.B. Generalized expectancies for internal versus external control of reinforcement. *Psychological Monographs*, 1966, *80*, (Whole no. 609).
Simon, J.G., & Feather, N.T. Causal attributions for success and failure at University examinations. *Journal of Educational Psychology*, 1973, *64*, 46-56.
Sears, P.S. *Correlates of need achievement and need affiliation and classroom management, self-concept and creativity*. Stanford University, Laboratory of Human Development, Unpublished manuscript, 1962.
Strong, E. *Vocational interests 18 years after college*. Minneapolis, Minnesota: University of Minnesota Press, 1955.
Trey, J.E. Women in the war economy—World War II. The *Review of Radical Political Economics*, 1972, *4*, 1-17.
Tyler, L. The antecendents of two varieties of interest patterns. *Genetic Psychology Monographs*, 1964, *70*, 177-227.
Tyler, F.B., Rafferty, J., & Tyler, B. Relationship among motivations of parents and their children. *Journal of Genetic Psychology*, 1962, *101*, 69-81.
U.S. Women's Bureau. *The myth and the reality*. Washington, D.C.: Printing Office, 1974.
Veroff, J. Social comparison and the development of achievement motivation. In Smith, C. (Ed.), *Achievement related motives in children*. New York: Russell Sage Foundation, 1969.
Weiner, B. *Theories of motivation*. Chicago: Markham, 1972.
Weiner, B. New conceptions in the study of achievement motivation. In B. Maher (Ed.), *Progress in Experimental Personality Research*, 1970.
Weston, P.J., & Mednick, M.T.S. Race, social class, and the motive to avoid success in women. In M.T.S. Mednick, S.S. Tangrie, & L.W. Hoffman, *Women and achievement: Social and motivational analyses*. New York: Hemisphere, 1975.
Women's Organization, Iran. *Occupation of women in positions of power and decision making*. Tehran, 1976.

Chapter 17

Women's Achievement and Career Motivation: Their Risk Taking Patterns, Home-Career Conflict, Sex Role Orientation, Fear of Success, and Self-concept*

HELEN S. FARMER
The University of Illinois

LESLIE J. FYANS, JR.
*The University of Illinois
and the Illinois State Board of Education*

A particular need in psychology is a female relevant model of achievement. This female relevant model would incorporate the relationship among some of the environmental, background, and psychological variables found by previous studies to influence women's achievement and career motivation. A dominant career motivation model represented by the work by Super (1957, 1976) defines the highly career motivated person in terms not unlike those of the achievement motivation model (i.e., persistence, independence, intrinsic motivation, self-esteem, etc.) Recently Super (1976) emphasized the point that a comprehensive theory of career motivation must take into account both situational and personal determinants and the ways in which these interact at various stages of individual development. Achievement motivation theory and research has seldom attempted to integrate or articulate with career motivation theory and research (Farmer & Backer, 1977). However, career motivation is an important dimension of the more specific achievement motivation and relates not only to post-schooling employment motivation, but also to the development throughout life of a range of work related interests,

---

*A smaller version of this chapter submitted for publication to *Psychology of Women Quarterly*.

This study was conducted in part with funding from the University of Illinois Research Board.

values and skills.

It has long been known that women do not achieve as much as men in the sciences, humanities, and the arts (Astin, 1967; Maccoby & Jacklin, 1974; Rossi & Calderwood, 1973). In spite of the fact that they represent over 40% of the professional labor force (Sexton, 1977) today, women represent less than 20% of the managers and administrators in the United States. What inhibits this achievement and productivity in women?

The model of achievement motivation developed by Atkinson (Atkinson & Raynor, 1974) and McClelland (1971) is fairly well established for middle class boys and men but not for girls. The model has identified several behaviors as typifying the high achiever: independence, persistence, preference for tasks of intermediate difficulty, high academic performance, and intrinsic motivation. This model does not hold up for girls (Bardwick, 1971; Horner, 1968) or for persons from other cultures and socioeconomic backgrounds (Maehr, 1974). Inconsistencies with the model are found for girls who obtain high scores on achievement motivation but do not show the predicted risk preferences for tasks of intermediate difficulty (Horner, 1968). Nor is girls' academic performance as high as that found for boys with the same achievement motivation scores (Horner, 1968).

A model more relevant to the achievement motivation of women might be one which includes the effects of *sex role socialization* practices and of present *discrimination and/or support systems* for females' achievement behaviors in their environment. The proposed model assumes that early sex role socialization practices lead to certain psychological predispositions such as risk preference, sex role orientation, self-esteem, and home career conflict and that these different dispositions produce different achieving behaviors depending on the support or lack of support a woman perceives to be present in her environment for her achieving behaviors.

One of the differences between achievement motivation and its behavioral manifestations and similar manifestations of career motivation is the striving for excellence in self-selected areas in the former (Atkinson & Raynor, 1974) whereas there is a striving for self-fulfillment in the latter (Super, 1963;

Holland, 1973). Career motivation has a lifetime, longterm dimension made up of a series of satisfying experiences, but ultimately ceasing only at the termination of life itself. Achievement motivation may be targeted to specific areas (i.e., grades in school or sports) and continually be transferred to new areas as accomplishments in one area no longer challenge the subject.

Atkinson and Raynor's (1974) view of the goal of cumulative achievement (i.e., career motivation) is limited to extrinsic rewards such as salary increases, promotions, status, fame and power, whereas career motivation theorists such as Super (1957, 1976) and Holland (1973) have stressed intrinsic rewards. Examples of intrinsic rewards include self-fulfillment, autonomy, and creativity. A view of career motivation as powered by both intrinsic and extrinsic goals seems more in keeping the view of female motivation as including a cooperative, contributory achievement style, and a valuing of contributions to the social welfare (Stein & Bailey, 1973).

Researchers on career motivation have documented differences in the career development of boys and girls other than those found by achievement motivation researchers (i.e., risk preference and level of achievement). Differences have been found in the career interests of boys and girls, men and women (Harmon, 1972). Differences in the timing of career choice have also been found (Ginsberg, 1966). In particular marriage plans have been found to have a moderator effect on the career motivation of girls greater than that found for boys (Psathus, 1968; Harmon, 1972).

To determine if women are motivated toward particular achievement tasks in ways similar to those they experience for cumulative achievement (i.e., career) both criteria variables (i.e., achievement and career motivation) were investigated.

A multiple level model (cf. Fyans, 1979) was the framework for defining the predictor variables of career and achievement motivation. This multiple level model incorporated *psychological* and *environmental* variables.

Psychological Variables

Self-Confidence

Maccoby and Jacklin's (1974) review of research of sex

differences indicated that little difference exists in general levels of self-confidence between girls and boys and men and women. However, differences emerge when self-confidence is studied in relation to specific tasks. These differences, although evidenced during adolescence, become significant in the college years. At this time women have less confidence than men in their ability to perform well on a variety of tasks (i.e., academic grades and anagram tasks) and have less sense of being able to control events that effect them. Women's self-confidence, however, is higher than men's in relation to social self-esteem (i.e., cooperation, empathy). Maccoby and Jacklin concluded that each sex appears to have higher self-confidence in areas of central ego involvement. It would seem important therefore to investigate self-esteem in relation to areas such as family, and social relations as well as academic self-esteem for purposes of understanding female achievement motivation.

Sex Role Orientation

Thelma Alper (1974), who has been studying the relationship of sex role orientation to achievement motivation in women for more than a decade, found that women with traditional female orientations, attitudes, and beliefs score lower on achievement motivation measures than women with non-traditional female orientations. Sandra Bem (1977) has advanced the notion of psychological androgyny, a sex role orientation balanced on traditional masculine and feminine traits. Bem found such women more "field independent" with respect to sex role stereotypes. Androgynous women were able to choose behaviors and activities for reasons other than sex role appropriateness. Other studies suggest that androgyny may be more characteristic of women who choose non-traditional careers (Tipton, 1976). Androgyny is a promising construct which deserves more study, especially on its behavioral correlates for both sexes, and its relationship to achievement.

Home-Career Conflict

Home-career conflict is understood to result for women

when they value both homemaking and career roles and at the same time view these roles as incompatible. This view of conflict is consistent with social role theory (Sarbin, 1954) and dissonance theory (Festinger, 1957). Many females are still socialized to believe that they cannot combine home and career and do them both well (Bem & Bem, 1973). Dissonance results for some women who wish to combine home and career but continue to believe that they can't without hurting: (a) the important relationship with their husband, or (b) the psychological health of their children.

Home-career conflict has been found to affect women's psychological outlook in a negative fashion. Early studies by Matthews and Tiedeman (1964), who examined the career choices of over 1,000 girls and college women, found that females who exhibited conflict felt guilty if they worked and depressed or frustrated if they stayed at home. Astin (1967) found that women feel depressed and hassled by having to both work and maintain a home. In a later study Astin (1976) found that reentry women (i.e., those returning to higher education after an absence) experience conflict and guilt about "leaving their homes and families to undertake a timeconsuming venture so personally fulfilling." In their recent review Tittle and Denker (1977) stated that home-career conflict continues to be an important mediator of career choice for women, however, its exact effect remains unclear.

Fear of Success

Horner (1968) did a study focused on unraveling some of the anomalies found by previous researchers on the female's motive to achieve. One of her hypotheses was that women's measured achievement motivation was inconsistent with that of men's because of conflict that some women experience between academic success and their feminine role.

Since 1968 several researchers have studied fear of success in men and women both in the U.S. and abroad (Feather, 1974; Gump & Rivers, 1975; Horner & Fleming, 1977; Monahan, Kuhn & Shaver, 1974; Tomlinson-Keasey, 1974; Zuckerman &

Wheeler, 1975). Horner and Fleming have extended the original measure to include not one but three story cues in an effort to establish greater validity in measuring this conflict. This new measure appears promising (Fleming, 1977). Shaver (1976) proposed that there is evidence that fear-of-success is not independent of fear-of-failure, particularly as conceived in the new scoring system developed by Horner and Fleming. The effect of this construct on achievement behavior both in school and in employed work needs to be clarified. Horner (1968) found that highly motivated women, high in fear-of-success, performed less well in competitive conditions compared to men of comparable ability. If competition with men reduces the performance of highly capable conflicted women, this construct merits investigation in any study of achievement and career motivation.

Risk Taking and Career and Achievement Behavior

Risk taking behavior has been found to be related to academic achievement and performance as well as career success by McClelland (1971). Boys who are highly motivated and high achievers in school demonstrate a moderate risk preference pattern in which uncertainty is not avoided and challenge is viewed positively. By taking on challenges boys are found to optimize their potential. Evidence from employed men supports this pattern for the career optimization of adults as well (McClelland 1971). Poor achievers are found to exhibit either a "high risk" pattern or a "low risk" pattern. This pattern, like the motivation model described earlier, does not hold for girls and women (Horner, 1968). High achieving girls with high motivation do not always exhibit the moderate risk pattern found for boys. Career motivation research too has found that risk taking ability is related to calculating the odds for and against a particular set of possible choices at each decision-point in a career (Super, 1976). The greater uncertainty women experience about the odds for and against success in a career may account for their less predictable risk patterns.

Environmental Influences

The Social Context

At least three aspects of the social context affect the achievement and career motivation of women (Edye, 1970). One aspect is the availability of resources which support the educational and career development of women. These include nondiscriminatory educational admissions and employment practices as well as the availability of child care and homemaking assistance. A second aspect is the expectations of significant others in the environment (i.e., parents, husbands, employers, teachers, etc.). A third aspect contributing to the motivation of women is supportive legislation related to providing equal opportunity and equal rights for all.

Harmon (1972) found it harder to predict the stability of career choice for college coeds compared to men, possibly due to the mediating effect of marriage plans. She found that women who aspired to high level careers in their freshman year often changed their choices to less demanding careers by the time they were college seniors. Harmon hypothesized that lack of reinforcement in the environment for their high career aspirations indirectly reinforced lower career motivation for these women. In support of Harman's finding, Hawley (1972) found that college women were influenced to raise or lower their aspiration level, depending on whether or not the attitudes of the men they knew toward working women were positive or negative. Tomilson-Keasey's (1974) finding that married, older, coeds had higher levels of motivation compared to unmarried coeds (as well as lower "fear of success" scores) also lends support to this thesis.

Early Socialization in the Family

The effect of early socialization patterns in the family has been found to effect motivation in boys differently from girls (Moss & Kagan, 1961). Results of studies are mostly limited to middle-class white families where boys of parents who encouraged certain attitudes and behaviors were found to be highly achievement motivated. Examples of such attitudes are "hard work is good," and "the real reward comes from knowing you have performed well." Examples of behaviors are showing initiative, being independent, and delaying gratification (Rubovits, 1975).

Rubovitz found that within the same families girls were found to receive different social learning experiences from boys. While boys were encouraged to be competitive, initiating, achieving, and independent, girls were often encouraged to be dependent, conforming, cooperative, and unconcerned about grades (Rubovits, 1975). When women raised in such households find themselves faced with the necessity to work, they are poorly prepared for the competitive marketplace or choose to work commensurate with their potential.

Purpose

The present investigation examines the relationship of environmental and psychological factors reviewed above to the career and achievement motivation of women returning to college after an absence to raise a family. This study was part of a larger study (Farmer, 1977) which included two other age groups: (a) college males and females (N=209) and (b) eleventh grade high school males and females (N=249). Where relevant, findings from this study are compared with findings from the larger study. An important contribution of this study was the extension of the model with the incorporation of sex role orientation as a classification variable. The relationship of sex role orientation to female achievement has been noted previously (Bem, 1977; Spence, 1978; Alper, 1974). It was hypothesized that sex role orientation could be employed as a classification variable, used in determining whether a student had an adrogynous, sex typed, or undifferentiated orientation. The question then became, what patterns of predictor variables would occur related to different levels of female achievement within each sex role group. Thus, this study was conducted from a *within-groups* framework, being guided by the following research question: Within each sex role group, which combination of variables relate to different levels of both achievement and career motivation when considered simultaneously?

It was hoped that such a within groups strategy would help define counseling and innovative programs to facilitate the achievement of females of differing sex role orientations.

Method

Subjects

Female subjects were 53 married mothers with a mean age of 37, who had returned to college. A state college in Illinois provided the site for these women.

Measures

The two criteria variables are described first, followed by the seven predictor variables. Reliability estimates for the measures are given in Table 1.

Table 1.
Criteria and Predictor Variables

Variable	Reliability[a]
Career Motivation	.71
Achievement Motivation	.91
Fear of Success	.90
Home-Career Conflict	.92
Self-Esteem Total	.92
Home Self-Esteem	.82
Academic Self-Esteem	.65
Family Self-Esteem	.66
Risk	.55
Sex Role Inventory	.86
Early Socialization	.86
Community Support	.93

[a] Reliability estimates are based on data collected for this study

Achievement motivation. Following the framework of Horner (1968), verbal leads were used to elicit imaginative stories. These verbal leads were based upon the research of Atkinson and Litwin (1960), French (1955) and Winterbottom (1953). Scoring for this measure was consistent with Atkinson (1958).

Career motivation. Career motivation was measured using Holland's (1973) and Roe's (1956) procedure for scoring occupational aspirations and the expressed level of career motivation of subjects. Predictive validity of this approach has been found by Holland (1973) to be .44.

Sex role orientation. The Bem Sex Role Inventory (BSRI) was selected to assess sex-role orientation (Bem, 1977). The BSRI consists of 60 items with three scales: masculine, feminine, neutral. Neutral scale scores were used to check for high social desirability response sets. Subjects were classified into one of four groups based upon their median scores on the masculine and feminine scales of the BSRI: (a) *sex typed* (if members of either sex had a score above the median on the scale for their sex), (b) *androgynous* (if subject's scores were higher than the medians on both sex scales), (c) *sex reversed* (if members of a particular sex had a score above the median on the scale opposite their sex), (d) *undifferentiated* (if subjects' scores were lower than the medians on both scales). Bem's (1977) medians were used for the Community College subjects.

Early socialization in the family. The measure tapping early socialization patterns of achievement and career motivation was developed by Fyans (1979). Factorial derived dimensions are as follows: Factor I: encouragement by parents of independence; Factor II: encouragement by parents of achievement behaviors; and Factor III: a warm-rewarding versus cold-rejecting home environment.

Perceived community resources. Three aspects of the social context which affect female achievement were assessed with this instrument. These aspects were: (a) availability of supportive resources (i.e., child care and nondiscriminatory admissions), (b) expectations of significant others in the environment, and (c) supportive legislation for equal rights and equal opportunity.

Self-concept. The Coopersmith inventory of self-esteem

(Coopersmith, 1967) was selected for this study. This Coopersmith scale can be divided into three independent self-concept scales: (a) in relation to social situations, (b) in relation to home environment, and (c) in relation to academic activities.

Risk taking tendency. The Kogan and Dorras (1975) adaptation of Kogan and Wallach's (1959) Choice Dilemnas Questionnaire (CDA) was used for this study. Six of the nine Kogan and Dorras items were used, eliminating three items found to be ambiguous (i.e., items about choices between physician and musician, and political risks in foreign countries). The remaining six items were balanced on work/school risks and on male/female actors. Preliminary analysis indicated, in line with the findings of Atkinson and Raynor (1974), a curvilinear relationship for this variable with achievement, Therefore the risk scores of each subject were squared to make them fit a linear scale for analysis.

Home career conflict. This measure was developed from previous work by Farmer and Bohn (1970) and Alper (1974). Alper reasoned that the projective format was justified based on Anastasi's (1976) argument that controversial values or attitudes are best assessed in this way. Four cues were developed representative of a range of home and work situations but ambiguous with respect to type of work or family.

Fear of success. The measure used in this study was similar to that suggested by Horner and Fleming (1977). Four new cues were developed, two with school settings and two with work settings. Cues were ambiguous with reference to field of study/work excluding specific fields such as chemistry, nursing, and medicine.

Construct Validity

An alpha factor structure was derived from an item-item analysis for all measures, exclusive of projectives (i.e., fear of success, home-career conflict and criteria measures). The successive differences between the eigenvalues after four factors were extracted was one-fifth as large as before and therefore no more factors were extracted. The four independent factors found were community support, sex role orientation, self-esteem and early socialization, providing some construct validity to these scales. The risk variable did not have a strong factor structure.

Administration of Measures

To control for order effects, a table of random numbers was used for ordering the tests each time a new group of women was tested. Examples of the random ordering of test presentation are available from the authors.

Analyses

Canonical analyses (Tatsuoka, 1971) were used because they permitted testing the effect of predictors on both achievement and career motivation variables simultaneously. For purposes of grouping for the canonical analyses, subjects' sex role scores (Bem, 1977) were used to identify feminine sex typed, sex reversed, androgynous, and undifferentiated groups. The canonical analyses treated career and achievement motivation as criteria variables and the nine dependent measures as predictor variables. Interpretation of significant canonical variates followed Tatsuoka: only the highest absolute values and those equal to about half or better are presented for discussion. Canonical analysis is useful in answering the question "What sort of self-concept variables are associated with what sort of motivation pattern?" Canonical analysis helps answer this question by determining linear combinations of the self-concept scales that are most highly correlated with linear combinations of the motivation measures. The technique has been compared to a doublebarrelled principal components analysis (Tatsuoka, 1971).

Results

Descriptive Data

Prior to the within groups canonical analyses, initial descriptive statistics were generated to explore the nature of the data. The dependent variable, achievement motivation, had a mean of 4.28 and standard deviation of 5.62. Achievement motivation scores ranged from -3 to 19. In employing the exploratory data analysis techniques of Mosteller and Tukey (1977) *three distinct clusters* of achievement motivation scores were found. What could be called the *high* achievement motivation group had scores which clustered around 4.35 on the scale. The *moderate* achievement motivation group had their scores centered around 1.69 on the achievement motivation scale. The scores for the *low* achievement group were clustered at the -.93 score on the scale.

Finally, it should be noted that the best predictions of achievement motivation were for students who scored at 5.8 and that the errors in prediction were higher above or below a score of 5.8.

Correlationally, the highest zero order associations with achievement motivation were early socialization (-.18), self-concept academic (.17), and risk-taking (.15), all nonsignificant. These correlations are based upon the total continuing education sample and moderation could occur when the sample is grouped by sex role or when other variables are accounted for.

The mean for the dependent variable, career motivation, was 4.53 with a standard deviation of 1.04 and a range from 0 to 6.

An exploratory data analysis (Mosteller & Tukey, 1977) provided *four distinct clusters* of career motivation scores. The scores for persons that could be called the *very high* career motivation group clustered around a score of 5.90 on the scale. The second group of *high* career motivation was centered around 4.80 on the career motivation scale. A group that could be defined as having *moderate* career motivation clustered about the 3.80 scale score. Finally, the *low* career motivation students were centered around a score of 3.00 of career motivation.

The highest zero order correlations with career motivation are with fear of success (.44, $p<.01$), perceived community support, (.16), and early socialization (.15). The latter two were non-significant. However, career and achievement motivation were significantly negatively correlated (-.34, $p<.01$) for this group of females.

Androgynous[1] Females

The first canonical equation (Equation I) relating career and achievement motivation with the independent variables for androgynous females was significant at the .001 level. The canonical R^2 between these sets was .78. Following the techniques of Tatsuoka (1971), the variables describing this canonical variate are given in Table 2.

The varibles described in Table 2 show that there is a group of androgynous females who have high career motivation (1.02) but also exhibit high fear of success (.86). In addition, Equation I shows these women to view themselves as socially unpopular (-.72). These androgynous women also displayed a tendency to take very high or very low risks (-1.42). Thus, intermediate risk taking, characteristic of high need achievement persons (Atkinson & Raynor, 1974) did *not* accompany androgynous women's high career motivation scores. Interestingly enough, for this androgynous group, career and achievement motivation were negatively correlated ($r = -.23$, $p<.05$).

The second canonical equation (Equation II) which was significant ($p<.025$) for androgynous females had an overall R^2 of .64. The varibles describing Equation II are given in

Table 2.
Canonical Variate One [a] for Androgynous Sex-Typed College Females

Predictor-Variable Set	Criterion-Variable Set
High Positive Weights	*High Positive Weights*
Fear of Success (.86)	Career Motivation (1.02)
High Negative Weights	*High Negative Weights*
(Risk)^2 -1.42	
Social Self-Esteem (-.72)	

[a] $R = .78***(p<.001)$
(61% of the variance)

---

[1] No sex reversed women were found in the study reported and only seven women were categorized as undifferentiated.

Table 3. The variables weights as described in Table 3 reveal a group of androgynous females who are characterized by high need achievement motivation but *low* career motivation. Likewise, these androgynous females appear to have very high self-esteem for academic pursuits (.56). However, the androgynous females described by Equation II had a decided *lack* of self-esteem for home-family affairs (-.95). Similarly, they felt a lack of support for achievement from either their early family socialization (-.60) or from their present community (-.41).

Sex Typed Females

The first canonical equation (Equation I) for sex typed females was significant at the .01 level. The overall R^2 for this equation was .72. The variables describing the relationship for sex typed females are given in Table 4.

The weighted pattern of variables in Table 4 reveal a group of sex-typed females who have high achievement motivation (.63) but low career motivation (-.60). This group of sex-typed women exhibited a high self-concept for academic pursuits (.93). However, this group also felt their early socialization environment was *unsupportive* of their achievement motivation (-.38). Similarly they perceived themselves as having low self-esteem in relation to home affairs (-.41). These females expressed low feelings of conflict between home and career

Table 3.
Canonical Variate Two [a] for Androgynous Sex-Typed College Females

Predictor-Variable Set	Criterion-Variable Set
High Positive Weights	*High Positive Weights*
Academic Self-Esteem (.56)	Need Achievement (.95)
High Negative Weights	*High Negative Weights*
Early Socialization (-.60)	Career Motivation (-.44)
Home Self-Esteem (-.95)	
Community Support (-.41)	

[a] $R = .64*$ (p<.025)
(41% of the variance)

Table 4.
Canonical Variate Two [a] for College Feminine Sex-Typed Females

Predictor-Variable Set	Criterion-Variable Set
High Positive Weights	High Positive Weights
Academic Self-Esteem (.93)	Need Achievement (.63)
High Negative Weights	High Negative Weights
Early Socialization (-.38)	Career Motivation (-.60)
Home Self-Esteem (-.41)	
Home-Career Conflict (-.47)	

[a] R = .72** (p .01)
(52% of the variance)

(-.47). As with androgynous females high in career motivation, these sex typed females expressed a negative correlation between achievement motivation (r = -.32, $p<.01$).

It is interesting to compare the results shown in Table 3 with those of Table 4. They both show results for females high in achievement motivation but low in career motivation, except one is for the androgynous females and the other for sex typed. As can be seen a major difference revolves around the feeling of lack of support from the community for androgynous females (-.41, Table 3) and the lack of home career conflict expressed by the sex typed females (-.47, Table 4).

The second canonical equation (Equation II) for sex typed females was also significant at the .01 level. This equation had an overall R^2 or .89. The variables defining Equation II are given in Table 5.

The results given in Table 5 indicate there is a group of sex-typed females who are high in both achievement (.87) and career motivation (.85). However, these doubly motivated sex-typed females also experience high levels of fear-of-success (.56). These women also felt strong community support for their achievement (.75). These sex-typed women expressed no preference for tasks of moderate risk (-1.40), but contrary to Atkinson and Raynor (1974) they chose tasks of very high or low riskiness. Further, they perceived themselves as socially

Table 5.
Canonical Variate One [a] for College Feminine Sex-Typed Females

Predictor-Variable Set	Criterion-Variable Set
High Positive Weights	High Positive Weights
Community Support (.75)	Career Motivation (.85)
Fear-of-Success (.56)	Need Achievement (.87)
High Negative Weights	High Negative Weights
(Risk)^2 (-1.40)	
Social Self-Esteem (-.98)	

[a] $R = .89^{**}$ ($p<.01$)
(79% of the variance)

unpopular (-.98). It is indeed interesting to note that the results just presented for the sex-typed females (Table 5) high in both achievement and career motivation almost mimic those for androgynous females who are high in only career motivation (Table 2). The major difference appears to center upon the inclusion of perceived community support for the sex-typed females.

Discussion

There are several important findings for female achievement and career motivation theory from the present study. Perhaps one of the more interesting ones concerns fear-of-success. Rather than being an inhibitor of achievement or career motivation, fear-of-success indeed appears to be a *co-terminus variable*. This finding holds for both androgynous and sex-typed females. Thus, it appears that as motivation in females increases, collaterally so does their fear that their actions will result in success with related losses in the area of heterosexual relations (Horner, 1968). When and where this conflict debilitates the tendency to achieve was not uncovered in this study. Previously Horner (1968) found high achievement motivated women with high levels of fear-of-success performed less well on academic tasks when they were competing with men of comparable ability. It may be that the debilitative effect of fear-of-success is an age-related phenomenon, since other data obtained from college students during the same time period indicated its inhibiting function there (Farmer, 1977).

Also interesting is the non-preference of motivated females of tasks of moderate risk. Noteworthy here is that this preference for either very high or low risks is found for both androgynous and sex-typed females. More revealing is that this risk preference is also expressed by the high fear-of-success women (see Tables 2 and 5). It is possible that the manifestation of high fear-of-success with high motivation influences females to choose careers (etc.) whose risk levels will decrease their fear-of-success. Thus they choose tasks on which they are either guaranteed sure success or sure failure and they completely avoid tasks in which there is less information about how the outcome will turn out (Weiner, 1974). This connection between female achievement, risk preference, and fear-of-success could be examined in future studies which focus upon exactly which levels of motivation are associated with which kinds of risks and which kinds of risks with fear-of-success sets. A path analytic diagram via LISREL (Joreskog, 1973) oriented research might be able to define these patterns more clearly in future studies.

Another interesting finding concerns the importance of perceived community spport for high achievement and career motivation in females. Apparently females of both sex role types (i.e., androgynous as well as sex typed) are dependent upon this perceived social sanction and support for their achievement activities (Tables 3 and 5). High school subjects (Farmer, 1977) exhibited a similar pattern. Sex role theory (Bem, 1977; Spence & Helmreich, 1978) would appear to predict differential responses to social support by androgynous versus sex typed females. However, in this study, the androgynous women who perceived the environment as unsupportive lacked career motivation, and sex-typed women who felt the community was supportive, aspired to high level careers. This apparent responsiveness of females of both sex types to social sanction is what could be expected from the social approval data generated from past achievement and attribution theory research (Feather, 1974). The measures, early socialization and community support, appear to be useful in picking up information on which women have inhibited career motivation. For example, women characterized by high achievement motivation and low career motivation (Tables 3 and 4) perceived themselves as having been raised in a family that did not support their achievement goals, and as living in an environment that was unsupportive as well. For these women their childhood experiences and the present context seem to be critical determinants of career motivation. Recently Parsons, Frieze and Ruble (1978) reported their findings that psychological variables were the best predictors of career motivation, accounting for 27% of the variance. In contrast the present study's use of canonical analyses permitted a look at factors inhibiting career motivation as well as those associated with high motivation. The percent of variance accounted for was also substantially higher (41, 52% for low motivation, 61, 79% for high motivation). These analyses revealed the important role environmental variables play relative to low career motivation.

A related and interesting result concerns the *social* self-concept of achieving females. Again both androgynous and sex-typed females who have high career motivation perceive themselves as socially unpopular with low social self-concepts (Tables 2 and 5). These women may feel that their high career aspirations are a "turn-off" to their peers.

Feminine sex-typed women had low levels of home-career conflict associated with low career motivation and high achievement motivation. For such conflict to occur women must be positively attracted to both home-making roles and career roles (Farmer, 1978, Festinger, 1957). Thus this finding is consistent with conflict theory.

The influence of academic self-concept in association with high achievement motivation can be seen for both androgynous and sex typed females in Tables 3 and 4. The finding, that people who feel a sense of academic competence have a high tendency to achieve, is a hallmark of past achievement motivation research.

The rather independent nature of career and achievement motivation was also highlighted by the present study. The results showed that is was not necessarily true that females who had high achievement motivation likewise had high career motivation. In fact, in two cases the results showed them to be negatively correlated. These variables should be tested as separate, though related, constructs and each researcher must define exactly what type of motivation their study is investigating. Perhaps more importantly the implications of these results are interwoven with the implications of all the other results of this study in that female achievement is not a simple, straightforward function of socialization as some have argued (Spence & Helmreich, 1978; Lipman-Blumen & Leavitt, 1976). Rather this study shows that female achievement is a process involving several variables from a *multiple level perspective*: cultural, sociological, personological. Further, veridical information on the nature of female achievement and subsequent counseling innovations based upon that knowledge cannot occur unless the researcher investigates female achievement from these cultural, sociological, and personological facets conjointly (cf. Fyans, 1979). This study is a first step in that direction. Bakan (1966), Bernard (1971) and Stein and Bailey (1973) have suggested including affiliative and nurturing values in the achievement motivation model to more adequately reflect women's achievement strivings. Using the Moreland et al.(1978) derived factors for the Bem Sex Role Inventory seems promising in this regard. Moreland et al. reviewed evidence from three studies, including their own, that the Bem Sex Role Inventory has two independent factors, one instrumental and the other expressive. These factors contain items from the masculine and feminine scales respectively but eliminate some items such as those asking respondents to indicate if they perceive themselves as "masculine" or "feminine." Other items from the neutral scale were found to load substantially on the expressive factor (i.e., helpful, friendly, likeable and happy). Perhaps achievement theorists can learn from career motivation theorists in this regard as well. The career assessment model has included values such as social service, and interest in working with persons vs. things for a long time (Strong, 1943; Super & Crites, 1963).

Studies that suggest increasing an androgynous orientation in women (Almquist, 1974; Burlin, 1976; & Tipton, 1976) in order to increase women's career motivation should be viewed cautiously in light of the findings in this study indicating that androgynous women may be low in career motivation (Table 4). Also, it should be remembered that the women studied were older than most college women and that the same motivation patterns may not hold for younger subjets. There is some evidence that a similar risk pattern (i.e., high and low) predicts high achievement motivated college women (Farmer, 1977). But the college women in the 1977 Farmer study did not have the same pattern of low social self-esteem associated with high motivation. These younger women appear to feel more support from their peers for their achievement strivings. More studies are needed comparing age groups on the factors contributing to achievement and career motivation.

Implications for practice suggest a variety of interventions rather than focusing change efforts in one area. McClelland's (1971) strategy of reeducating adults and adolescents could be applied to girls and women to help change their values, attitudes and self-concepts in the direction of greater achievement motivation. Atkinson's (Atkinson & Raynor, 1974) strategy of changing the environment to optimize motivation could be applied to family education, teacher education, teachers, employers, legislators and policy makers. A third approach suggested some years earlier by Cronbach (1957) is to match the individual and the environment in some manner to optimize achieving behavior. For example, a sex-typed girl may be highly achieving given adequate support for achieving behavior at home and school, whereas an androgynous girl may thrive in a somewhat different environment. This third approach is not free from difficulty, since matching individual differences to environments assumes that these differences are relatively unchanging and stable. All three approaches alone are seen as having limitations, namely relying on change in either the individual or the society to solve the problem. Change strategies seem called for at all levels of society, individual, institutional, and societal. Perhaps most important are change strategies which address both sexes, since change in one sex cannot help but effect change in the other.

The study described in this paper provides some tentative

directions for theory, research, and practice relative to the achieving behavior of girls and women. The fact that conflict was found to be highest in highly motivated women suggests that much remains to be done before women's full potential is unleashed both for their personal benefit and that of society.

REFERENCES

Almquist, E. Sex stereotypes in occupational choice—The case for college women. *Journal of Vocational Behavior*, 1974, 5, 13-22.
Alper, T. Achievement motivation in women: Now-you-see-it-now-you-don't. *American Psychologist*, 1974, 29, 194-203.
Anastasi, A. *Psychological Testing*, (4th ed.), New York: Macmillan, 1976.
Astin, H. Factors associated with the participation of women doctorates in the labor force. *Personnel and Guidance Journal*, 1967, 46, 240-246.
Astin, H. A profile of the women in continuing education. In H. Astin (Ed.), *Some action of her own: The adult woman and higher education*. Lexington, Mass.: Lexington Books, 1976.
Atkinson, J. (Ed.). *Motives in fantasy, action, and society*. Princeton: Van Nostrand: 1958.
Atkinson, J., & Litwin, G. Achievement motive and test anxiety conceived as motive to approach success and motive to avoid failure. *Journal of Abnormal and Social Psychology*, 1960, 60, 52-63.
Atkinson, J., & Raynor, J. *Motivation and Achievement*. Washington, D. C.: Winston & Sons, 1974.
Bakan, D. *The duality of human existence*. Chicago: Rand McNally & Co., 1966.
Bardwick, J. Androgyny and humanistic goals, or goodbye cardboard people. In M. McBee & K. Blake (Eds.), *The American woman: Who will she be?* New York: Macmillan, 1974.
Bem, S. On the utility of alternative procedures for assessing psychological androgyny. *Journal of Consulting and Clinical Psychology*, 1977, 45, 196-205.
Bem, S. & Bem, D. Training the woman to know her place: The social antecedents of women in the world of work. Harrisburg: Pennsylvania State Department of Education, Bureau of Pupil Personnel Services, 1973. (ERIC Document Reproduction Service No. ED 082 098).
Bernard, J. *Women and public interest*. Chicago: Aldine-Atherton, 1971.
Burlin, F. Locus of control and female occupational aspiration. *Journal of Counseling Psychology*, 1976, 23, 126-129.
Coopersmith, S. *The antecendents of self-esteem*. San Francisco: Freeman, 1967.
Cronbach, L. The two disciplines of scientific inquiry. *American Psychologist*, 1957, 12, 671-684.

Edye, L. Eliminating barriers to career development of women. *Personnel and Guidance Journal.* 1970, *49*, 24-36.

Farmer, H. What inhibits achievement and career motivation in women. In L. Harmon, L. Fitzgerald, J. Birk, & M. Tanney (Eds.). *Counseling women.* Monterey, California: Brooks Cole, 1977.

Farmer, H. *Career and family present conflicting priorities for married women returning to college.* Paper presented at AERA Annual Meeting, Toronto, March, 1978.

Farmer, H., & Backer, T. *New career options for women: A counselor's sourcebook.* New York: Human Sciences Press, 1977.

Farmer, H., & Bohn, M. Home-career conflict reduction and the level of career interest in women. *Journal of Counseling Psychology,* 1970, *17,* 228-232.

Feather, N. Fear of success in Australian and American student groups: Motive or sex-role stereotype? *Journal of Personality.* 1974, *42*, 190-201.

Festinger, L. *A theory of cognitive dissonance.* Stanford, Calif.: Stanford University Press, 1957.

Fleming, J. Comment on "Do women fear success?" by David Tresemer, *SIGNS,* 1977, *2*, 706-717.

French, E. Some characteristics of achievement motivation. *Journal of Experimental Psychology,* 1955, *50*, 232-236.

Fyans, L.J., Jr. A new multi-level paradigm for conducting cross-cultural and socio-cultural psychological research. In M. Cole and W. Hall (Eds.) *Laboratory for Comparative Cognitive Development Quarterly.* University of California at San Diego, September, 1979.

Ginsberg, E. & Associates. *Life styles of educated women.* New York: Columbia University Press, 1966.

Gump, J., & Rivers, L. The consideration of race in efforts to end sex bias. In E. Diamond (Ed.), *Issues of sex bias in interest measurement.* Washington, D. C.: U. S. Government Printing Office, 1975.

Harmon, L. Variables related to women's persistence in educational plans. *Journal of Vocational Behavior.* 1972, *2*, 143-153.

Hawley, P. Perceptions of male models of feminity related to career choice. *Journal of Counseling Psychology.* 1972, *19*, 308-313.

Holland, J.L. *Making vocational choices: A theory of careers.* Englewood Cliffs, N.J.: Prentice-Hall, 1973.

Horner, M. *Sex differences in achievement motivation and performance in competitive and non-competitive situations.* Unpublished doctoral dissertation, University of Michigan, Ann Arbor: 1968, No. 69-112, 135.

Horner, M., & Fleming, J. *Revised Scoring Manual for an Empirically Derived Scoring System for the Motive to Avoid Success.* Available from Jacqueline Fleming, Barnard College, New York City, 1977.

Joreskog, K. A general method for estimating a linear structural equation system. In A. Goldberger, and O. Duncan, (Eds.), *Structural equation models in the social sciences.* New York: Seminar Press, 1973.

Kogan, N., & Dorros, K. *Sex differences in risk taking and its attribution.*

Paper presented at the American Psychological Association Annual Meeting, Chicago, 1975.
Lipman-Blumen, J., & Leavitt, H. Vicarious and direct achievement patterns in adulthood. *The Counseling Psychologist*, 1976, *6*, 26-32.
Lowell, E. The effect of need for achievement on learning and speed of performance. *Journal of Psychology*, 1952, *33*, 31-40.
Maccoby, E. & Jacklin, C. *The psychology of sex differences.* Stanford, California: Stanford University Press, 1974.
Maehr, M. *Sociocultural origins of achievement.* Monterey, California: Brooks-Cole, 1974.
McClelland, D. *Assessing human motivation.* New York: General Learning Press, 1971.
Matthews, E. & Tiedeman, D. Attitudes toward career and marriage and the development of life style in young women. *Journal of Counseling Psychology*, 1964, *11*, 375-384.
Monahan, L., Kuhn, D., & Shaver, P. Intrapsychic versus cultural explanations of the "fear of success" motive. *Journal of Personality and Social Psychology.* 1974, *29*, 60-64.
Moreland, J., Gulanick, N., Montagues, E., & Harren, V. Some psychometric properties of the Bem Sex-Role Inventory. *Applied Psychological Measurement.* 1978, *2*, 249-256.
Moss, H., & Kagan, J. Stability of achievement and recognition seeking behaviors from early childhood through adulthood. *Journal of Abnormal and Social Psychology.* 1961, *62*, 504-513.
Mosteller, F., & Tukey, J. *Data analysis and regression.* Reading, Mass.: Addison-Wesley, Inc., 1977.
Parsons, J., Frieze, I., & Ruble, D. Intrapsychic factors influencing career aspirations in college women. *Sex Roles*, 1978, *4*, 337-347.
Psathas, G. Toward a theory of occupational choice for women. *Sociology and Social Research*, 1968, *52*, 253-268.
Roe, A. *The psychology of occupations.* New York: Wiley and Sons, 1956.
Rossi, A., & Calderwood, A. (Eds.), *Academic women on the move.* New York: Russell Sage Foundation, 1973.
Rubovits, P. Early experience and the achieving orientations of American middle-class girls. In M. Maehr & W. Stallings (Eds.), *Culture, child, and school.* Monterey, California: Brooks/Cole, 1975.
Sarbin, T. Role theory. In G. Lindzey (Ed.) *Handbook of social psychology* (Vol. 1). Cambridge, Massachusetts: Addison-Wesley, 1954.
Sexton, P. Women and work: R & D Monograph 46. U. S. Department of Labor, Washington, D. C.: U. S. Government Printing Office, 1977.
Shaver, P. Questions concerning fear of success and its conceptual relatives. *Sex Roles*, 1976, *2*, 305-320.
Spence, J., & Helmreich, R. *Masculinity and feminity: Their psychological dimensions, correlates, and antecedents.* Austin, Texas: University of Texas Press, 1978.

Stein, A. & Bailey, M. The socialization of achievement orientation in women. *Psychological Bulletin*, 1973, *80*, 345-364.

Strong, E. *Vocational interests of men and women*. Stanford: Stanford University Press, 1943.

Super, D. *The psychology of careers*, New York: Harper and Row, 1957.

Super, D. Self concepts in vocational development. In D. Super, R. Starishevsky, N. Matlin, & J. Jordaan. *Career development self-concept theory*. New York: College Entrance Examination Board, 1963.

Super D. Vocational guidance: Emergent decision-making in a changing society. *Bulletin of the International Association for Educational and Vocational Guidance*, 1976, *29*, 16-23.

Super, D., & Crites, J. *Appraising vocational fitness*. New York: Harper and Row, 1962.

Tatusoka, M. *Multivariate analysis: Techniques for educational and psychological research*. New York: Wiley, 1971.

Tipton, R. Attitudes towards women's roles in society and vocational interests. *Journal of Vocational Behavior*, 1976, *8*, 155-165.

Tittle, C., & Denker, E. Re-entry women: A selective review of the educational process, career choice, and interest measurement. *Review of Educational Research*, 1977, *47*, 531, 584.

Tomlinson-Keasey, C. Role variables: Their influence on female motivational constructs. *Journal of Counseling Psychology*. 1974, *21*, 232-237.

Weiner, B. *Achievement motivation and attribution theory*. Morristown, N.J.: General Learning Press, 1974.

Winterbottom, M. *The relation of childhood training in independence to achievement motivation*. Unpublished doctoral dissertation, University of Michigan, 1953.

Zuckerman, M., & Wheeler, L. To dispel fantasies about the fantasy-based measure of fear of success. *Psychological Bulletin*, 1975, *82*, 932-946.

VI
Teacher Expectations and Achievement Motivation

Chapter 18

Teacher Expectations and Student Learning

MARGARET C. WANG
Learning Research and Development Center
University of Pittsburgh

WARREN J. WEISSTEIN
Learning Research and Development Center
University of Pittsburgh

The potentially adverse influence that teacher perceptions may have on student achievement and achievement motivation has been an increasing concern of educational researchers. This concern has heightened with the passage of the Education for All Handicapped Children Act (PL 94-142) and the current movement to mainstream exceptional children into regular classrooms. One type of perception which is thought to be potentially detrimental to student achievement and achievement motivation is a teacher's perception of the achievement level of students. It has been suggested that teacher perceptions of student achievement levels can be detrimental, in that they may influence the teacher's expectations. These expectations in turn affect the way a teacher interacts with his or her students and, consequently, may cause students to exhibit patterns of behavior and academic performance that conform to teacher expectations. Furthermore, conformity to teacher expectations is more likely to have an adverse impact on students whom teachers perceive to be low in their expectancy to achieve. The underlying assumption here is that the behavior of students is more or less directed toward certain expectations (the teacher's as well as their own), and the extent to which these expectations influence their behavior will depend upon a multitude of situational and psychological factors.

Although results from experimental studies which attempt

to manipulate and induce teacher expectancy effects appear to be somewhat inconsistent and contradictory (e.g., Dusek, 1975; Good and Brophy, 1978), they suggest that teacher expectations may be self-fulfilling and that teacher expectations are related to teacher behaviors (Brophy and Good, 1974; Jeter and Davis, 1973). To thoroughly understand how and why teacher expectations may influence teachers to behave in ways that tend to be self-fulfilling, however, we need detailed information on the process and situation variables that influence student conformity to those expectations. Research on this topic has been quite scanty and limited in scope.

Good and Brophy have pioneered some interesting research in this area (1978). They developed a model that provides a framework for investigating the processes and conditions through which teachers might influence student behaviors to conform with their teacher expectation. According to Good and Brophy, teacher expectations tend to be fulfilled when the following five conditions have been satisfied:

(1) The teacher expects specific behavior and achievement from particular students.
(2) Because of these different expectations, the teacher behaves differently toward different students.
(3) This treatment tells the student what behavior and achievement the teacher expects from them and affects their self-concept, achievement motivation, and level of aspiration.
(4) If this treatment is consistent over time, and if the students do not resist or change it in some way, it will shape their achievement and behavior. High-expectation students will be led to achieve at high levels, while the achievement of low-expectation students will decline.
(5) With time, students' achievement and behavior will conform more and more closely to that originally expected of them (Good & Brophy, 1978, p. 72).

The fourth condition, as described in the Good and Brophy model, seems particularly relevant to studies which aim to examine how teachers influence student learning in a way which is consistent with teacher expectations of their achievement level. According to the model, if children are to display a level of achievement which conforms to teacher expectations, they must be situated in a learning environment where they are not able to resist teacher attempts to control their learning behavior. Such a learning environment is one in which teachers can consistently control reinforcements which shape the learning behavior of children as they interact with them.

In a study designed to test the validity of their model, Brophy and Good (1970) examined: how often teachers called on students to answer questions related to material they read; whether the responses given to teacher-posed questions were correct or incorrect; and the type of feedback given by teachers to student responses (i.e., criticism, praise, no feedback, providing the answer for students, and repeating or rephrasing questions). They found that while no differences appeared to exist in the number of interactions which teachers initiated with students they perceived to be high and low achievers, differences did exist in the form those interactions took. On occasions when children in each group gave correct responses to questions posed by teachers, the teachers appeared to praise correct responses more often when they were made by students whom teachers expected to be high, as opposed to low, achievers. Conversely, teachers criticized incorrect responses more frequently when those responses were made by students who were perceived to be low achievers. Also, when the low achievers made incorrect responses, they were less likely to get a second chance to respond. This situation did not exist with respect to those students identified as high achievers. Teachers gave these students a second opportunity to respond to questions under such circumstances. They also rephrased or repeated questions to increase the probability that these students would get them correct. Overall, teachers tended to criticize or ignore students they perceived to be low achievers while they praised students they perceived to be high achievers.

The results of this particular study suggest that the pattern of reinforcement teachers used to control student learning behavior may have influenced the amount of personal effort students displayed on learning tasks and the learning outcomes they exhibited. In terms of personal effort, students whom teachers perceived to be low achievers initiated fewer instructional interactions with teachers than students whom teachers perceived to be high achievers. In terms of learning outcomes, the low-expectancy students gave more incorrect responses to teacher-posed questions and attained lower scores on the standardized achievement test. These findings seem to support Good and Brophy's notion that teachers can influence children to display a level of academic performance which conforms with teachers' expectations of them.

Findings from a study by McDermott (1977), based on his observational data on teacher and student behaviors in reading classes, suggest that teachers can influence the achievement motivation of children whom they perceive to be high and low achievers merely by ignoring the children in the latter group. The teacher in the McDermott study was able to influence student learning in this fashion by limiting the amount of time he or she listened to students read when working with a low, as opposed to a high skill reading group. According to McDermott's analysis, the low reading group received less listening time for two reasons. The first related to the way the teacher organized turn-taking in reading. The teacher seemed to organize turn-taking in the low reading group randomly. This lack of systematic pattern for turn-taking generally consumed a large portion of the group's reading instruction time. Much of the teacher's time was involved in determining who would read next. In contrast, when the teacher worked with students in the more advanced reading group, he or she went around the reading table in order. Secondly, it was found that the teacher tended to allow more group disruptions, such as other children entering the reading group to ask for teacher help, when the children within that group were low in reading skills. Overall, children in the low reading group received only one-third of the reading time received by children in the high skill reading group. This situation appears to have adversely influenced the motivation that children in the low skill reading group displayed in their desire to learn to read. Such a lack of teacher attention can serve as a negative reinforcer which is likely to reduce students' motivation to learn.

One of the most interesting series of investigations designed to replicate the Good and Brophy study was conducted by Cooper (1977) and Cooper and Baron (1977). According to Cooper, while students who are perceived to be low achievers would like to initiate instructional interactions with teachers in order to gain knowledge about how to do learning tasks correctly, the knowledge they seek is inconsistent with the pattern of behavior teachers ascribe to these students. Since such student-initiated interactions with teachers appear inconsistent with behavior which teachers perceive to be associated with low achievers, Cooper hypothesized that three events should occur: (1) teachers should perceive that they have greater control over the learning behavior of high than of low achievers; (2) teachers should attempt to gain control over the learning behavior of students they perceive to be low achievers by negatively reinforcing (i.e., criticizing) student-initiated interactions with them; and (3) as the amount of teacher criticism of student-initiated interactions increases, the amount of personal effort children perceive they contribute to their own learning outcomes decreases. Data from their studies seem to support these hypotheses. A high positive correlation was found between teacher perceptions of students as high, moderate, or low in their expectancy to achieve and the amount of influence teachers perceived they had over the learning behavior of these students. Furthermore, negative reinforcement from teachers was found to have a dampening effect on the number of student-initiated interactions which ocurred among children who tended to be among the most criticized in the classroom (i.e., students perceived to be low achievers).

Based on evidence from the few studies we have cited in support of the Good and Brophy Model, it appears that teachers' expectations may cause students to display a level of academic performance which conforms with teacher perceptions of them. Furthermore, teacher expectations appear to have their most adverse impact on students whom teachers perceive to be low achievers. It seems that such adverse effects are most likely to occur in learning environments where *teachers have consistent control over reinforcements which shape student behaviors.*

To examine the Good and Brophy model from another

perspective, one can also interpret the model to suggest that teachers who are situated in learning environments designed to minimize their exclusive and/or consistent control of reinforcements which shape students' learning behavior may have a more difficult time influencing the learning behaviors of their students. Such a learning environment would be one in which *students are permitted to assume some responsibility for controlling those reinforcements* which shape their own learning behavior. We refer to such a setting as a student-centered and self-managed learning environment.

It is from this perspective that the study reported in the present chapter was designed. We were particularly interested in studying the extent to which low achievers in a self-managed learning environment can overcome the potentially damaging effects that teacher expectations may have on their academic performance. Specifically, the study was designed to document differences in the nature and frequency of interactions between low and high achieving students and their teachers and peers in such an environment as well as document the extent to which these differences may contribute to overcoming teacher expectation effects. We were interested in studying: (1) the nature and frequency of interactions between teachers and students, and the extent to which patterns of teacher interactions with high and low achieving students differ; (2) the extent to which high and low achieving students differ in their classroom behaviors and learning habits (i.e., the frequency of observed on-task behaviors, the setting in which they choose to work, etc.); (3) the extent to which the nature and frequency of peer interactions differ between the high and low achieving groups; and (4) the extent to which student task completion rates and student perceptions of achievement responsibility differ among high and low achievers.

The Study

Setting

The study was carried out in a non-graded, primary classroom designed with a multi-aged and heterogeneous ability grouping pattern that aims to minimize the importance of group differences (i.e., low achievers, high achievers, chronological age, etc.), and emphasize the strengths and learning needs of the individual child. The instructional program is designed to provide learning environments that are effective in adapting learning experiences to individual differences. It aims to develop in students the competencies required to achieve mastery of the basic academic skills while fostering in them the ability to take increasing responsibility for regulating and managing their own learning in their social and physical surroundings. The program's emphasis on the development of student self-responsibility for learning assumes that self-managed classroom environments may overcome the adverse influence that teacher expectations can have on a child's motivation to learn. It further assumes that the adverse effect that teacher expectations of student academic performance can have on the classroom learning processes and student learning outcomes may be minimized through instructional design.

Therefore, explicit in the design of this program, known as the Adaptive Learning Environments Model (Wang, 1979), is the establishment of a classroom setting where children are given the opportunity to carry out their learning tasks with some independence from teachers. Such independence requires the acquisition of skills in: (1) planning a schedule for completing their daily assignments; (2) selecting learning tasks related to an instructional objective which they were assigned to master; and (3) using previously acquired skills to successfully complete learning tasks with minimal teacher assistance. This last skill is of particular importance in the context of the current investigation. Knowledge of prerequisite skills on which a learning task is based helps children determine their own capability to successfully perform the task. Information about their own capability to learn a new task, or complete a task with previously acquired skills, serves as a self-administered reinforcement which, in turn, may influence their motivation and learning behavior.

The Adaptive Learning Environments Model is designed to facilitate acquisition of these skills through three related program components. They include: a prescriptive learning component geared to teaching basic skills; an exploratory learning component which encourages students to assume some responsibility for structuring and defining their learning tasks and applying their skills and knowledge in a variety of situations; and an instructional-learning management system which is designed to assist teachers in helping students gain increasing capacity to self-regulate and self-manage their learning in school.

The prescriptive learning component. This component is designed specifically to teach basic skills. It includes a series of individualized learning experiences which are diagnostically tailored, in accordance with a student's competencies and learning needs. The series of learning experiences are based on a continuum of instructional objectives which have been hierarchically sequenced and empirically validated (e.g., Wang, Resnick, and Boozer, 1971; Resnick and Wang, 1969). Structured around these empirically validated hierarchies, the prescriptive learning component of the program is designed to ensure that acquisition of simple skills will provide a foundation for mastering more complex skills.

An integral feature of the prescriptive learning component is the diagnostic and skill mastery testing program. This testing program provides the teachers and students with a built-in system which gives them evaluative feedback that they can use to monitor and reinforce student learning progress. The testing program, when well implemented, is particularly reinforcing because it is designed to take advantage of the fine-grained incremental steps in the curricular hierarchies which form natural checking points. These incremental steps and checking points provide opportunities for frequent formal feedback to students about their progress.

Another important feature of the prescriptive learning component is its built-in implementation flexibility to systematically individualize student learning experiences. A large variety of learning options and instructional methods have been developed in order to adapt to individual differences in students. Instructional approaches and materials used to design a learning task can be arranged in a wayf which permits the learning task in question to vary from a highly independent learning activity to one which requires extensive tutorial assistance from teachers and/or peers. While some students may require tutorial assistance, it is expected that most students functioning under the Adaptive Learning Environments Model should be able to successfully complete prescriptive learning tasks with minimal teacher or peer assistance. One reason for this contention is that the diagnostic and skill mastery testing program helps teachers place students at levels within the curriculum where they should be able to use their prerequisite skills to complete a learning task with little difficulty.

The exploratory learning component. Within this more open-ended curricular component, students are encouraged to assume responsibility for structuring and defining their learning tasks by expanding upon skills which both interest them and are among their strengths. Unlike the prescriptive learning tasks, exploratory learning tasks are generally selected and designed by students themselves, with teacher assistance when needed. In order to assist students in assuming responsibility for structuring and defining their learning tasks, they are taught the skills of choosing, planning, and executing those tasks independently, with minimal amounts of teacher assistance and supervision. As they work on exploratory learning tasks, students are encouraged to plan, explore, make mistakes, and apply their previously acquired skills and knowledge in new situations.

Learning tasks included in the exploratory learning component can be explorations in topics related to subject areas such as reading, math, science, social studies, creative writing, block construction, creative arts, perceptual skills, music, and sociodramatic play. The range of different exploratory learning tasks made available to students is largely determined by student interest, teacher expertise, and material and space constraints.

The instructional-learning management system. One of the unique features of the Adaptive Learning Environments Model is its instructional-learning management system, known as the Self-Schedule System (Wang, 1974a). The Self-Schedule System was designed to help teachers and students become more proficient at functioning in learning environments where student self-management is not only a desired outcome but a requirement for successful implementation of the program. This aim is achieved through a multi-faceted approach: (1) developing teacher competencies to effectively take advantage of the design features of the program in order to maximize the use of school time for instructional purposes; (2) teaching students skills they need in order to assume increasing self-responsibility for planning and carrying out their learning activities; and (3) helping students gain confidence in their own ability to control reinforcements which shape their learning behavior.

Under the Self-Schedule System, students may be found working simultaneously on free choice and teacher-prescribed activities, either individually or in small groups. As children work in all areas of the classroom, teachers circulate among them to check their work and to provide reinforcements and/or instructional and management assistance as needed. Students are permitted to move freely about the classroom. They can do their learning tasks independently, with minimal teacher assistance. However, students can also be called together into small groups for tutoring purposes, for testing, or for special activities such as art, music, and physical education. Along with the freedom students have to flexibly schedule learning activities under the Self-Schedule System, they must assume certain responsibilities. For example, students are expected to take the responsibility to budget their time to make sure that they complete all the teacher-assigned prescriptive learning tasks and at least two self-selected learning tasks each day. They are also expected to take the initiative to ask for teacher or peer assistance when needed.

The Adaptive Learning Environments Model is designed in a way that fails to satisfy a condition which Good and Brophy (1978) believe is necessary to fulfill the negative expectations which teachers may formulate regarding the level of students' academic achievement. Therefore, the Adaptive Learning Environments Model is though to be potentially effective in helping students resist the adverse effects of teacher expectations. By nature of its design, the Adaptive Learning Environments

Model requires the transfer of a large amount of the control teachers generally have over students' learning behavior to students. Thus the learning behavior of students is no longer exclusively contingent upon teacher reinforcements. Through explicit instructions, an increasing amount of teacher control is transferred to students to help them assume increasing responsibility for controlling intrinsic reinforcements which shape their learning behaviors.

Subjects

All the children enrolled in this particular primary classroom served as subjects for the current study. The class included 46 children, (ages 5 through 8), two teachers and one instructional aide. Based on I.Q. scores and achievement test scores, high and low achieving students in the classroom were identified and served as a sub-sample for intensive study. Children with I.Q. scores of 85 and above, who performed below the 40th percentile on a standardized achievement test in math and/or reading, were identified as low achievers. Students who scored above the 95th percentile on a standardized reading and/or math achievement test were identified as high achievers. Based on these criteria, four of the children in the classroom were identified as low achievers and eight were identified as high achievers. Students who did not fit into either the high or low achieving categories were referred to as others. None of the identities of the children were revealed to the teachers.

Measures

Intellectual Achievement Responsibility Questionnaire (IAR). The IAR (Crandall, Katkovsky & Crandall, 1965) was administered to obtain information on students' perceptions of the locus of control of their intellectual and school achievements. The IAR was designed to assess the extent to which children assume responsibility for the successes or failures they experience in their learning environment. According to the authors, internal perception is defined as the belief that "rewards and punishments are dependent upon their own (subjects') instrumental behavior, and external perception is defined as "the perception that those events (which cause rewards and punishments) occur at the whim or discretion of some agent other than the individual" (Crandall & Lacey, 1972, pp. 1123-1124). The IAR consists of 34 forced-choice items. Each item describes either a positive or a negative achievement experience which routinely occurs in the child's daily life.

Narrative recordings of classroom processes. Narrative recordings of observed classroom processes were used to analyze classroom interaction patterns. The recordings consist of verbatim reports of teacher and student behaviors and the setting in which the behaviors occurred. The records were used in the present study to obtain information regarding the interactions between students and teachers and the interactions among peers. We were particularly interested in studying the pattern of interactions, that is, information about with whom a specific child interacted most frequently, the reason for each interaction, and the nature of the situation in which the interaction occurred.

Student Behavior Observation Schedule (SBOS). The SBOS (Wang, 1974b) was used to obtain systematic information on student learning processes. It was specifically designed to obtain information on: (1) the frequency of student-teacher interactions and their purposes (e.g., whether an interaction occurred for instructional or management purposes); (2) the frequency and purposes of interactions among students (i.e., whether they were initiated for communicating ideas, for sharing, or for disruptive purposes, etc.); (3) the percentage of time students spent working in group interactive, group parallel, or individual settings; and (4) the extent to which children exhibited on task or distracted behaviors. The SBOS has been used in several previous classroom process studies. Its validity and reliability have been reported elsewhere (Wang, 1976; Wang & Stiles, 1976).

Task completion rates. Detailed records on the number of prescriptive tasks assigned by teachers and the number of prescriptive and exploratory tasks completed by students were kept on a daily basis through a computerized classroom information system (Wang & Fitzhugh, 1978). Task completion rates were calculated by dividing the number of tasks completed by the number of tasks assigned.

Analysis of Classroom Observation Data

When analyzing the data of one particular group (i.e., low or high achievers), observed frequencies obtained from the SBOS data were summarized in terms of mean percentages. The mean percentages reported in Table 1, for example, were calculated by dividing the total number of observed frequencies emitted by a given group of students by the total number of behaviors observed for that particular group.

To overcome the problem of comparing the results of two unequal groups of subjects, we have adopted a scoring procedure that controls for the differences in group size by taking into account the distribution of frequencies actually observed (the observed score) for each child in both groups and the distribution of observed frequencies one would expect by chance (expected score) for each child in a given group. The expected score is an estimation of expected frequencies based on the size of the two groups. It was reasoned that by subtracting the proportion of those frequencies one would expect to obtain by chance from the distribution of the observed frequencies, the resulting differences would reflect the actual differences between the groups beyond chance. Basically, difference scores control for group size by determining the extent to which the frequency of observed interactions between teachers and a group of students is greater or less than one would expect by chance, taking into account the number of students in the group. Positive difference scores indicate that there are more interactions than one would expect to occur by chance between teachers and a group of students of given size. Negative difference scores suggest that there are fewer interactions than would be expected to occur by chance between teachers and a student group of given size. The difference scores were used to compare the differences in observed interactions between the teachers and the high and low achieving students (Weisstein & Wang, 1978).

Table 1
Frequency of Interactions Between Teachers and Students

Behavior Categories	Mean Percent of Observed Frequencies		
	Total Class	High Achievers	Low Achievers
Interactions initiated by:			
Student	63%	43%	36%
Teacher	18%	54%	63%
Unknown	18%	1%	0%
Purpose:			
Instructional	64%	76%	52%
Management	27%	13%	47%
Unknown	9%	10%	0%

Results

Patterns of Classroom Interaction Between High and Low Achievers and Their Teachers

The SBOS data was first examined to determine the overall frequency and purpose of interactions between teachers and students in a classroom environment designed to adapt to the learning needs of the individual students. Table 1 provides a summary of mean percentages observed within each behavior category relating to teacher-student interactions. The first column shows the mean percentages of observed frequencies for all students enrolled in the class within each of the behavior categories. The second and third columns report the distribution of observed frequencies for high and low achieving students across the categories.

As shown in Table 1, 63% of all interactions were initiated by students. Eighteen percent were initiated by teachers. Sixty-four percent of all interactions between teachers and students were for instructional purposes. Twenty-seven percent were for management purposes. When separating the data into high and low achievement levels, some contrasting trends in teacher-student interaction patterns were observed. The mean percentage of observed frequencies reported in columns 2 and 3 indicate that while 76% of all interactions teachers had with high achieving students were for instructional purposes, *only 52% of their interactions with low achieving students were for instructional purposes.* The data further indicates that the percentage of management interactions between teachers and low achievers appears to be much higher (47%) than management contacts between teachers and high achievers (13%) or between teachers and the class average (27%).

To test the validity of the frequency data, we attempted to control for the effects of differences in the sample sizes of the high and low achieving groups. The observed frequencies between teachers and students in the two groups were further analyzed, using the procedures outlined under the "Analysis of Classroom Observation Data" section of this paper. We calculated the differences between the expected frequencies (taking into account the sample size) and the observed frequencies for each group and statistically tested the differences for the discrepancies between expected and observed frequencies for the two groups. Table 2 reports the results of the t-test comparison of these differences.

Interactions between teachers and high and low achievers regardless of who initiated them. In examining the first set of comparisons as listed in Table 2 (i.e., interactions between teachers and students regardless of who initiated them), statistically significant differences were found in the frequencies of teacher instructional and management interactions between high and low achieving students. Teachers were observed to spend less than the expected amount of instructional interactions with low achieving students (mean difference score = -2.157) and more than the expected amount of instructional interactions with high achievers (mean difference score = 1.091). This difference is statistically significant beyond the .001 level.

Teacher-initiated interactions with low and high achievers.

Table 2
Comparison of Frequency and Nature of Teacher Interactions with
Low and High Achievers

Comparison	Discrepancy scores between expected and observed frequencies				df	t	2 tail prob
	Low Achiever		High Achiever				
	X	S	X	S			
A. Interactions between teachers and students, regardless of who initiated them.							
1. Instructional interactions	-2.157	0.863	1.091	0.643	10	7.41	p<.001
2. Management interactions	1.170	0.910	-0.445	0.287	10	4.80	p<.001
B. Teacher-Initiated Interactions							
1. Total number of interactions, without regard to purpose	-0.324	0.242	0.173	0.123	10	4.85	p<.001
2. Instructional interactions	-1.382	0.978	0.750	0.704	10	4.38	p<.001
3. Management interactions	1.502	1.197	-0.747	0.753	10	4.04	p<.002
C. Student-Initiated Interactions							
1. Total number of interactions, without regard to purpose	-0.994	0.367	0.505	0.374	10	6.58	p<.001
2. Instructional interactions	-0.494	0.135	0.253	0.207	10	6.49	p<.001
3. Management interactions	-0.165	0.191	0.850	0.117	10	2.85	p<.01

The data reported in Table 2 also suggest that teachers tend to initiate more than the expected amount of interactions with high achievers (.173) and less than that expected with low achievers (-.324). The differences in the frequencies of teacher-initiated interaction between the two groups were statistically significant beyond the .001 level. The data indicates that the teachers initiated more than the expected amount of instructional interactions with high achievers (.750) and less management interactions with them (-.747). They initiated more management interactions (1.502), and less instructional interactions (-1.382), with low achievers. These differences were also statistically significant.

Student-initiated interactions with teachers. In examining the nature and frequency of teacher-student interactions that have been initiated by students (Table 2), we found that low achievers in general initiated less contacts than were expected

(-.994), as opposed to those expected of high achievers (.505). The data also clearly suggests that while low achieving students initiated interactions with teachers for both instructional and management purposes, they did so less often than expected. High achievers, on the other hand, sought teacher help more frequently than expected for both instructional and management purposes.

Patterns of Classroom Interaction Between High and Low Achievers and Their Peers

The narrative recordings of student classroom behaviors were used as our basic data set for investigating the peer interaction patterns. We wanted to find out not only how frequently students interacted with one another in an environment where the opportunity for spontaneous interaction was provided, but also who tended to interact with whom and for what purpose.

The data on peer interactions was analyzed using a procedure somewhat comparable to the one used to analyze teacher-student interactions. For example, since there are twice as many high achievers as low achievers, one would expect a low achiever to have more chance interactions with high achieving students than with other low achievers. Hence, a scoring system similar to the difference score procedure described earlier was used to determine whether the frequency of observed interactions which one group of students had with a given peer group was greater or less than expected, taking into account the size of the peer group. In terms of peer interactions, chi-square tests were used to determine whether the number of observed interactions which one group of students had with another significantly differed from the number of expected interactions. Table 3 provides a summary of these chi-square results. The first column lists the nature of comparisons. The second, third and fourth columns indicate with whom (the type of students) the interaction took place. The plus (+), minus (-) and zero (0) signs indicate whether a particular interaction was observed more, less, or about as frequently as would be expected, when taking into account group size. The

last column reports the results of the chi-square analyses.
Interactions between low achievers and their peers. Listed under Section A of Table 3 are results of four sets of chi-squre comparisons that were carried out to analyze the interactions between low achievers and their peers. We were interested in discovering whether low achievers chose one group of students significantly more than another when initiating peer contacts, and whether they differentiated between the types of students with whom they interacted based on the purpose of their interactions.

The data reported in Table 3 suggests that low achievers seemed to initiate interactions with high achievers significant-

Table 3
Comparison of Discrepancy Scores on the Frequencies
and Nature of Peer Interaction

Interactions	Low-Achievers	Others	High-Achievers	Chi-Square Results
A. Interactions initiated by low achievers with their peers				
1. Regardless of Purpose				
a. with others vs with high achievers		−	+	$p<.05$
b. with other low achievers vs with others	0	0		NS
c. with other low achievers vs high achievers	−		+	$p<.01$
2. For Instructional Purposes				
a. with others vs with high achievers		0	0	NS
b. with other low achievers vs with others	0	0		NS
c. with other low achievers vs high achievers	−		+	$p<.05$
3. For Management Purposes				
a. with others vs with high achievers		0	0	NS
b. with other low achievers vs with others	0	0		NS
c. with other low achievers vs with high achievers	0		0	NS
4. For Personal Reasons				
a. with others vs with high achievers		0	0	NS
b. with other low achievers vs with others	0	0		NS
c. with other low achievers vs with high achievers	+		−	$p<.05$

Table 3 continued

Interactions	Low-Achievers	Others	High-Achievers	Chi-Square Results
B. Interactions initiated by high achievers with their peers				
1. Regardless of Purpose				
a. with others vs with other high achievers		0	0	NS
b. with low achievers vs with others	+	–		p<.05
c. with low achievers vs with other high achievers	0		0	NS
2. For Instructional Purposes				
a. with others vs with other high achievers		0	0	NS
b. with low achievers vs with others	0	0		NS
c. with other low achievers vs with high achievers	0		0	NS
3. For Management Purposes				
a. with others vs with other high achievers		0	0	NS
b. with low achievers vs with others	0	0		NS
c. with low achievers vs with other high achievers	0		0	NS
4. For Personal Reasons				
a. with others vs with other high achievers		0	0	NS
b. with low achievers vs with others	0	0		NS
c. with low achievers vs with other high achievers	0		0	NS

Note: 0 = No discrepancy between expected and observed frequencies
+ = More than expected
– = Less than expected

ly more frequently (beyond the expected level) than with other groups of peers in their class. It is interesting to note that low achievers tended to interact significantly more frequently than expected with other low achievers for personal reasons, while they interacted significantly more than expected with high achievers for instructional purposes. This result is particularly interesting. The data seems to suggest that in the particular environment where the study was conducted, low achieving students tended to use their high achieving peers as a source of instructional assistance. They seemed to be aware of children who were most likely to be able to provide them with the help they needed on instructional matters.

Interactions between high achievers and their peers. Overall, no major differences were observed in the interaction patterns between high achievers and their peers. High achievers interacted as frequently with low achievers for instructional, management and personal reasons as they did with other peer groups.

The Effects of Teacher and Student Behavior Differences on Student Learning Processes and Outcomes

To investigate the extent to which the observed differences in interactional patterns between teachers and students and among peers may influence the student learning processes and learning outcomes, we examined classroom observation data obtained from the SBOS, and the task completion rates of students. Our main question was whether the impact of differences in teacher and peer behaviors toward low and high achieving students on student learning processes and outcomes could be significantly minimized in a learning environment that had been explicitly designed to adapt to the learning needs of individual children.

Student learning processes. The SBOS data was analyzed to determine the extent to which the learning processes of low achieving and high achieving students differed. We compared the frequency of observed behaviors of these two groups of students to determine: (1) whether low achieving and high achieving students differed in the frequency with which they engaged in exploratory and prescriptive tasks; (2) whether one group of students tended to work more in one particular instructional setting than the other; and (3) the extent to which children in the two groups differed in the manner in which they worked on learning tasks.

In general, the data did not show major behavioral differences between the two groups. The SBOS data, as shown in Table 4, suggests that students did not differ much with respect

Table 4
Student Classroom Behavior Patterns

Categories	Mean Percent of Observed Frequencies		
	Total	High Achievers	Low Achievers
Activity Types:			
Prescriptive	74%	67%	61%
Exploratory	15%	17%	9%
Other	11%	16%	29%
Setting:			
Group: interactive	19%	60%	28%
Group: parallel	3%	11%	7%
Individual	78%	29%	64%
Manner:			
On task	70%	72%	76%
Waiting for teacher's help	7%	11%	12%
Distracted	24%	17%	12%

to the amount of time they were observed engaging in prescriptive or exploratory activities. Although high achievers were observed to have engaged more frequently in both prescriptive (67%) and exploratory (17%) activities than did the low achievers (61% and 9% respectively), the differences did not appear to be large. The observed differences between the two groups in terms of on-task behavior also appeared to be minimal. Overall, the low achievers were observed to be a little more on task and slightly less distracted.

The only difference between the two groups related to the type of instructional settings in which they tended to work. The low achievers were primarily found to work on assignments in individual settings (64%). In contrast, high achievers worked in individual settings only 29% of the time. High achievers appeared to work in group interactive settings much more frequently (60%) than low achievers (28%). In terms of the type of setting which students within the class preferred, it is interesting to note that the low achievers seemed to be more similar to the total group pattern than were the high achieving students. Based on this data and that shown in Tables 2 and 3, the SBOS data seems to suggest that although there were some significant differences in the interaction patterns between high and low achievers and their teachers and peers, these differences do not seem to have any consistent effects on student learning processes in the particular environment in which the study was carried out.

Student learning outcomes. Because the nature of teacher-student interactions between high and low achievers was found to be different, and because students were not observed to display large differences in the learning processes they used in this particular classroom, it seemed worthwhile to determine whether the differential interactions between teachers and students affected such learning outcomes as student task completion rates. It is our hypothesis that if the amount of on task behavior is highly related to task completion rates, student task completion rates between the two groups probably would not differ. On the other hand, if task completion rates are highly dependent on the amount of instructional interaction between teachers and students, then we would expect to find higher task completion rates among high achievers.

While the pattern of interaction between teachers and low achieving students may not have been as conducive to learning (i.e., more management and fewer instructional interactions) as the pattern that emerged between teachers and high achieving children, it appeared that children who were low achievers in reading and math were able to successfully complete a similar number of tasks as high achieving students. Interestingly, they completed more tasks than the teachers expected them to do. To compare task completion rates between the two groups, the rate for each group was calculated by tallying the total number of tasks assigned and the total which was actually completed in a given subject. To control for differences in group size, the total number of tasks assigned and completed was divided by the number of students in the group for which the figures were calculated.

No large differences in task completion rates were found between the two groups. The task completion rate for low achieving students in reading was 115%, suggesting that low achieving students completed all the reading tasks they were expected to finish plus 15% more. Those who were low achievers in math had a 125% task completion rate. They completed 25% more math tasks than teachers expected. The mean task completion rate for high achievers in math and reading, on the other hand, was 125% for math and 107% in reading.

These results indicate that low achieving students were able to successfully complete as many tasks as comparable groups of high achieving students. Furthermore, and perhaps one of the more interesting results, task completion rates did not seem to relate as much to teacher interactions as we tended to believe. It is our observation that one possible explanation for why low achievers in this class completed more tasks than teachers expected was their ability to plan and manage their own learning behaviors. That is, when teachers failed to provide them with instructional assistance, they figured out some other means of getting the assistance that was needed to correctly complete their tasks.

To further investigate the relationship between academic achievement and a student's perception of his or her self-management skills, we examined interview results obtained from IAR. We predicted that since all students in the program were taught to function independently in carrying out routine assignments, low achievers should be as competent as high achievers in managing their learning. Consequently, low achieving students should also perceive themselves to be as competent as high achievers in carrying out their task completion responsibilities. We further postulated that it is through this sense of locus of control over how learning takes place in schools that low achieving students in this program were able to complete as much work as they were required to do, without depending on constant intervention from the teacher. This postulate runs contrary to data obtained on low achieving students in most programs.

Task completion rates of low achieving students and their IAR scores seem to support our hypothesis. *The mean total IAR scores for low achieving students was almost identical to that of high achievers.* The low achieving students obtained a mean score of 23 while the mean for high achievers was 22.3.

Discussion

A major implication of this study is that the adverse influence which teacher expectancies might have on the achievement motivation of children can be minimized by designing learning environments that explicitly teach students to control their own learning behavior. In spite of teacher attempts to interact with low achievers in a way which may have influenced their academic performance to conform with teacher expectations, the data in our study seems to suggest that teacher expectancies did not adversely affect student motivation. For example, even though teachers initiated fewer interactions with low achieving students for instructional assistance, the low achievers were able to complete their learning tasks correctly either on their own or by actively seeking out teacher or peer assistance. In other words, teachers functioning under the Adaptive Learning Environments Model *were not able to influence the behaviors of low achieving students* in ways that could adversely affect the learning processes and/or learning outcomes of these students.

In terms of student learning processes, there did not appear to be large differences between high and low achievers with respect to: (a) the frequency of task behavior; (b) the frequency of times both groups waited for teacher assistance; and (c) the frequency of times they were distracted when working on learning tasks. There also did not appear to be much difference between high and low achievers with regard to learning outcomes such as task completion rates, or the amount of personal effort that they perceived they contributed to their academic performance as measured by the IAR.

It is our speculation that the reason low achieving students in this study were able to overcome the influence that teacher expectations can have on achievement motivation is related to the program feature that places students in control of their own learning behavior. The student's ability to manage his or her own learning is likely to lessen the amount of teacher control over reinforcements that may adversely influence student behavior and achievement motivation. It is for this reason that we believe teachers in our study were not able to satisfy a condition which Good and Brophy (1978) believe is necessary if teacher expectations regarding student academic performance are to be fulfilled.

While the findings from our investigation remain speculative, due to the nature of the sample and the small number of children included in the study, our formal and informal observations seem to suggest that students who are high in self-management skills are less susceptible to the influence that teachers may have on their learning behavior. These observations seem to be consistent with results from other research studies on this topic. Findings from a study conducted by Pines and Julian (1972) represent a particularly interesting case in point. Pines and Julian's data indicates that students who are high in self-management are able to independently discover how to use previously learned concepts to solve problems without the presence of a teacher. Such an outcome could explain how low achieving students in our investigation completed more learning tasks than expected, even though teachers failed to give them an amount of instructional assistance equivalent to that given to high achieving students. This observation was also supported by Davis and Phares (1967). They found that under ambiguous conditions, where experimental subjects possessed no knowledge of the prerequisite skills needed to solve a problem, individuals who were high in self-management seemed to more actively seek information needed to work on a learning task than persons who were low in this trait. Such a situation may have occurred in our study. When they lacked knowledge of the prerequisite skills necessary to correctly complete a learning task, and when teachers failed to provide the necessary instructional help, low achievers actively searched for other sources of information. In our case, the additional information source that the low achievers selected was high achieving students. Interestingly, our study revealed that low achievers were also sought after by their peers as valued information resources.

It appears that self-management skills in the context of the present study have helped children, particularly the low achievers, to successfully function in classroom environments where the potentially adverse effects of teacher expectations on achievement levels were overcome. We believe this has been achieved both by teaching students to use prerequisite concepts to independently complete learning tasks and by teaching them to actively seek teacher and student assistance in order to meet their learning needs. It is important to note that we also suspect that acquisition of self-management skills is likely to have little influence on the amount of personal effort that students believe they contribute to their learning outcomes unless those skills are acquired in a classroom environment which is conducive to learning those skills.

One classroom environment which may influence students' perceptions of themselves in this manner is a child-centered and self-managed learning environment which encourages students to assume some responsibility for managing their own learning behavior. This position is supported by the results of several recent investigations. In one study of this nature, Arlin and Whitley (1978) attempted to determine whether the perceptions students formulate regarding their opportunity to use self-management skills can influence the extent to which students perceive that their learning outcomes are a product of personal effort. Arlin and Whitley's study was carried out in two schools. One school contained open classrooms which encouraged self-managed learning and individualization of instruction. The second school was composed of traditional self-contained classrooms in which individualization of instruction was not the predominant mode of learning. The study found that perceptions students formulate regarding their opportunity to manage their own learning behavior may have a greater influence on the amount of personal effort students perceive they contribute to their learning outcomes when students are situated in open as opposed to traditional classroom environments. Consequently, the teaching of self-management skills may have a more positive influence on the achievement motivation of low achieving students in open classroom environments similar to the one on which the current investigation was based.

Findings from studies similar to Arlin and Whitley's also suggest the potential for the use of such instructional intervention strategies as teaching self-management skills to improve and maintain achievement motivation, particularly in low achieving students, who are more susceptible to teacher expectancy effects. In spite of the pilot nature of such studies as the one described in this paper, we believe that their implications for minimizing teacher expectancy effects on student achievement and achievement motivation through instructional design are particularly promising.

REFERENCES

Arlin, M., & Whitley, T. W. Perceptions of self-managed learning opportunities and academic locus of control: A causal interpretation. *Journal of Educational Psychology*, 1978, *70*(6), 988-992.

Brophy, J., & Good, T. Teacher's communication of differential expectations for children's classroom performance: Some behavioral data. *Journal of Educational Psychology*, 1970, *61*, 365-374.

Brophy, J., & Good, T. Changing behavior related to teacher expectations. In J. Brophy & T. Good (Eds.), *Teacher-student relationships: Causes and consequences*, New York: Holt, Rinehart and Winston, Inc., 1974.

Cooper, H. M. Controlling personal rewards: Professional teachers' differential use of feedback and the effects of feedback on the student's motivation to perform. *Journal of Educational Psychology*, 1977, *69*(4), 419-427.

Cooper, H., & Baron, R. Academic expectations and attributed responsibility as predictors of professional teachers' reinforcement behavior. *Journal of Educational Psychology*, 1977, *69(4)*, 409-418.

Crandall, V. C., Katkovsky, W., & Crandall, V. J. Children's belief in their own control of reinforcements in intellectual-academic achievement situations. *Child Development*, 1965, *36*, 91-109.

Crandall, V. C., & Lacey, B. W. Children's perceptions of internal-external control in intellectual-academic situations and their embedded figures test performance. *Child Development*, 1972, *43*, 1123-1134.

Davis, W. L., & Phares, E. J. Internal-external control as a determinant of information-seeking in a social influence situation. *Journal of Personality*, 1967, *35*, 547-561.

Dusek, J. Do teachers bias children's learning? *Review of Educational Research*, 1975, *45*, 660-684.

Good, T., & Brophy, J. *Looking into classrooms*. New York: Harper and Row, 1978.

Jeter, J., & Davis, O. *Elementary school teachers' differential classroom interaction as a function of differential expectations of pupil achievement*. Paper presented at the annual meeting of the American Educational Research Association, 1973.

McDermott, R. P. Social relations as contexts for learning. *Harvard Educational Review*, 1977, *47*(2), 198-213.

Pines, H. A., & Julian, J. W. Effects of task and social demands on locus of control differences in information processing. *Journal of Personality*, 1972, *40*, 407-416.

Resnick, L. B., & Wang, M. C. *Approaches to the validation of learning hierarchies*. Paper presented at the Eighteenth Annual Western Regional Conference on Testing Problems, San Francisco, May, 1969.

Wang, M. C. *The rationale and design of the Self-Schedule System*. Pittsburgh: University of Pittsburgh, Learning Research and Development Center, 1974. (a)

Wang, M. C. *The use of direct observation to study instructional-learning behaviors in school settings*. Pittsburgh: University of Pittsburgh, Learning Research and Development Center, 1974. (b)

Wang, M. C. (Ed.). *The Self-Schedule System for instructional-learning management in adaptive school learning environments*. Pittsburgh: University of Pittsburgh, Learning Research and Development Center, 1976.

Wang, M. C. *Mainstreaming exceptional children: Some instructional design and implementation considerations*. Paper presented at the meeting of the American Educational Research Association, San Francisco, April, 1979.

Wang, M. C., & Fitzhugh, R. J. Planning instruction and monitoring classroom processes with computer assistance. *Educational Technology*, February, 1978, 7-12.

Wang, M. C., Resnick, L. B., & Boozer, R. The sequence of development of some early mathematics bahaviors. *Child Development*, 1971, *42*, 1767-1778.

Wang, M. C., & Stiles, B. An investigation of children's concept of self-responsibility for their school learning. *American Educational Research Journal*, Fall, 1976.

Weisstein, W. J., & Wang, M. C. *An investigation of classroom interactions between academically gifted and learning disabled children with their teachers*. Paper presented at the meeting of the American Educational Research Association, Toronto, April, 1978.

VII
Achievement Motivation: A Look Toward the Future

Chapter 19

From Single-Variable to Persons-in-Relation*

DAVID E. HUNT
Ontario Institute for Studies in Education

It has been more than 25 years since the original work on *The Achievement Motive* (McClelland et al., 1953) appeared, and therefore I consider directions for future research in achievement motivation with special reference to earlier chapters. My comments will attempt to incorporate the methodological suggestions proposed by Walberg and Uguroglu and the proposal by Parsons and Goff to shift emphasis from an agentic to more communal perspective. My remarks are especially influenced by deCharms' proposal to study *persons-in-action* rather than single variables, and the first section is a variation and extension of his theme. I next consider how conceptualizing a person informs methodological decisions, and finally, I propose the study of persons-in-relation as the appropriate phenomenon for psychological inquiry.

On the Concept of a Person

If future research on achievement motivation is to contribute to psychological understanding, it must be conceptually interwoven into a conception of a person (deCharms chapter) and it must be methodologically synthesized into a multiple causal framework (Walberg and Uguroglu chapter). Questions need to be framed in a form more comprehensive and complex than the typical x-causes-y fashion. Atkinson's chapter proposes incorporating ability *and* achievement, yet his argument still rests on an x-causes-y argument, i.e., behavior (f) motivation.

As deCharms suggests, the problem is one of part-whole

relations to which Gestalt psychologists contributed but did not provide concrete solutions. The counterargument to deCharms' proposal for a person (whole) context in which achievement motivation (part) is viewed is that it requires solving all the questions before we begin. A quotation from Sarason (1976) clarified this issue:

> how you approach and deal with the part is influenced mightily by where you see it in relationship to the whole, that is, what you hope to do and the ways in which you go about it are consequences of how you think it is imbedded in the larger picture.

Stated in terms of Walberg and Uguroglu's framework, research on one-factor relations (part) must always be conceptualized within a multiple causal (whole) framework which requires at least a provisional concept of the person.

With the exception of a few die-hard reductionists, most personality psychologists would generally agree that a conception of person is necessary. Disagreement occurs about *how* and *when* such a conception of a person is to be made explicit. deCharms' major point in proposing the Ossorio "gameboard" is to emphasize that some whole concept of person which tentatively organizes the parts in relation to one another is essential, even though the conception will be revised. In short he argues for the necessity of *some* conception of a person to remind us of the context within which the single variable part operates. Better an imprecise attempt to characterize the whole person than to ignore the incomplete nature of single variable part descriptions. Most personality researchers hold implicit conceptions of a person within which they study single variables so that the first step in following deCharms' suggestion is for researchers to explicate their implicit conceptions. I assume that deCharms uses Ossorio "parts" to illustrate one way of conceptualizing a person, though not necessarily the only, or even best, way since deCharms initially mentions four other "parts": motivation, perception, communication, and learning.

Let us consider an example of incomplete description, and then turn to some not very successful strategies to go from part to whole. Until recently the *Annual Review of Psychology* chapter on personality alternated each year between motivation and structure which is typical of the reductionistic compartmentalization which has prevented comprehensive understanding of the person. How are these compartments to be broken and a whole created? One approach is exemplified by the chapter by Lipman-Blumen et al., which extends the traditional definition of achievement to vicarious achievement. Such an extended-part strategy is less incomplete than the original single variable, but is unlikely to produce an adequate concept of a person. As Parsons and Goff observe, altruism and social support are important in their own right (perhaps even more so from a communal view) as distinct from achievement. Another attempt to move from single-variable part to whole has been through hyphenation: social-cognition, cognitive-social learning, humanistic-behaviorism, etc. Although hyphenation is very fashionable at the moment I find it unsatisfactory because it fails to acknowledge *that the whole is more than the sum of its parts.* In simply adding another part, hyphenation misses the point of the relation among the parts.

deCharms implies that following a pragmatic paradigm, i.e. trying to make a difference, is likely to force a conception of a person. One might think that field research might be more likely to force an articulation of part-whole relations than continued single-variable laboratory research yet as recent work has shown, it is as easy to continue to do single variable work in the field as it is in the laboratory. However, if one works in collaboration with practitioners (as in deCharms, 1976), the likelihood of conceptualizing work in a context of a person is much greater.

The "something extra" in the sum of the parts is the *relation* among the parts, and this will not be solved by simply applying multiple regression formulae. One advantage, hinted by deCharms, in a visual conception of a person is that it requires considering how the parts relate to one another, i.e., not just how motivation affects behavior, but how a combination of intention, knowledge, and know-how combine to influence action. More than forty years ago, E. C. Tolman (1938) proposed what he called a "schematic sow bug" which characterized how the "parts" of the sow bug (sensory capacity, motor capacity, learning capacities, demands, etc.) operate interdependently to determine behavior at a choice point. One thinks of computer simulation when considering such complex models, and this might be useful. However, the *conception* of the interrelation among parts will not come directly from the computer but from the researcher.

Other examples of earlier conceptions of persons are Heider (1958) and Murray (1938). It is especially ironic that Heider proposed a comprehensive conception of a person using a "naive analysis of action" (1958) yet his current influence is manifest in the incomplete single variable description of attribution. I have been critical of single variable work yet much of my own research has been on the role of one personality variable, Conceptual Level (Hunt, 1978a). Having been aware of the limits of understanding provided by a single variable I proposed that persons be characterized in terms of four "accessibility characteristics": *cognitive orientation* (or CL), *motivational orientation, value orientation*, and *sensory orientation* (Hunt, 1971). I have thus viewed CL in this context although our research has not often dealt simultaneously with other accessibility characteristics.

I think that personality researchers would be more likely to understand the necessity for an adequate conception of a person in the area of person perception than in personality characteristics. If we consider our own implicit conception of others, we quickly realize that they are organized in wholes or persons, not in terms of single variables. To put achievement motivation or any single variable into a concept of a person requires a shift in our ways of thinking. Perhaps the most concrete basis for recommending such a shift is to consider how this way of thinking affects methodological decisions.

Methodological Implications

All elements of methodology—measurement, research design, and analysis—should flow from our conception of a person. Since psychology is concerned with understanding persons, methodological decisions cannot be based on arbitrary principles used in some sciences and agriculture.

Measurement

The method of measuring a personality characteristic should be determined by the theorist/researcher's conception of a person and the nature of the personality characteristic. One would hardly measure repressed hostility by a direct self-report, e.g., "Do you have hidden feelings of concealed anger?" yet the rationale for using self-report questionnaires is often as ill-advised. Much of the criticism of the TAT-type measure of achievement motivation (e.g., Entwistle, 1972) overlooks the rationale of the method to the construct. As deCharms puts it elsewhere (1976), a "thought sample" is required. Measurement issues are blurred by notions of projective-non projective or internal consistency. Direct report and preference choice have not proven as useful as "thought samples" in measuring achievement motivation because they are less relevant to the characteristic being measured. It is surprising that after 25 years of research in achievement motivation in which most of the measurement has been by "thought samples" that the work is still criticized because it does not meet arbitrary criteria of measurement in the nonpersonal disciplines, e.g., internal consistency.

Second, if we consider the testing situation in terms of deCharms' gameboard scheme, then we must ask what does the person being assessed want, or more to the point, why should he reveal himself? Viewing personality assessment in a context of the person's intentions is critical of conducting investigations which have meaning to the world of human experience. If all the person wants is to finish the task as soon as possible, then this will influence his response. Increased requirements for ethical standards such as informed consent have forced researchers to become more sensitized to the person's perspective in being assessed. Psychologists laugh nervously about personality research being a study of the college sophomore, yet it has been so, and what is more it is psychology of persons in a constrained, enforced condition (Argyris, 1968).

Third, considering measurement in light of one's conception of a person reminds us that some personality characteristics may be more relevant than others for some persons. Research in achievement motivation began with a prototype of persistent, striving person driven by the Protestant Ethic, and there are a certain number of such persons in the population for whom this type is an excellent way to understand them. There may be others for whom it is quite irrelevant, yet we continue to study "highs" and "lows." More vivid examples come from the "Authoritarian Personality" or "Machiavellian Personality." In each case, the prototype may convey a wealth of meaning of a person, yet the clarity is diluted when everyone is arrayed on a scale of Authoritarianism or Machiavellianism. To say a person is low in authoritarianism or need achievement only poses the problem to look further. Each single personality characteristic probably provides considerable understanding for a small proportion of the population, and measurement should be designed to take account of such differential salience. So should the design of research to which we now turn.

Design

One way to consider alternative research approaches is as follows: (1) "If you want to understand a phenomenon, control all variables except one" (experimental), (2) "If you want to understand a phenomenon, observe how it changes over time" (developmental-longitudinal), and (3) "If you want to understand a phenomenon, try to change it" (action). As discussed earlier, it seems more likely that the necessity for a conception of a person will be required when one conducts either developmental or action research. It is currently fashionable to describe such work as *ecologically valid,* but it is important that one not simply transplant single variable research into field situations. I believe that psychological inquiry should deal with phenomena which *generalize* over settings and over settings and over time (cf. Fyans and Maehr chapter). That such phenomena are complex and do not lend themselves to highly controlled studies does not disqualify them from study. Psychologists must focus on the phenomena—the nature of human experience in settings over time—no matter how difficult. Perhaps the most glaring example of remote reductionism is the suggestion by Cronbach & Furby in the paper, "How should we measure 'change' or should we?" (1970), who conclude that, because of psychometric difficulties in meeting arbitrary assumptions, change should probably not be measured at all!

Several chapters show a commendable emphasis on longitudinal investigation (deCharms, Hill, Maehr, Ruble, Raynor). One consequence of longitudinal investigation is that it almost always requires revision of single variable approaches, and ultimately a more comprehensive conception of the person.

Analysis

If the unit of inquiry becomes the person rather than a variable or an abstraction, then the analysis should reflect the shift. Parametric analysis of average scores on a dimension or scale ignores whether the principle applies to individual persons, e.g., the principle of stimulus generalization first espoused by Hovland (1937) on the basis of analysis of decreasing mean scores, applied to only one of the 20 subjects (Hunt & Sullivan, 1974). Parametric analysis is based on the assumption that the inquiry is about abstractions or variables rather than persons. A simple Fisher exact test may be more appropriate to the inquiry about persons as well as the level of precision of our present knowledge. An inspection of a 2×2 matrix quickly reveals the specific *persons* who were accounted for ("hits") and those who were not ("misses"), surely a more relevant estimate than the percentage of variance.

Finally, the methods of analysis should be based on what Walberg and Uguroglu called "causal multiplicity" and "multiple learnings." I realize that it is difficult to square this recommendation for multiple cause analysis with the use of the Fisher exact text. Yet we need methods which will flow from our conception of a person, especially the interrelations among constructs, and to know whether they apply appropriately to particular persons or not.

Persons-in-Relation

Parsons and Goff use Bakan's (1966) distinction between agentic and communal modes to point out the individual-centered nature of achievement motivation and to point to the need for a communal perspective. They also maintain that a communal perspective "points to the need to consider more than one motive system." Several years before Bakan's book, the philosopher John Macmurray made a similar distinction in "The Self as Agent" (1957) and "Persons in Relation" (1962). If one takes a "persons-in-relation" perspective seriously, then the unit of analysis becomes persons and the relation between them.

deCharms showed how a "gameboard" of five features could characterize a person-as-agent. Let us imagine that we have two (or more) persons, each with features of identity, intention, knowledge, know-how, and action, that we are concerned with how the two persons relate to one another (cf. Hunt, 1977; 1978b; 1978c). Person-in-action becomes persons-in-relation, and the further question of the adequacy of the conception of a person is the degree to which the features characterize dyadic interchange. For example, it becomes essential to add some perceptual feature when conceptualizing persons-in-relation so that the persons have ways of sensing and becoming aware of each other.

Like Lipman-Blumen et al., Parsons and Goff are concerned with the inadequacy of achievement motivation for characterizing women in present-day society. Unlike Lipman-Blumen et al., who simply extend achievement motivation to include altruism and social support, Parsons and Goff argue for a broader perspective, i.e., communal or persons-in-relation within which to view altruism, achievement motivation, and other systems as separate but interrelated motives.

Overemphasis on the individual has characterized North American cultural values in the 70's: the "New Narcissism," the "Me generation" and "Do your own thing," and therefore a shift to communal perspective will not occur quickly. However, it is interesting to consider the central role of achievement motivation in this perspective of cultural values.

Sampson (1977) comes to the conclusion that a "persons-in-relation" perspective is necessary in his very valuable paper, "Psychology and the American ideal," yet he reaches it through a consideration of historical, dialectical, and interdependent perspectives. In a similar vein, Sarason (1974) attributes the failure of community psychology movement to its non-community, or individualistic perspective.

If it is difficult to explicate our concepts of a person, then it is even more difficult to construe such conceptions in an interdependent fashion. Achievement motivation is certainly not the only individual, non-interpersonal construct. I have recently become aware that our work on Conceptual Level had become very individually oriented with its emphasis, for example, on self-directed learning even though in the original Harvey-Hunt-Schroder statement (1961) the most valued stage was that of interdependence.

Taking on a "persons-in-relation" way of thinking requires not only a larger unit (of at least two persons and their relation) but new constructs to describe such relation. I have recently (1979) proposed what I called the New Three R's of persons-in-relation: *responsiveness, reciprocality,* and *reflexivity* as beginning constructs to conceptualize the process of mutual adaptation.

This may all seem a long way indeed from achievement motivation, but I think not. The point is that the construct, achievement motivation, must be reconceptualized first as a part of a whole person and then as a part of a person interacting with another person.

REFERENCES

Argyris, C. Some unintended consequences of rigorous research. *Psychological Bulletin,* 1968, *70,* 185-197.
Bakan, D. *The duality of human existence,* Chicago: Rand McNally, 1966.
Cronbach, L. J., & Furby, R. How should we measure "change" or should we? *Psychological Bulletin,* 1970, *74,* 68-80.
deCharms, R. *Enhancing motivation: Change in the classroom,* New York: Irvington, 1976.
Entwisle, D. E. To dispel fantasies about fantasy-based measures of achievement motivation. *Psychological Bulletin,* 1972, *77,* 377-391.

Harvey, O. J., Hunt, D. E., & Schroder, H. M. *Conceptual systems and personality organization*, New York: Wiley, 1961.

Heider, F. *The psychology of interpersonal relations*, New York: Wiley, 1958.

Hovland, C. I. The generation of conditioned responses: I. The sensory generalization of conditioned responses with varying frequencies of tone. *Journal of General Psychology*, 1937, *17*, 125-148.

Hunt, D. E. *Matching models in education: The coordination of teaching methods with student characteristics*. Toronto: Ontario Institute for Studies in Education, 1971.

Hunt, D. E. Theory-to-practice as persons-in-relation. *Ontario Psychologist*, 1977, *9*, 52-62.

Hunt, D. E. Conceptual Level theory and research as guides to educational practice. *Interchange*, 1978, *8*, 78-90 (a).

Hunt, D. E. Theorists are persons, too: On preachng what you practice. In C. Parker (Ed.), *Encouraging student development in college*. Minneapolis: University of Minnesota Press, 1978, 250-266. (b)

Hunt, D. E. In-service training as persons-in-relation. *Theory into Practice*, 1978, *17*, 239-244. (c)

Hunt, D. E., & Sullivan, E. V. *Between psychology and education*. Hinsdale, Illinois: Dryden, 1974.

Macmurray, J. *The self as agent*, London: Faber, 1957.

Macmurray, J. *Persons in relation*, London: Faber, 1962.

McClelland, D. C., Atkinson, J. W., Clark, R. A., & Lowell, E. L. *The achievement motive*, New York: Appleton-Century-Crofts, 1953.

Murray, H. A. *Explorations in personality*, New York: Oxford University Press, 1938.

Ossorio, P. G. Never smile at a crocodile. *Journal of Theory of Social Behaviour*, 1973, *3*, 121-140.

Sampson, E. E. Psychology and the American ideal. *Journal of Personality and Social Psychology*, 1977, *35*, 767-782.

Sarason, S. B. *The psychological sense of community*, San Francisco: Jossey Bass, 1974.

Sarason, S. B. Community psychology, networks, and Mr. Everyman. *American Psychologist*, 1976, *31*, 317-323.

Tolman, E. C. *Prediction of vicarious trial and error by means of the schematic sowbug*. Paper presented to American Psychological Association meeting, 1938. (Available in Tolman, E. C. *Behavioral psychological man*. Berkeley: University of California Press, 1958, pp. 190-206).

Author Index

A

Abelson, N.D., 37, 63, 95
Abramson, Lyn Y., 100, 101, 108, 111, 137, 164, 313, 314, 316, 318, 319, 320, 321, 324, 335, 342
Acland, H., 93
Adelson, J., 356, 372
Adorno, T.W., 138, 164
Advisory Panel on SAT Score Decline, 188
Allport, G.W., 137, 164
Almquist, E., 409, 411
Alper, T.G., 140, 164, 385, 387, 393, 397, 400, 411
Amaral, P., 20
American College Testing Program, 16, 19
Amitai, A., 223
Amsel, A., 97, 111
Anastasi, A., 400, 411
Andrews, C.R., 249, 263, 264
Andrews, G., 315, 342
Angelini, A.L., 140, 164
Angoff, W.H., 16, 19
Argyris, C., 451, 455
Arlin, M., 442, 443
Armstrong, W., 357, 373
Astin, H., 366, 372, 391, 394, 411
Atkinson, John W., i, 3, 4, 9, 10, 11, 12, 13, 14, 15, 16, 18, 19, 20, 42, 43, 45, 61, 72, 105, 111, 112, 136, 137, 164, 167, 175, 188, 192, 193, 198, 216, 219, 223, 224, 227, 235, 242, 243, 250, 264, 314, 342, 350, 351-355, 356, 359, 360, 368, 372, 374, 387, 388, 391, 392, 399, 400, 403, 405, 410, 411, 456

B

Backer, T., 380, 383, 388, 390, 412
Baer, D.J., 340, 343
Bailey, M.M., 356, 361, 373, 392, 409, 413
Bailey, R.C., 309
Bakan, D., 355, 365, 366, 372, 409, 411, 454, 455
Bales, R.F., 164, 167
Baltes, P., 179, 189
Bandura, A., 133, 137, 164, 250, 264

Bane, M.J., 93
Bardwick, J.M., 374, 375, 387, 391, 411
Baron, R., 421, 443
Bar-Tal, D., 290, 297, 310, 385, 387
Bass, B.M., 155, 165
Battle, E.S., 228, 242, 361, 372
Becker, Ernest, 180, 188
Beery, R., 333, 334, 342
Belmont, J.M., 337, 342
Bem, D., 394, 411
Bem, Sandra L., 140, 164, 374, 382, 383, 384, 385, 387, 393, 394, 397, 399, 401, 407, 409, 411
Benesh-Weiner, M., 225, 244, 314, 323, 333, 340, 345
Berliner, D.C., 249, 264
Berlyne, 127
Bernard, J., 140, 164, 409, 411
Bialer, I., 234, 242
Bies, Robert J., 135
Birch, D., 9, 10, 11, 12, 19, 20, 175, 188, 216, 219, 223
Birk, J., 380, 387
Blake, R.P., 138, 164
Blaxall, M., 140, 164
Blumen, J. (Lipman), 140, 141, 142, 165 (see also Lipman-Blumen)
Bloom, B.S., 127, 133
Bodine, Richard J., 79, 92
Boggiano, A.K., 225, 227, 235, 242, 244
Bohn, M., 400, 412
Bolles, R.C., 128, 133
Bongort, K., 10, 11, 12, 19
Bontzin, R.R., 100, 112
Boozer, R., 424, 444
Botwinick, J., 176, 188
Bowers, K.S., 52, 92
Brehm, J.W., 100, 101, 108, 111, 113
Brennan, M., 46, 83, 92
Brewster Smith, M., 133
Broadhurst, P.L., 12, 20
Bromley, D.B., 233, 243
Brophy, J., 418, 419, 420, 421, 426, 441, 443
Brown, F.G., 165
Brown, J.S., 97, 111, 193
Brown, M., 192, 224
Bruner, J.S., 323, 325, 328, 342
Burger, J.M., 290, 310

457

Burlin, F., 409, 411
Burnham, D.H., 150, 167
Butterfield, E.C., 46, 63, 95, 337, 342

C

Calderwood, A., 391, 413
Callaway, H., 377, 388
Campbell, D.F., 388
Campbell, D.T., 299, 310, 380
Campbell, E.O., 92
Campbell, E.Q., 250, 264
Carnegie Commission on Higher Education, 349, 372
Cattell, R., 133
Catton, N.R., Jr., 330, 342
Child, D., 133
Child, I.L., 230, 245
Christie, R., 139, 140, 165
Clark, R., 388
Clark, R.A., 112, 136, 167, 227, 243, 456
Clark, R.W., 3, 4
Clifford, 4
Cohen, D., 93
Cohen, M.W., 32, 33
Cohen, S., 327, 332, 342, 344
Coleman, J.S., 45, 92, 173, 188, 250, 264
College Board, 17, 20
(College Entrance Examination Board)
Condry, J., 355, 372
Cook, R.E., 290, 311
Coons, A.E., 138, 167
Cooper, H., 421, 443
Cooper, H.M., 290, 310, 443
Cooper, J., 380, 387
Coopersmith, S., 333, 342, 399, 400, 411
Coser, R.L., 140, 165
Covington, M., 333, 334, 342
Crandall, V.C., 3, 4, 42, 43, 92, 228, 229, 230, 231, 234, 235, 242, 250, 252, 264, 269, 271, 282, 286, 361, 372, 385, 388, 427, 428, 443
Crandall, V.J., 229, 230, 234, 235, 243, 252, 265, 277, 286, 361, 385, 388, 427, 443
Crites, J., 409, 414
Cronbach, L.J., 38, 92, 257, 264, 410, 411, 453, 455
Crumrine, J., 140, 166
Csikszentmihalyi, M., 178, 180, 188

D

D'Andrade, R., 229, 244
Darling, M., 383, 388
Darom, E., 290, 310
Davidson, K.S., 94
Davidson, N., 264
Davidson, W., 286
Davis, Keith K., 112
Davis, O., 418, 443
Davis, W.L., 441, 443
Deaux, K., 268, 286, 289, 293, 308, 310
Debus, R.L., 249, 263, 264, 315, 342
deCharms, Richard, i, 3, 4, 26, 27, 28, 29, 32, 33, 137, 165, 182, 188, 319, 342, 447, 448, 449, 451, 453, 454, 455
Deci, E., 127, 133, 137, 165, 319, 342
Denker, E., 394, 414
Denmark, F., 356, 372
deVilliers, P.A., 11, 20
Diaz-Guerrero, R., 46, 93
Diener, C., 337, 342
Dinner, S.H., 235, 242
Dodson, J.D., 12, 20
Dolan, L., 131, 133
Dollard, N.E., 136, 167
Dorros, K., 383, 388, 400, 412
Dougherty, 28
Douvan, F., 356, 372
Downs, R.M., 324, 342
Drisha, S.D., 134
Dunteman, G., 155, 164
Dusek, J.B., 43, 58, 92, 418, 443
Dweck, C.S., 36, 42, 43, 44, 57, 63, 92, 103, 111, 137, 165, 240, 242, 249, 263, 264, 268, 271, 272, 273, 286, 313, 315, 316, 323, 336, 337, 342, 385, 388
Dyer, S., 355, 372

E

Eaton, W.O., 52, 53, 54, 55, 92, 93
Edward, A., 176, 188
Edwards, D.W., 272, 286
Edye, L., 396, 412
Egbert, R.I., 172, 174, 182, 185, 186
Egeland, B., 236, 245
Eisen, M., 233, 243
Ekehammar, B., 52, 92
Elig, Timothy W., 289, 290, 295, 298,

308, 309, 310
Eller, S.J., 103, 113, 320, 344
Emrich, A.M., 32, 33
Endler, N.S., 52, 92, 176, 188
English, L.D., 224
Enna, B., 264, 286
Entwisle, D.R., 11, 20, 284, 286, 451, 455
Epstein, C.F., 165
Erikson, E.H., 213, 223
Evans, R., 185, 188
Eysenck, H.J., 12, 20

F

Farber, S.M., 329, 342
Farmer, Helen, 380, 383, 388, 390, 397, 400, 407, 408, 410, 412
Farris, E., 289, 293, 308, 310
Feather, N.T., 3, 4, 9, 19, 42, 61, 72, 91, 106, 111, 175, 188, 198, 223, 228, 242, 264, 293, 310, 315, 343, 385, 389, 394, 407, 408, 412
Feld, S.C., 44, 92
Feldman, N.S., 235, 237, 242, 244, 250
Fencil-Morse, E., 112, 336, 343
Festinger, L., 104, 111, 394, 408, 412
Fiedler, F.E., 138, 165
Fiedler, M.L., 32, 33
Fisher, J., 286
Fiske, D.N., 299, 309, 310
Fitzhugh, R.J., 429, 444
Fleming, J., 140, 165, 394, 395, 400, 412
Fleishman, E.A., 138, 139, 165
Ford, Gerald, 157, 165
Franklin, B., 157, 165
Frazier, E.F., 375, 388
Freire, 127
French, E., 140, 165, 399, 412
Freud, S., 126, 144, 165
Friend, R.M., 250, 264
Frieze, Irene Hanson, 5, 95, 232, 242, 267, 268, 286, 287, 289, 290, 295, 297, 298, 308, 309, 310, 311, 345, 351, 357, 362, 372, 373, 385, 387, 408, 413
Fuller, R., 357, 373
Furby, R., 453, 455
Fyans, L.J., Jr., i, 3, 4, 36, 37, 43, 44, 46, 71, 83, 84, 92, 94, 251, 265, 390, 392, 399, 409, 412, 452

G

Gabriel, R., 341, 344
Garland, T.N., 357, 373
Geer, J.H., 99, 113
Geis, F.R., 138, 139, 165
Gergen, Kenneth J., 112
Gezi, K., 45, 92
Gienapp, J.C., 172, 174, 182, 185
Gilbert, J.P., 116, 133
Gillmore, G., 237, 244, 251
Gilmore, G., 167, 265
Ginsburg, E., 392, 412
Gintis, H., 93
Ginzburg, E., 392, 412
Gladstone, R., 309
Glass, D.C., 326, 343
Glasser, William, 96, 111
Gleser, G.C., 264
Godfrey, R., 341, 343
Goetz, T.E., 240, 242, 249, 264, 268, 271, 286, 313, 315, 342
Goff, 4, 447, 449, 454
Golding, S.L., 257, 264
Goldstein, M., 267, 293, 311
Golins, G., 331, 344
Good, T., 418, 419, 420, 421, 426, 442, 443
Gregory, R.L., 323, 343
Groves, D., 99, 113
Gulanick, N., 413
Gump, J., 394, 412

H

Haertel, G.D., 114, 117, 132, 134
Halperin, M.S., 230, 235, 242
Handley-Isaksen, Alice, 135
Haner, C.F., 97, 111
Hannah, T.E., 102, 112
Hannsa, B.H., 286
Harmon, L., 380, 388, 392, 396, 412
Harmon, P., 341, 343
Harren, V., 413
Harris, R., 142, 166
Harter, S., 42, 43, 95, 235, 236, 242, 319, 343
Harvey, O.J., 455, 456
Hastings, J.T., 39, 93
Hawley, P., 396, 412
Hayduck, L.A., 284, 286

Hays, N.L., 256, 264
Hebb, D.O., 12, 20, 324, 343
Heckhausen, H., 112, 227, 234, 243, 290, 311
Heider, F., 27, 33, 127, 290, 310, 450, 456
Heins, M., 364, 372
Heller, K.A., 238, 243
Helm, B., 309
Helmreich, R.L., 140, 369, 373, 407, 409, 413
Helper, M.M., 134
Herrnstein, R.J., 11, 20
Heyne, B., 93
Hibschuser, J., 113
Hieronymous, A.N., 29, 33
Hill, Kennedy T., 3, 4, 34, 36, 37, 39, 42, 43, 44, 45, 46, 47, 52, 53, 54, 55, 56, 57, 58, 59, 60, 61, 63, 67, 72, 79, 83, 84, 92, 93, 94, 165, 453
Hiroto, D.S., 99, 112
Hobson, C.J., 92, 250, 264
Hochschild, A.R., 140, 165
Hodges, K.E., 243
Hodges, K.L., 287
Hoffman, L.W., 140, 165, 356, 373
Holland, 392, 399, 412
Holtzman, W.H., 46, 93
Horner, M., 137, 140, 141, 142, 166, 354, 355, 356, 357, 360, 373, 391, 394, 395, 399, 400, 406, 412
Horner, M.S., 10, 20, 375, 385, 386, 388
Horner, T., 3, 4
Horowitz, R., 132, 134
Houston, C.S., 329, 343
Houts, P.L., 38, 93
Hovland, C.S., 453, 456
Hubner, J.J., 134
Huesmann, L.R., 324, 343
Hull, C.L., 126, 136, 167
Hunt, David, i
Hunt, D.E., 447, 450, 453, 454, 455, 456
Hutt, M.L., 62, 93

I

Inhelder, B., 108, 112
Inkeles, A., 157, 166

J

Jacklin, C.N., 140, 167, 271, 282, 286, 391, 392, 393, 413
Jacobs, P.D., 100, 113
James, W., 125, 126, 326, 343
Jencks, C., 45, 93
Jeter, J., 418, 443
Johnson, D.J., 134
Johnson, P., 372
Jones, Edward E., 102, 112
Jones, S.L., 97, 102, 112
Joreskog, K., 407, 412
Julian, J.W., 441, 443

K

Kagan, J., 127, 175, 188, 230, 231, 243, 396, 413
Kaplan, R., 325, 330, 343
Kaplan, S., 324, 325, 326, 343
Karabenick, J.D., 238, 243
Katkovsky, W., 234, 242, 250, 252, 264, 372, 385, 388, 427, 443
Katz, I., 250, 264
Keen, P., 163, 166
Kelley, H.H., 314, 322, 323, 324, 343
Kelly, F.J., 340, 341, 343
Kelly, J.R., 176, 188
Kermis, M.D., 58, 92
Kessel, L.J., 281, 286
Kievit, M., 383, 388
Kleiber, D., 4, 171, 179, 189
Klein, D.C., 103, 112, 336, 343
Klein, R., 176, 189
Kliman, A.S., 337, 343
Klinger, E., 107, 112, 175, 189
Klosson, E., 242
Kluender, M., 182, 185, 188
Koenigs, S.S., 32, 33
Kogan, N., 383, 388, 400, 412
Kohlberg, L., 229, 243, 272, 286
Kozak, M.J., 99, 113
Kravitz, R.M., 140, 166
Kruglanski, A.W., 200, 223
Kubal, L., 100, 112
Kuhn, D., 394, 413

Kukla, A., 5, 95, 102, 106, 112, 113, 177, 178, 189, 227, 228, 243, 250, 261, 262, 264, 267, 268, 283, 286, 290, 311, 333, 345
Kun, A., 225, 238, 243, 244, 313, 314, 322, 323, 333, 340, 344, 345

L

Lacey, B.W., 428, 443
Lange, 126
Lanzetta, J.T., 102, 112
Lazarus, M., 45, 93
Leavitt, H., 4, 409, 413
Leavitt, H.J., 135, 142, 157, 163
Lee, Jongsook, 103, 104, 105, 108, 112
Lefcourt, 127
Lekarczyk, D.T., 63, 93
Lens, W., 18, 19
Lerman, D., 314, 316, 317, 340, 345
Lesser, G.S., 140, 166
Levine, A., 140, 166
Levinson, D.J., 157, 166
Lewin, Kurt, 125, 126, 138, 166, 219, 221, 223, 350
Lewis, J., 44, 92
Light, R.J., 116, 133
Lighthall, F.F., 94
Likert, R., 138, 139, 167
Lindquist, E.F., 29, 33
Linn, Robert L., i, 83, 93
Lipman-Blumen, J., 4, 135, 140, 141, 142, 157, 166, 385, 388, 409, 413, 449, 454
Lippit, R., 138, 166
Litwin, G.H., 106, 111, 399, 411
Livesley, W.J., 233, 243
Lockheed, M., 166
Loebl, J.H., 244
Lowell, E.L., 3, 4, 112, 136, 167, 227, 243, 388, 413, 456
Lynch-Sauer, J., 261, 264

M

McCandless, S., 356, 372
McClelland, D.C., 3, 4, 105, 112, 126, 136, 144, 150, 151, 157, 175, 189, 226, 227, 229, 243, 351, 374, 375, 388, 391, 395, 410, 413, 447, 456
McClelland, L., 357, 373
Maccoby, E.M., 140, 167, 271, 282, 286, 391, 392, 393, 413
McDermott, R.P., 420, 443
McHugh, M.C., 286, 289, 291, 292, 310, 311
McKeachie, W.J., 128, 134
McKenna, B., 38, 94
McMahon, I.D., 293, 311, 385, 388
Macmurray, J., 454, 456
McNelley, F.W., 175, 189
McPartland, J., 92, 250, 264
McPhail, A., 79, 92
Magnusson, D., 52, 92, 176, 188, 299, 310
Maehr, M.L., i, 3, 4, 36, 37, 43, 46, 71, 72, 83, 92, 93, 94, 134, 167, 172, 174, 175, 176, 177, 179, 180, 182, 185, 188, 189, 237, 241, 249, 250, 251, 262, 263, 264, 280, 283, 285, 286, 287, 353, 375, 388, 391, 413, 452, 453
Maier, S.F., 99, 112, 113, 312, 344
Margolin, B. 223
Markus, G., 181, 189
Markus, H., 284, 287
Maslow, A., 127, 137, 144, 167
Massad, P., 97, 112
Masters, J.C., 43, 94, 239, 243
Matthews, E., 394, 413
Mednick, M.T.S., 140, 167, 376, 388, 389
Mehrabian, A., 3, 4,
Mergler, N.L., 58, 92
Mettee, D.R., 334, 343
Meyer, W.U., 290, 292, 311
Michaels, E.J., 228, 244
Michelson, S., 93
Miller, A.G., 157, 167
Miller, I.W., 100, 101, 103, 108, 112
Miller, J., 136, 167
Mischel, W., 137, 167
Monahan, L., 394, 413
Montagues, E., 413
Montemayor, R., 233, 243
Mood, A.M., 92, 250, 264
Moreland, J., 409, 413
Moss, A., 175, 188

Moss, H., 230, 230, 243, 396, 413
Mosteller, E., 116, 133, 402, 413
Moulton, R.W., 206, 223, 227, 243
Mouton, J.S., 138, 164
Murray, H.A., 450, 456
Murray, R.A., 136, 137, 144, 167

N

Nadelman, L., 261, 264
Nanda, H., 257, 264
Nation, J.R., 97, 112
National School Boards Association, 38, 94
Neal, J.M., 250, 264
Neale, J., 287
Nelson, S., 264, 286
Newman, Richard S., 4, 312
Nicholls, J.G., 4, 174, 175, 176, 177, 180, 189, 233, 235, 236, 238, 243, 249, 250, 262, 264, 266, 267, 268, 269, 270, 271, 272, 275, 276, 280, 283, 285, 287,
Nierenberg, R., 112, 267, 293, 311
Nisbett, R., 324, 332, 337, 344
Norman, W.H., 100, 101, 103, 108, 112
Nottelmann, E.D., 58, 94
Nowlie, V., 167
Nygard, R., 106, 112

O

O'Malley, P.M., 18, 19
Ossorio, P.G., 23, 26, 27, 33, 448, 456
Overmier, J.B., 99, 112

P

Packard, R., 140, 166
Padia, W.L., 341, 344
Panciera, L., 113
Papanek, H., 140, 167
Parham, I., 175, 189
Parsons, J.E., 238, 240, 242, 243, 244, 272, 287, 351, 357, 372, 373, 408, 413, 447, 449, 454
Parsons, T., 140, 167
Patterson, Kerry J., 135
Peevers, B.H., 233, 243
Peter, N.V., 237, 245
Peterson, P.L., 132, 134

Pettigrew, L.A., 389
Pettigrew, T.F., 332, 344, 375
Pfeffer, J., 137, 147, 167
Phares, E.J., 441, 443
Piaget, J., 108, 112
Pines, H.A., 441, 443
Plan Organization—Iran, 376, 378, 388
Plass, J.A., 55, 56, 57, 58, 59, 60, 67, 72, 94
Poloma, M.M., 357, 373
Preston, A., 372
Postman, L., 325, 328, 342
Price, L.H., 10, 11, 12, 19
Price-Williams, D.R., 45, 94
Psathas, G., 392, 413
Puryear, G.R., 140, 167

Q

Quinto, F., 38, 94

R

Rafferty, J., 383, 389
Rajartanam, J., 257, 264
Ramirez, M., III, 45, 94
Raynor, J., 391, 392, 400
Raynor, J.O., i, 3, 4, 5, 10, 14, 16, 18, 19, 20, 42, 43, 45, 61, 91, 107, 112, 137, 164, 175, 188, 190, 192, 197, 199, 215, 219, 223, 224, 243, 244, 250, 264, 356, 391, 392, 400, 403, 405, 410, 411, 453
Raynor, J.W., 20
Reagan, B., 4, 140, 164
Reed, L., 5, 95, 345
Reed, Witt., 165
Repucci, N.D., 103, 111, 165, 313, 315, 323, 342, 385, 388
Resnick, L.B., 424, 444
Rest, S., 5, 95, 102, 112, 345
Revelle, W., 12, 20, 228, 244
Rholes, N.S., 233, 242, 244
Riter, A., 223
Rivers, L., 394, 412
Rodin, J., 327, 344
Roe, A., 399, 413
Rokeach, M., 360,
Rokoff, G., 140, 165
Rosellini, R.A., 99, 113
Rosen, B., 182, 183, 189
Rosen, B.C., 229, 244

Rosenbaum, R.M., 5, 95, 295, 311, 345
Rosenberg, M., 332, 335, 338, 339, 344
Rosenshine, B.V., 249, 264
Ross, J., 244
Rossi, A., 391, 413
Roth, S., 100, 112
Rotter, J.B., 126, 316, 318, 344, 385, 389
Roussel, J., 97, 111
Ruble, Diane N., 4, 71, 74, 76, 225, 227, 233, 235, 238, 242, 243, 244, 287, 351, 357, 372, 408, 413, 453
Rubovits, P., 397, 413
Ruebush, B.K., 36, 43, 45, 94
Ruhland, D., 357, 373
Russell, D., 314, 316, 317, 340, 345
Ryle, G., 26, 33

S

Salancik, G.R., 137, 147, 167
Salili, Farideh, 4, 71, 94, 157, 167, 237, 244, 251, 264
Sampson, E.E., 455, 456
Sarason, I.G., 50, 64, 94
Sarason, S.B., 36, 37, 42, 43, 45, 57, 61, 93, 94, 180, 189, 448, 455, 456
Sarbin, T., 394, 413
Sawusch, J.R., 11, 20
Schachter, S., 332, 344
Schaefer, S., 142, 166
Schaie, K., 175, 179, 189
Schechner, R., 157, 167
Schiller, D., 114, 117, 132, 134
Schott, J., 341, 344
Schroeder, H.M., 455, 456
Schultze, J.R., 327, 344
Sears, P.S., 167, 383, 389
Sears, R.R., 167
Secord, P.F., 233, 243
Seitz, V., 37, 63, 95
Seligman, Martin, E.P., 99, 111, 112, 113, 137, 164, 167, 312, 313, 315, 316, 331, 336, 342, 343, 344
Sexton, P., 391, 413
Shabtai, L., 233
Shaklee, H., 238, 244
Shantz, C.N., 233, 234, 244
Shavelson, R.J., 134
Shaver, P., 394, 395, 413

Shea, B.M., 45, 94
Sheehy, G., 213, 224
Sherrod, D.R., 326, 332, 344
Shiffler, N., 261, 264
Shusterman, L., 113
Simon, H.A., 146, 167
Simon, J.G., 293, 310, 385, 389
Singer, J.E., 326, 343
Skinner, B.F., 126
Slater, P., 139, 164
Small, A., 243, 287
Smith, C.P., 3, 5, 95, 225, 230, 244
Smith, E.R., 289, 290, 293, 311
Smith, M., 42, 93
Smith, M.B., 128, 134, 365, 373
Smith, M.L., 341, 344
Sohn, D., 250, 264, 267, 287
Solomon, R.L., 99, 113
Sorenson, R.L., 71, 94
Spence, J., 126, 397, 407, 409, 413
Spence, J.T., 140, 167, 369, 373
Spielberger, C.D., 42, 45, 61, 95
Stallings, W.M., 71, 94
Stanton, G.C., 134
Stea, D., 324, 342
Steihman, W.M., 165
Stein, A., 392, 409, 413
Stein, A.M., 356, 361, 373
Stiles, B., 428, 444
Stipek, D., 317, 318, 345
Stogdill, R.M., 138, 167
Strong, E., 380, 389, 409, 414
Sullivan, E.V., 453, 456
Super, D., 390, 391, 392, 395, 409, 414
Sutton, R., 272, 280, 287
Swartz, J.D., 46, 93

T

Tangri, S., 356, 372
Tanney, M., 380, 387
Tatsuoka, Maurice, i, 401, 403, 414
Teasdale, John D., 111, 137, 164, 313, 342
Tennen, H., 103, 113, 320, 344
Theil, 120, 122, 134
Thomas, J.C., 328, 344
Thorndike, R.L., 11, 16, 20, 126
Thornton, J.W., 100, 113
Tickamyer, A., 140, 166
Tiedeman, D., 394, 413

Tipton, R., 393, 409, 414
Tittle, C., 394, 414
Tolman, E.C., 125, 126, 324, 344, 350, 449, 456
Tomlinson-Keasey, C., 394, 396, 414
Triandis, Harry, i
Tresemer, D., 140, 167
Trey, J.E., 376, 387, 389
Tukey, J., 402, 413
Turiff, S., 20
Turner, E.V., 134
Tyler, B., 383, 389
Tyler, F.B., 383, 389
Tyler, L., 380, 389

U

Uguroglu, M.E., 3, 5, 114, 130, 134, 448, 453
U.S. Women's Bureau, 378, 389

V

Valins, S., 332, 344
Valle, V.A., 267, 286, 287, 295, 311
Van Bergen, A., 235, 244
Veroff, J., 43, 71, 78, 95, 175, 189, 230, 231, 232, 234, 235, 236, 239, 240, 244, 268, 287, 356, 357, 366, 373, 383, 389
Vroom, V.H., 137, 138, 167

W

Waite, R.R., 94
Walberg, H.J., 3, 4, 5, 114, 117, 130, 132, 134, 447, 448
Wallach, 400
Walsh, V., 331, 344
Wang, M.C., 417, 423, 424, 425, 428, 429, 444
Webb, E.J., 163, 166
Weiner, B., 3, 5, 43, 44, 61, 95, 102, 106, 107, 112, 113, 167, 225, 226, 227, 228, 231, 232, 235, 237, 240, 242, 243, 244, 245, 249, 264, 267, 287, 289, 290, 293, 294, 310, 311, 313, 314, 315, 316, 317, 318, 319, 320, 322, 323, 330, 333, 340, 344, 345, 385, 389, 407, 414
Weinfeld, F.D., 92, 250, 264
Weinstein, 4

Weiss, J., 134
Weissner, H.J., 140, 167
Weisstein, W.J., 417, 429, 444
Weisz, J.R., 317, 318, 345
Weston, P.J., 389
Wheeler, L., 395, 414
White, R.W., 137, 138, 166, 167, 319, 333, 345, 373
Whiting, J.W.M., 167
Whitley, T.W., 442, 443
Williams, J.P., 45, 58, 64, 65, 66, 72, 95
Wine, D., 176, 188

Fyans/author index/psj/4-2816B-m

Wine, J., 58, 95
Winter, D.B., 3, 4
Winter, D.C., 175, 189 356, 373
Winterbottom, M.R., 3, 5, 229, 245, 399, 414
Wirtz, W., 38, 95
Witryol, S.L., 280, 287, 288
Wittrock, M.C., 128, 134
Wohlwill, J., 326, 345
Women's Organization—Iran, 377, 378, 380, 389
Wortman, C.B., 100, 101, 102, 103, 108, 113
Wurtz, 28

Y

Yerkes, R.M., 12, 20
Yetton, P.W., 138, 167
York, R.L., 92, 250, 264
Young, E., 236, 245

Z

Zajonc, R.B., 181, 182, 189
Zaksh, P., 223
Zandar, A., 357, 373
Zellman, P., 372
Zigler, E., 37, 42, 43, 46, 63, 95, 230, 245
Zimbardo, P.G., 43, 94
Zuckerman, M., 394, 414

Subject Index

A

Ability
 as causal element, 290
 attributions of, 272
 concept of, 234
 in educational outcomes, 117
Achievement, 250
 agentic, 355
 assessment of, 38
 behavior, defined, 284
 changes in concepts of, 233
 choices, 251
Achievement ethic, the, 180
 and aging, 180
Achievement level
 perceptions of (teacher), 417
Achievement motive
 defined, 374
 an ethnocentric conception, 375
Achievement motivation, 96, 137
 and aging, 175
 and culture, 177
 and values, 349
 attribution theory of, 268
 developmental influences on, 241
 in career choices, 350
 major determinants of, 229
 sexual differences in, 143, 267, 269
 theories of, 226, 227, 242
Achievement motivation, women, 375
 black, 375
 contextual factors in, 375
 determinants, 386
 early socialization in, 376
 situational factors in, 375
Achievement Orientation Scale, 370, 371
 subscale (Parsons, Goff), 371
Achievement values
 cross-cultural differences, 158
Achievement - women, Iran, 374
 and vocational behavior, 374
Achieving styles, 135
 collaborative relational, 145
 dimensions of, 156
 direct, 135, 149, 150
 etiology, 145
 flexibility in, 159
 intensity, 159
 other assumptions in, 147
 power direct, 145
 range of, 157
 relational, 135, 149, 150, 151
 reliant relational, 145
Adaptive Learning Environments
 Model, 423, 426, 440
 exploratory learning, 425
 instructional learning
  management, 425
 prescriptive learning, 424
Adult personal functioning
 theory of, 197
Adult personality
 motivational determinants, 191
Affiliation as success, 356
Agency
 Bakav, 365
 defined, 365
Agentic achievement, 355
Agentic domain, 355
Alternative approaches
 to development of achievement
  motivation, 230
Analysis
 parametric, 453
Androgyny, 393, 401, 403, 406, 409
A.S.I., 162
Assessment
 modified motivational, 368
 values, 370
Atkinson's mathematical model, 352
Attainment value, 3
Attitudes
 self-discriminatory, 382
Attribution, 96, 102, 108
Attributions
 causal (questionnaire), 252
 emotional reaction, 317
 external, 318
 style, 4
Attributional framework, 341
Attribution measures, 294
Attributional style, 250
 assessment of, 253

causal, 267

B

Biases
 individual differences in, 251
 in task selection, 251
Bipolar ratings
 in assessments, 290, 293
BSRI: Bern Sex Role Inventory, 399, 408

C

Career behavior
 women's, 395
Career motivation
 in women, 390
 or cumulative achievement, 392
 community support for, 407
Career striving, 204
 and aging, 210
 in an open contingent path, 213
Causal attributions
 instruments for measuring, 289
 measuring, 289
 sex differences in, 273, 275, 284
Causal elements, 290
 ability, 290
 effort, 290
 luck, 290
 task difficulty, 290
Causal schemata, 314
 and attribution, 314
Causation
 personal, 27, 183
CDA - Choice Dilemmas Questionnaire, 400
Challenge
 concepts of, 236
Choice of one major cause
 in assessments, 289
Cognitive clarity, 322, 325
 in wilderness, 330
Cognitive confusion, 325
Cognitive deficits, 322
Cognitive maps, 324
 theory, 324
Cognitive test demands, 68
Communion
 Bakan, 365
 defined, 365
 perspective, 367
Competence
 perceived, 317, 318
Competence motivation theory, 319
Concepts
 ability, 234
 challenge, 236
 masculine vs. feminine, 268
 success/failure, 235
Congruence, 251
 of attribution with selections, 256
Consistency, 251
Constructs
 in achievement motivation
 theories, 227
Controllability
 of causal attributions, 314
Cultural values
 central role of achievement
 motivation, 454
 individual, 201

D

Deficits
 alleviation of, 321
 cognitive, 322
 emotional, 318
Depression
 therapeutic strategies for, 321
Developmental framework, 232
 cognitive components of, 232
 social-environmental components, 239
Developmental perspective
 on achievement motivation, 226
Difficulty value, 198
Direct domain, 145
Direct relational model, 142
Discriminating attitudes, 380
Discrimination
 employment, 349
 sexes, 349
 women, 391
Dynamics of action, 9, 10, 11

E

Educational testing policy, 34, 52
 programs, 39

Subject Index

Effort as causal element, 290
Emotional deficit, 318
Enactment, 160
Environmental influences
  on women, 395
  socialization, 396
Epistemic paradigm, 26, 34
Esteem-income, 197
  and action, 219
  and value, 198
  attaining, 217
  maintaining, 217
  time-linked sources, 221
Etiology of achieving styles, 145
Evaluation, 34
  anxiety, 3
  practices, 39, 40
  situations, 41
Executive functioning, 337
Expectancy of success, 228
Expectancy value, 3
Expectancy-Value theory of motivation, 314
  persistence behavior as support, 315
Expectations
  and achievement motivation, 417
  as guides, 138
  minimize adverse effects, 440, 443
  of teachers, 417
  success, 138
Extrinsic value, 201

F

Failure, 96
  effects of, 107
  fear of, 42, 46, 61
Fear of failure, 42, 46, 61
Fear of success, 9, 354, 357
  women's, 390, 394, 406
Feedback
  peer, 433
  success, 99
  teacher, 274
Female achievement, 1, 140
Forms of measurement, 3
Freedom to succeed, 101
Frustration, 96, 97, 98

G

"Gameboard", 449
  de Charms, 452, 454
  Ossorio, 449
Globality
  and deficits, 320
  in attributional schemata, 314
"Graying of America", 172
  and age consciousness, 181
  field for research, 188
  global issue, 187
  motivational factors in, 183
  socialization changes because of, 182, 185

H

Helplessness
  see learned helplessness
  chronic, 316
  personal, 324
  transient, 316
  universal, 319
Home-career conflict
  women's, 390, 393

I

IAR
  Intellectual Achievement Responsibility Questionnaire, 253, 385, 427
Identity crisis, 214
Incentive value, 229, 355
  and globality, 362
  omission of, 359
Independent measures
  vs. ipsative, 292
Independent ratings
  in assessment, 289
Individual culture value, 201
  consensual, 201
  individual, 201
Individual variance variable, 227
Induced incongruity, 325
Informational content
  in wilderness, 326
Informational processes
  in wilderness, 328
Instrumental value, 200

Inventory of self-esteem (Coopersmith), 399
Involvement, 329
Ipsative measures
 defined, 292

L

Lack of control
 learned helplessness, 104
Law of Effect, 11
Leadership styles, 139
 studies, 139
Learned helplessness, 63, 96, 100, 138, 312
 and causal attribution, 313
 and mastery orientation, 313
 deficits, 320
 from lack of control, 104
 personal vs. impersonal nature, 318
 strategies, 321
Learning environments
 student-centered, 421
 teacher-controlled, 420
Levels of reinforcement, 97
 partial, 98
Life choices
 and attainment value, 361
 as task choices, 350
 differences, men's/women's, 352
 relative worth, 357
 vs. wife-mother role, 361
Locus of causality, 314, 317
 and affect, 317
Locus of control, 316
Luck as causal element, 290

M

Manipulation
 attributions, 320
 incentives, 13, 14
 variables, 33
Mastery orientation, 313
 vs. learned helplessness, 313, 315
Measurement, 43, 368, 452
 instruments, 289
 of achieving styles, 162
 scale method, 308
 self, 43
 techniques, 289

Methodology
 from conception of person, 451
Minimal competency testing, 39, 51
Model
 Adaptive Learning Environments, 423, 426, 440
 Atkinson's mathematical, 352
 direct relational, 142
 motivational, 350
Model of achievement, female
 environmental variables in, 392, 395
 needed in psychology, 390
 one proposed, 391
 psychological variables in, 392
 sex-role orientation in, 393
Model of agency and communion, 365
Model of motivation
 affective arousal, 137
 direct relational, 142
Motivation
 achievement, 3, 9, 10, 23, 41
 intrinsic/extrinsic, 138
 negative, 3, 36, 48, 76
 optimize, 40
 over-, 16
 positive, 36, 37, 82
 strength of, 12, 14
 theory of, 9, 125, 126, 127
Motivation, and
 ability, 19
 academic achievement, 130
 educational productivity, 114, 117
 home environment, 131
 learning, 130
 open education, 132
 point of no return, 121
 test performance, 47
Motivational factors, 37, 38
 in test performance, 45
 negative, 39
 variables, 48
Motivational test bias, 36, 37, 50, 82, 83
 causes, 39
 minimize, 41, 76
Motivational models, 350
Motive to avoid failure, 105
Motive to succeed, 105
Motives
 as traits, 138

Subject Index

N

N-ACH (need for achievement), 3, 137
Need to achieve, 3, 101, 106, 137
Need power, 356
Needs
 egoistic, 147
 physical, 147
 social, 147
 synonymous with motives, 148
Negative motivation, 3, 36, 48, 76

O

Open-ended attributions, 298
Open-ended responses, 284, 289
 advantages, 291, 292
 in assessment, 289
 poorer reliability, 308
 poorer validity, 308
 vs. structured responses, 293, 294
Origins, 27
 training, 29, 32, 319
Outward Bound, 330

P

Parametric analysis, 453
Past success, 211
 value, 211
Pawns, 27, 29, 319
Peer interactions, 433, 434
Perceived competence, 317, 318
 vs. perceived contingency, 319
Perceived contingency, 317
 and perceived competence, 319
Perception
 defined, 323
Performance, 100
 control of, 100
Persistence
 and attainment value, 361
Person
 concept of, 448
 characterization of, 451
Personal causation, 27, 183
Personal helplessness, 319
Personal styles, 139
Persons-in-action, 22, 23, 26, 33, 448

Persons-in-relation, 454
 vs. persons-in-action, 454
 necessary perspective, 455
Population density
 behavioral effects, 327
Population shift, 174, 176
 and changing motivation, 186
 and changing achievement, 186
Pragmatic paradigm, 27, 28, 33
Predictive associations, 257
Predictors
 of learning outcomes, 114
Preference, 263
 guided by attribution, 263
 when evaluation minimized, 263
Problem-solving, 333
 in Outward Bound, 333
Productivity
 and causality, 115
 and cost effectiveness, 115
 theories of, 117, 118
Psychological careers, 202
Psychological variables, 392
 self confidence, 392
 sex role orientation, 393
 risk-taking behavior, 395

Q

Quality
 of instruction, 118

R

Reactance, 96, 101
Re-entry women, 394
Referrent individuals, 332
Reinforcement
 influence on effort, 420
 levels of, 97, 98
 peer, 433, 434
 social, 384
 teacher control of, 418
 value, 229
Relational domain, 145
Report cards, 44, 73
Risk-taking, 3
 patterns, 390
 women's, 395

S

SBOS-Student Behavior Observation Schedule, 428
Self-concept, 195
  low vs. high, 316
  means of self-identity, 201, 218
  senses of self, 197, 198
  women's, 390
Self-determination theory, 319
Self-discriminating attitudes, 382
Self importance
  and future importance, 193
Self-management skills, 441
Self Schedule System (Wang), 425
Sex differences
  in achievement motivation, 4, 374
  in achievement orientation, 385
  in anxiety, 57
  in causal attributions, 268, 270, 278
  in task value, 363
  on ASI, 163
Sex role orientation, 383
  women's, 390, 393
Sex role socialization
  re women, 391
Sex roies, 141
  stereotypes, 141
  relation to achievement, 142
Single-variable, 448
Social learning, 138
Social reinforcement
  need for—female, 384
Sociocultural factors in school achievement, 45
Stability, 315
  in causal attributions, 314
SAT,
  standardized achievement test, 34, 40
Student-centered learning environments, 421
  influence self-perceptions, 442
Styles
  leadership, 139
  personal, 139
Success/failure, 235
  causal attribution for, 167
  concepts of, 235
  influence on congruence, 253
  in testing, 80
  perception of, 250
Success feedback, 99
Support systems
  for women's achievements, 391

T

Task difficulty
  as causal element, 290
Task selection, 250, 251
  biases in, 251
Task value
  defined, 361
  sex differences in, 363
Teacher criticism
  of low achievers, 419
  related to lowered effort, 421
Teacher expectations, 4
Teacher perceptions
  adverse?, 417
  how they influence, 418
  of achievement level, 417
  self-fulfilling?, 418
Test anxiety in children, 42, 44, 46
  changes, 55
  correlation with performance, 51
  effects, 50
  measure, 43
Test demands
  intellectual, 69
  non-content, 81
Test performance, 16, 37
  to optimize, 41
Testing
  achievement, 44
  dual, 86
  educational, 34, 52
  individual vs. group, 72
  minimal competency, 39, 51
  modification, 84
  new procedures, 41
  pressure in, 61, 78
  programs, 39
  success/failure experiences, 80
Tests
  of ability, 9
  of educational achievement, 9
Test-taking skills, 36, 41, 79, 82

Test-taking strategy, 35
  maladaptive, 61d
Theory, Expectancy Value, 314
  theory of
    achievement motivation, 226, 227, 242
    adult personal functioning, 194
    cognitive maps, 322
  competence motivation, 319
  productivity, 117, 118

  productivity, 117, 118
  self-determination, 319
Time
  limits, 52
  of instruction, 118
  pressure, 52, 55, 59, 78, 80, 84
Traits
  as motives, 138

U

Universal helplessness, 319

V

Values
  extrinsic, 201
  instrumental, 200
  vs. motives, 364
Vicarious achievement ethic, 141

W

Women in Iran, 376
  education, 377
  employment, 378
  vocational aspirations, 379
Women, re-entry, 394
Women's achievement and career
  motivation, 390

DATE DUE

MAY 5 1982

MAY 1 4 1982

APR 2 6 1988

MAY 6 1988

OCT 1 7 1989

APR 2 7 1993